In the Kingdom of Gorillas

In the Kingdom of Gorillas

The Quest to save Rwanda's Mountain Gorillas

Bill Weber and Amy Vedder

AURUM PRESS

First published in Great Britain
2002 by Aurum Press Ltd
25 Bedford Avenue, London WC1B 3AT

First published in the US by Simon and Schuster 2001

A catalogue record for this book is available from the British Library.

Maps copyright © 2001 Mark Stein Studios
Book design by Ellen R. Sashara

All photographs, unless otherwise credited, are by Bill Weber and Amy Vedder.

ISBN 1 85410 839 5

1 3 5 7 9 10 8 6 4 2
2002 2004 2006 2005 2003

Printed by MPG Books, Bodmin

Acknowledgments

O N A SUMMER EVENING in 1992, we described our work in Rwanda to Bill McKibben and Terry Tempest Williams as we sat around an Adirondack fire. They both insisted that we tell our story of mountain gorillas, human characters, and lethal conflicts to a larger audience. We were grateful for the interest of two such accomplished writers but failed to follow their advice. As we continued to focus on our conservation work, the Rwandan story grew more complicated, then exploded into civil war and a bloody genocide. Bill McKibben again encouraged us to write a book. His agent, Gloria Loomis, called the next day to say she wanted to meet with us. Bill has given further invaluable advice as the book has progressed, and Gloria has been a staunch supporter of the need to tell the full story: gorillas, people, progress, and despair. We are deeply indebted to their advice, assistance, and steady encouragement.

The Wildlife Conservation Society has supported our work from the beginning. In 1977, WCS gave us a grant to study mountain gorillas and the problems that confronted them in Rwanda. It was a risk to sponsor a young couple with no advanced degrees; it was a bigger risk to endorse our plan to look at the needs of both the gorillas and local human populations. The result was the framework for the successful Mountain Gorilla Project. In later years, WCS backed our decision to expand our work beyond gorillas to focus on a series of more neglected forests, species, and people across central Africa. But the Rwanda story remained untold. George Schaller, WCS Director for Science and our first inspiration to work with mountain gorillas, added his encouragement. In 1999, WCS Vice President John Robinson approved a flexible work schedule to allow us to write this book.

Our work with the Mountain Gorilla Project would not have been possible without the initial support of The Fauna and Flora Preservation Society, under the direction of John Burton. Sandy Harcourt was especially supportive of our ideas and involvement in his role as principal advisor to FFPS. Jean-Pierre von der Becke, Conrad Aveling, Rosalyn Aveling, Roger Wilson, Jeff Towner, Mark Condiotti, and Craig Sholley all contributed greatly over the years to the success of the Mountain Gorilla Project in the field. Eugène Rutagarama, José Kalpers, Katie Frohart, and Anecto Kayitare have continued this work in recent years. Beginning in 1985, we received financial assistance from the U.S. Agency for International Development for a series of more comprehensive conservation projects in Rwanda, with particular thanks to Gene Chiavaroli for his initial support. Rwandan partner institutions included the Office of National Parks and Tourism, the Ministry of Education, the Ministry of Agriculture, and the National University of Rwanda. Alain Monfort, Nicole Monfort, Jean Pierre van de Weghe, and Thérèse Abandibakobwa opened their lives and homes to us and dispensed valued counsel throughout our early years in Rwanda. In addition to her role as director of Karisoke Research Center, Liz Williamson has provided friendship, updates on the status of gorillas and people, and the incomparable opportunity to revisit the gorillas we first came to know almost a quarter century before.

Martha Schwartz, Susan Lenseth, Marion Vedder, Louise Fox, and Kent Redford read early draft chapters of this book and offered the encouragement needed to continue with what seemed like a monumental task. Bill McKibben, George Schaller, Kay Schaller, Josh Ginsberg, Stacey Low, and David Watts read the completed first draft and provided invaluable advice. David also drew on his own extensive work with the mountain gorillas and our shared experiences to confirm events and fill in many missing facts and incidents. Craig Sholley contributed additional information based on his experiences in Rwanda. Kurt Kristensen revisited painful memories to open a personal window on the terrible first days of the 1994 Rwandan genocide.

It has been a pleasure to work with Bob Bender as our principal editor at Simon & Schuster. He has allowed us to tell our story in the manner we felt best, while improving the final product in many important ways. Johanna Li of S&S helped prepare the final manuscript. Ingrid Li of WCS adapted traditional Rwandan art for use in the book.

Finally, we thank the Rwandan field staff—past, present, and future—of the Parc National des Volcans and the recently proposed Parc National de Nyungwe. Their dedication and commitment to the protection of these world-class sites throughout the most difficult of times is a source of personal inspiration and a model for global conservation.

*To our parents—Marion, Chuck, Mary, and George—
who raised us to follow our hearts and minds, however far away
they might lead us, and to our sons—Noah and Ethan—
whose passion for the wild gives us hope for the future*

Contents

List of Illustrations

Guide to African Words, Names, and Places

In the Kingdom of Gorillas contains names of places and people, as well as certain expressions, in Kinyarwanda and Swahili. These are Bantu languages spoken in Rwanda. Kinyarwanda is spoken by all Rwandans, whereas Swahili is a trade language spoken by many northern Rwandans and those who work in the commercial sector. Key pronunciation points are as follows.

The letter *a* has a broad *ah* sound as in *father* or *mama.*
The letter *e* is usually pronounced like *ay,* as in *say* or *play.*
The letter *i* is pronounced like *ee,* as in *feel* or *steel.*
The letter *o* is pronounced like *oa,* as in *oats* or *goats.*
The letter *u* is pronounced like *oo,* as in *moo* or *boot*
   (e.g., Uganda is *ooganda,* not *youganda*).

The combination *ny* as in Nyungwe is pronounced like the *ñ* in *señor.*
The combination *ng* is like the *ng* in *singer* (unless you're from Long
   Island). (Say *singer,* then say *inger,* then say *nger:* you got it.)
The combination *cy* is pronounced like *sh,* as in *shadow.*
The combination *cu* is pronounced like *chew.*

All Swahili words are accented on the second to last syllable. This system
   works reasonably well for Kinyarwanda, too.

Rwandans usually have two personal names, one Western (e.g., Jonas,
   Thérèse) and one African (e.g., Nemeye, Rwelekana). Use of either is a
   matter of personal preference.

13

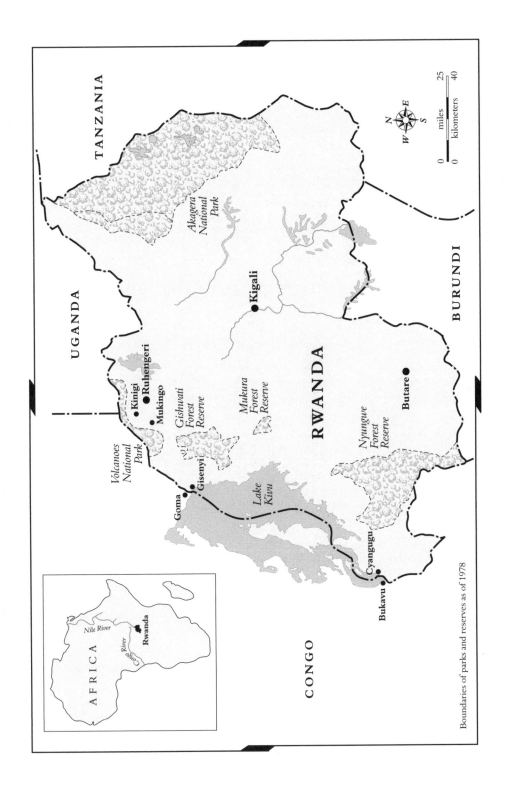

Boundaries of parks and reserves as of 1978

Section One

Under the Gun

Chapter One

The View from Bukavu

G O AHEAD. SIT NEXT TO HIM. Adrien deSchryver's suggestion was part tease, part dare. Amy hesitated, then smiled as she began to crawl toward Casimir, a massive silverback scowling at us from beneath a tree about thirty feet away. Surprised by her eager response, deSchryver grabbed Amy by the belt and pulled her back to his side. For the next thirty minutes, we watched Casimir and his family of gorillas from a respectful distance of fifteen to twenty yards. Thick bamboo limited our views to isolated body parts. The gorillas tolerated our presence, but several stayed completely out of sight and all were clearly nervous. Occasional screams ripped the still mountain air. Powerful smells and strange plants enhanced the sensory stew. We were elated.

Our pygmy guide, Patrice, was calm—seemingly bored—throughout our time with the gorillas. But on our hike back to park headquarters he grew more animated when another creature caught his eye. Patrice stalked his tiny prey until it led to an invisible target. Locating the entryway to the bee's nest, he ignored repeated stings as he ripped open the rich ground. Within a minute, he returned to our group with a wide grin on his face and large chunks of dripping honeycomb in his hands. DeSchryver grabbed some moss from a tree and showed us how to soak up the abundant honey like a sponge; Patrice preferred to eat the comb, larvae and all. Soon our faces were smeared with an indescribably exotic mix of flavors and substances. On that late summer day in 1973, we entered the land of gorillas and honey.

Five years later, we would experience the wonder of sitting peacefully among mountain gorillas in Rwanda—and the awesome responsibility of trying to save their population from extinction at the hands of humans. But in 1973 we were Peace Corps volunteers in eastern Congo, with much more to learn before we could make any meaningful contribution to conservation.

WE MET IN 1969 at Swarthmore College and married three years later. Two kids from small towns in upstate New York, we shared the best and worst of the late 1960s and early 1970s. The King and Kennedy assassinations, Vietnam, Kent State, acid rock, Earth Day, Women's Lib, the Generation Gap, and seemingly endless cultural conflict. Through much of that turbulent time, Swarthmore's Quaker tradition was a calming influence. While other campuses went up in flames, our passions were doused with a smothering blanket of Quaker understanding—and the admonition to use our learning and experience to go forth and make the world a better place. Following graduation, our budding interest in conservation and an urge for adventure led us to Africa via the Peace Corps. We weren't qualified for specialist positions in parks or wildlife management, so we joined more than one hundred other volunteers to be trained as the first teachers sent to Congo, which was then known as Zaire.

Our arrival in Africa was more eventful than expected. En route from London to begin our Peace Corps training in eastern Congo, we were scheduled to land in the Rwandan capital of Kigali and change planes for a short hop over the border to Bukavu. An overnight coup in Rwanda had closed the airport, however, and we were forced to make an unannounced landing at Uganda's Entebbe airport. This was not a reassuring prospect at a time when Field Marshal, General, and Dictator-for-Life Idi Amin Dada was ruling the former "Black Pearl of Africa" with an increasingly erratic and ruthless hand. Worse, Amin himself was due at the airport at any moment to welcome President Bongo of Gabon. After quickly refueling, we took off and headed west toward Congo. As Lake Victoria shimmered below in the early morning sun, our mood relaxed—until the pilot announced that we had been told to return to the airport. We later learned that Amin had seen our plane lift off and scrambled two MiGs to reinforce the order to return. Back on the ground, we watched as our suitcases, trunks, backpacks, and guitars were thrown from the cargo hold onto the runway. Ugandan security agents then collected our passports and ordered us to join our belongings on the hot pavement below. There, we were surrounded by at least twenty khaki-clad soldiers bearing AK-47s and persistent scowls. Parental premonitions about the foolhardiness of our African adventure began to seem all too real.

After two hours of exposure to the equatorial sun, we were ordered inside the small bunkerlike building that passed for Uganda's international airport. Within three years, its layout would be printed on the front page of every major newspaper in the aftermath of the Israelis' "Raid on Entebbe" to rescue a planeload of their own citizens seized by terrorists. Our experience was far less dramatic. We spent more than fifty hours of tension-filled boredom, talking among ourselves, nibbling at a suspect mix of green beans, green French fries, and green meat of unknown origin, while drinking rationed supplies of warm Bell lager beer. Ugandan airport workers cast furtive glances our way and rarely spoke. The few who dared to break the silence asked us not to judge their country by this incident.

Our curious captivity included the right to watch the Ugandan national news on the airport bar TV each evening. There we learned that we were "mercenaries bound to destabilize Rwanda." This was a strange charge to pin on a group of 112 young Americans, two-thirds of whom were women. To document the threat, Amin appeared in person early on the second morning to take a picture of our increasingly ragtag band with an Instamatic camera that looked ridiculously small in his beefy hands. It fell to President Mobutu Sese Seko of Congo to convince Amin that he was holding nothing more than a group of volunteer schoolteachers. Early on our third day, we were released and flown to the Congolese capital, Kinshasa.

Meanwhile, the July 5 coup in Rwanda proved to be bloodless, with no outside interference and apparently little internal resistance. Diplomats would later describe the event as a peaceful transition of power from a stagnant clique of southern Rwandan Hutu to their more dynamic cousins from the northern volcanoes region. We paid little attention to such details at the time. We would learn much more about the regional politics of Rwanda in the years to come.

BY SOME STROKE OF FATE, we were assigned to teach at a small school along the shores of Lake Kivu, partway between the town of Bukavu, where we had received our Peace Corps training, and the Kahuzi-Biega National Park, where we had first seen wild gorillas. We shared a house in Bukavu with another volunteer named Craig Sholley. Sitting on our front porch most afternoons, we watched dramatic storms sweep across the vast expanse of Lake Kivu, its picturesque islands framed by imposing mountains on all sides. The pulsing red glow of the Nyiragongo volcano dominated the evening sky, about seventy-five miles to the north. To the east, the Congo–Nile Divide rose to ten thousand feet to form the rugged backbone of Rwanda. The border with Rwanda was ten minutes from our house, but in the immediate wake of the coup it was rarely open. Congolese and Rwandan residents crossed back and forth between the border towns of Bukavu and Cyangugu, but foreigners were generally not welcome. When Rwanda did open its doors, visitors were required to be clean-shaven, short-haired, and well dressed. We joked that this would assure that no one but CIA agents could enter. Rumors held that ancient tribal animosities and other dark secrets were shielded by the closed border and Rwanda's perpetual cloud cover. Most young travelers who reached Bukavu while wandering across Africa were left to imagine Rwanda's attractions—including the exotic Nyungwe Forest, where an especially potent form of marijuana was cultivated by the mountain forest dwellers for whom Pygmy Thunder was named. Far to the north, a reclusive American woman was said to live alone among the mountain gorillas that made their home in the Virunga volcanoes. In 1974, we received our only firsthand look at the country during a short transit drive through Rwanda to neighboring Burundi, in the company of an American diplomat. Compared to Congo, it seemed a much less developed country, a simple society where earthen huts with thatched roofs dotted an overwhelmingly agrarian landscape. On that Sunday, though, Rwanda's dirt roads were filled with people: drab men in secondhand Western suits, accompanied by women in flowing robes of dazzling colors. We wondered if they were Hutu or Tutsi.

Adrien deSchryver was chief warden of the Kahuzi-Biega park. Six years earlier, when Congo's bloody civil war raged around Bukavu, deSchryver was a one-man force for law and order within the park. Poachers had already killed his brother and narrowly missed Adrien himself. Born in Congo (when it was a Belgian colony), with a Congolese wife and family, he was single-mindedly committed to the park and its gorillas. These were eastern lowland gorillas, or Grauer's gorillas as they were sometimes known from their scientific name *Gorilla gorilla graueri*. They were one of three recognized subspecies of gorilla. The lowland gorillas of west-central Africa were much more numerous, whereas the mountain gorillas on the other side of Lake Kivu were considered the most endangered of the three. But no one really knew how many eastern lowland gorillas existed in the Kahuzi-Biega park, let alone across their range in the surrounding Kivu highlands. DeSchryver wanted a census of the gorillas. He also wanted assistance with his effort to habituate Casimir's group and other gorillas for tourism viewing, which he felt was the only way to save the park from intensive poaching. We had discussed our interest in conservation during earlier visits, but were pleasantly shocked when he asked if we would help with the census and tourism. We jumped at the chance. The Peace Corps was supportive, too, agreeing to fund our positions in Kahuzi-Biega as long as we finished our original two-year commitment as teachers.

In many ways, we were fortunate that we began our work in Africa as teachers. If we had started in conservation, with strong pressure to save some park or species, we might have been quickly pulled into adversarial positions with local people and government officials. Instead, teaching brought us into constant contact with Africans and their view of the world. We saw how our students learned and came to understand reasoning and values that shaped their perceptions. We gained firsthand experience working with the dysfunctional Congolese education bureaucracy—and saw how pervasive corruption could crush individual initiative at a very young age. We became fluent in French and learned Swahili, a regional Bantu language that opened up a rich and rewarding world of contact with the large majority of local people who spoke no European language. Most of all, we were able to take our time and absorb the African way of life and culture that surrounded us. We tried to follow the advice of a Jesuit priest who had addressed our Peace Corps group toward the end of our formal training. *You will see many strange and different things over the next two years,* he said. *Always keep a question mark in front of your eyes and ask "why" before you judge something you see as wrong just because it is different.* It was excellent advice.

During the summer break between school years, we traveled overland by truck, boat, train, and bus to Tanzania and Kenya. There we saw the incomparable wildlife spectacles of the East African plains. We traveled and camped in ten different parks in habitats ranging from mountains to deserts and savanna grasslands to coral reefs. Our commitment to the cause of conservation grew even stronger. Yet our East African experience also reinforced our perceptions from Congo that local Africans gained little from tourism, while absorbing almost all of the direct costs of conservation. They were prohibited from hunting and other forms of traditional use on lands declared as parks, yet most conservation-related jobs went to people from outside the local area. Park revenues flowed straight into central government coffers. We wondered how Africa's vast biological wealth could survive in the face of overwhelming human poverty and growing pressures for economic development. It was a haunting concern. Like any important challenge, it was also an opportunity. We had originally thought that only Amy, with her degree in biology, could do serious conservation science. Faced with pressing questions of local perceptions and the true costs and benefits of conservation, Bill saw how his social science background was equally relevant. All that was left was to convince others that a multidisciplinary approach to conservation—one that considered the needs of both people and wildlife—was worthy of support. Such an approach was certainly needed around Kahuzi-Biega, and we felt that it would be of value everywhere we had been during our African travels.

⁓⌒⌒⌒⌒⌒

IN MID-1975, we returned to the United States to secure an affiliation with an appropriate graduate school. We chose the University of Wisconsin because of its strong tradition in conservation that stretched back to John Muir and Aldo Leopold, and its excellent field biology program, which had produced the world's foremost gorilla expert at that time, George Schaller. Wisconsin also had initiated a radical new program that offered an interdisciplinary Ph.D. in applied conservation science, perfect for Bill's interests. We intended to take a few courses in our respective programs, absorb as much as we could from faculty and fellow students, and return to Congo after one semester. Complications arose between the Peace Corps and the Congolese park service, though, and our return was delayed. Fortunately, a fellow grad student named Tag Demment had discussed our background and interests with Richard Wrangham, a rising star at that time in the field of primatology. Richard had just returned from a short stay at the Karisoke Research Center in Rwanda, where Dian Fossey was studying endangered mountain gorillas. Richard called to tell us that our applied work on gorillas and the problems confronting them would be of great value if carried out in Rwanda. That there was a need didn't surprise us. But his contention that Dian Fossey would welcome our work didn't fit with what we had heard about the strange recluse of Rwanda. Richard acknowledged her "eccentricities"—an extreme understatement, we would learn, even by his British standards—but reiterated the need.

When Wrangham secured a written statement of interest from Fossey, we modified our proposal to fit Rwandan circumstances and sought financial support. In 1977, there were only two U.S.-based conservation groups with significant international programs: the World Wildlife Fund and the Wildlife Conservation Society (then known as the New York Zoological Society). At the time, WWF was considered to have a more applied focus and we approached them first. In a private meeting in Madison, however, a senior WWF official regretfully told us that he had discussed the gorillas' plight with Fossey and concluded that "the situation was hopeless." This was a serious blow to our funding prospects, but also a spur to get to Rwanda before it truly was too late to help the gorillas.

We were not confident that our proposal to the Wildlife Conservation Society would be well received. WCS was certainly interested in gorillas: they had funded George Schaller's pioneering study almost twenty years earlier and had given some initial support to Fossey in the late 1960s. Part of our research would appeal to them, with its biological focus on feeding ecology, habitat use, and population dynamics: all essential factors in the mountain gorilla's survival equation. The "people" side of that equation was another matter, however. While we were convinced that social and economic factors were just as important as those in the biological realm, this was not a widely held belief in the world of conservation in the 1970s. Fortunately, WCS proved itself not bound by tradition, nor daunted by the task at hand. In September 1977, we received our full request of $11,850 for an eighteen-month project. All we needed was final clearance from Fossey and the Rwandans.

In October, we met Dian Fossey for the first time, at a hotel restaurant in Chicago. We were joined by another prospective researcher named David Watts. It was an awkward dinner, as Dian made small talk and watched the three of us eat while barely touching her own main course of grilled fish. She did, however, unwrap and consume at least a dozen pats of butter during the meal. She also dumped the entire bowl of sugar cubes into her handbag on leaving the table. Back in her room, Dian stretched out her six-foot-two frame on her bed, put on her reading glasses, and began to read our proposal out loud. She stopped occasionally to comment on minor points but raised no substantive concerns—except to note in passing that ecology was boring and that work with local people was hopeless. At the end of the evening, we set a tentative arrival date and asked if there was anything else we needed to do, such as requesting authorization for our work from the Rwandan park service. Dian said she would take care of everything.

Chapter Two

Why Are You Here?

Why are you here without authorization? Could Rwandan scientists go to Yellowstone, build a research station, and install themselves for as many years as they want? Without permits? Could they establish their own rules and violate your laws?

— DISMAS NSABIMANA, Director, Rwanda National Parks

If you see anyone in the park, shoot them.

— DIAN FOSSEY

ON FEBRUARY 3, 1978, Walter Cronkite began the CBS evening news with a report that Digit, a young silverback gorilla, had been killed by poachers in Rwanda. Few in the listening audience knew anything about Rwanda or its people, but millions had seen Digit reach out to touch Dian Fossey in a recent National Geographic film special. For them this was almost like a death in their extended family. For Dian, it was a death in her only family.

For us, Digit's death was a deep shock and a dark warning. We were still in the U.S., making final preparations to go to Rwanda to study and help protect its rare and endangered mountain gorillas. In less than a month we would have been sitting with Digit in his forest home. Instead, we were heading into a storm. And Cronkite's broadcast showed that when it came to gorillas that we knew by name, the world was paying very close attention to their fate.

Our arrival in Rwanda's capital of Kigali was uneventful. Tall *Eucalyptus* trees planted under the Belgian colonial regime still shaded its quiet streets, leaving the curious scent that we associated with cough drops in the air.

Somewhere, nearly 130,000 residents were going about their daily affairs, yet the urban bustle of a modern capital city was nowhere to be seen. Perhaps this reflected the fact that Rwanda in 1978 had fewer than three thousand registered motor vehicles in a country of almost five million people. One of those vehicles was an old Peugeot pickup truck doing double-duty as a *taxi brousse,* or bush taxi. After negotiating a rate—and declining the preferred seating offered to Amy inside the cab—we joined almost twenty other passengers for the five-hour northerly trek toward the Virunga volcanoes. Standing in the back and swaying together as we clung to a lattice frame of metal pipes, we passed the time in conversation, reactivating the "backcountry Swahili" we remembered from our Peace Corps days in neighboring Congo. As we talked about our planned work, we realized that our fellow travelers did not share the Western world's awareness of, nor concern for, the last of the mountain gorillas. Digit's death was not broadcast on Radio Rwanda.

The slow pace and 360-degree exposure from the back of the pickup gave us time to absorb the changing landscape. The initial stretch was unusual not only for its level paved road, but also for the intensive conversion of the Nyabugogo wetlands, which it bordered. The neatly ordered patterns of well-tended rice paddies completely filled the former marsh. Crowned cranes, black-necked herons, and a few sacred ibises appeared to enjoy full use of the altered habitat, though countless other species no doubt had been eliminated by the wetlands' conversion. At the end of the valley, an ugly concrete structure proclaimed itself home to a combined pig-raising and molasses-producing project. Funded by the United Nations Development Program, it produced an unforgettable, if indescribable, mixture of odors.

Leaving the Nyabugogo, our dirt track wound up a steep series of switchbacks before leveling off on an extended ridge. To our left, Rwanda's major river, the Nyabarongo, flowed red with a heavy load of sediments. Yet the surrounding countryside was surprisingly dry: a mosaic of pale browns and greens in which small plots of already harvested sorghum, beans, and corn were juxtaposed with permanent stands of coffee and bananas. Freshly tilled fields awaited the first rains. Halfway to our destination, we slipped into the narrow canyon of the Base River, where a few giant tree ferns stood in silent testimony to a former, more diverse forest community. But beyond that point, nature again yielded to the much greener farmlands of the rugged Ruhengeri highlands, where the rainy season had already begun.

Rounding a long, descending curve about four hours into our ride, we saw the Virunga volcanoes reveal themselves one by one, from west to east, towering over the northern landscape. Our fellow travelers helped to confirm the names which we had read in the works of George Schaller and Dian Fossey: *Karisimbi*—"the cowrie shell"—its often snow-capped summit rising to 14,797 feet; *Visoke*—"the watering hole"—with its magnificent crater lake; *Sabyinyo*—its jagged ancient rim evoking "the teeth of the old man"; *Gahinga*—"the hoe"—classically flat-topped like Visoke, with a distinctive breach on its western flank; and *Muhavura*—"the guide"—its 13,540-foot perfect cone visible from all corners of the country. Standing shoulder to shoulder, in camouflage cover of mixed light and clouds, the volcanoes that day seemed like solemn sentinels, guarding the last mountain gorillas that roamed their forested slopes.

In the outpost town of Ruhengeri, we bought some food and negotiated a ride to the park border with the driver of another pickup truck. His load of potatoes provided some relief from the bone-jarring grind up a primitive track of lava rock that led to the base of Mt. Visoke. This was the end of the road. Behind us, the land was cleared of all natural cover, replaced by well-maintained fields that abutted the park along an almost perfectly straight line of the kind that humans like to impose on nature. Above us loomed the more alluring chaos of the remnant forest of the Parc National des Volcans: our home for the next several years.

The foot trail to the Karisoke Research Center began with a narrow cut through a lava wall, where bits of rubbed-off hide testified to the recent passage of elephants—perhaps returning from a crop-raiding mission to nearby fields. After a brief, precipitous climb, the trail skirted the steepest slopes as it wound up and around the southern flank of Visoke. Along the way, it crossed a half-dozen streams and passed through an equal number of open glades. These grassy clearings attracted large numbers of Cape buffalo, whose ample dung testified to their presence. Gorillas avoided the clearings, though, perhaps as much because of the mud, which sucked at our boots, as the open exposure. Finally, after about an hour's climb and short of breath, we reached the research camp. Karisoke was named for its location in the saddle between Mts. Karisimbi and Visoke. The fantastic shapes of one-hundred-foot-tall *Hagenia* trees dominated the generally marshy environment and dwarfed the camp's six buildings. These were simple structures made of raised wooden platforms and frames, over which corrugated tin siding and roofs were nailed. The result was protection from rain, but not from the constant humidity and cold, which, at ten thousand feet, dropped to freezing on many nights. Tired and ecstatic to finally reach our new home, we felt no discomfort as we fell asleep under several sleeping bags that first evening.

The next morning, we set off to meet the gorillas. Dian Fossey was still in

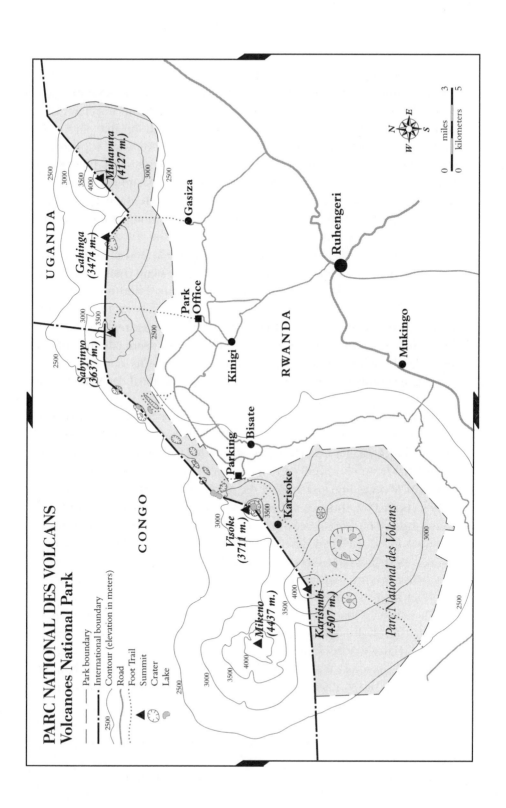

PARC NATIONAL DES VOLCANS
Volcanoes National Park

Park boundary
— ·· — International boundary
——— Contour (elevation in meters)
2500
——— Road
········· Foot Trail
▲ Summit
⬭ Crater
◦ Lake

UGANDA

CONGO

RWANDA

Muhavura
(4127 m.)

2500
3000
3500
4000
3000
2500

Gahinga
(3474 m.)

● Gasiza

Sabyinyo
(3637 m.)

3000
3500
3500
2500
2500

Park
Office

● Kinigi

● Bisate

Parking

Visoke
(3711 m.)

● Karisoke

3000
3500

3000

Mikeno
(4437 m.)

3500
4000
3000
3500
2500

Karisimbi
(4507 m.)

4000
3500

3000

2500

Parc National des Volcans

● Ruhengeri

● Mukingo

N
W · E
S

0   3 miles
0   5 kilometers

the U.S. to set up a fund in Digit's memory, so our guide for this first contact was Ian Redmond, a young British biologist who was acting camp manager in her absence. After barely passing Ian's first test—traversing a ten-foot log bridge across Camp Stream—we followed the edge of an extensive marsh west of camp, then climbed a series of small hills on the eastern flank of Karisimbi. There we quickly rejoined the trail where Ian had left the gorillas the day before. Tracking from that point was a straightforward procession through the gorillas' nesting site of the night before and on to that morning's fresh trail. The trail itself was a minor highway of trampled vegetation, punctuated by bowl-like depressions and neatly discarded plant matter where individuals had stopped to feed. Focused on the trail sign, and trying not to trip over the grasping vegetation, we failed to notice when we caught up with the group. There was no sound at all from the gorillas; certainly nothing like the repeated barks from Casimir that greeted our arrival when we followed his family almost five years earlier in Congo's Kahuzi-Biega park. Instead, we practically stumbled over a small juvenile who was sitting calmly to one side of the trail.

This was Group 4, one of several families named in the order in which they were discovered by Dian. Within minutes of our kneeling down where we could see most of the group, a two-year-old named Kweli approached. He was beautiful, with deep brown eyes that reflected both light and human kinship. His name meant "Truth." Kweli looked away shyly as he moved to inspect the new white apes more closely. His mother, Macho, sat nearby, calmly observing her infant with the dazzling eyes that earned her her Swahili name. On this first encounter, Kweli gingerly touched the hem of Amy's jeans, then twirled slowly away in a comic series of spins that brought him back to his mother's side. We looked at each other with wide eyes and full smiles. Inside, our hearts were pounding as our minds raced to place the experience in some kind of context. But there was no precedent for the wonder of such direct contact across species lines.

There was a sobering side to that first meeting and others over the next few days. Group 4 was Digit's group. Did the other gorillas remember the attack and Digit's death? His piercing screams must have accompanied the hail of blows from spears and machetes, and reached the ears of Kweli and the others fleeing the poachers' attack. How did they distinguish us from poachers? Why did they still tolerate any humans at all?

These questions hung in the air like the perpetual Virunga mist. Yet they didn't lessen the magic of sitting among wild gorillas, as if we were at a forest gathering with friends. On our second day, Kweli's seven-year-old brother, Augustus, sauntered up to Bill, paused only inches away, then pushed him gently backward and knuckle-walked across his chest before sitting down to feed on some choice celery nearby. He was clearly bolder than Kweli, and unfazed by his human observers. For four days, we lost ourselves in the Virunga wonderland. We saw gorillas every morning, and began to learn their individual personalities as well as their family relationships. We walked around Camp Meadow each afternoon and watched black-fronted duiker and bushbuck browsing along the forest fringe. Late afternoon was a good time to work on our Swahili and learn from the Rwandan trackers and other camp staff. At night we would relive the day's events and experiences. It was a heady time. But it came to an abrupt end with the arrival of a message that we were to return immediately to Kigali to meet with the director of the Office Rwandais du Tourisme et des Parcs Nationaux: the Rwandan national park service, or ORTPN, as it was commonly known.

D ISMAS NSABIMANA was a striking presence. Tall and trim, with a handsome face, he seemed comfortable and self-confident. On this day, he certainly had the upper hand. *Pourquoi êtes-vous ici sans autorisation?* he queried. The question of our presence and who had authorized it was fraught with potential, all of it bad. Dian Fossey could surely explain our apparent lack of authorization to work in Rwanda, we thought, but she was not with us. She had just returned to Rwanda from the U.S. the day before and was resting at the U.S. ambassador's residence. Still, we assured Mr. Nsabimana that she could account for our status. The problem, according to Nsabimana, was that none of her arrangements had been made with Rwandan authorities.

"Could Rwandan scientists go to Yellowstone, build a research station, and install themselves for as many years as they want? Without permits? Could they establish their own rules and violate your laws?" Nsabimana's anger grew. In part, he was getting back for the adverse international publicity heaped on Rwanda in the wake of Digit's killing, and we were much easier targets than the famous Dr. Fossey. But he was also right. No government authorization, in fact, had ever been requested or granted for our work. This was a matter of principle for Dian, as we would learn, who only reluctantly recognized a limited form of Rwandan sovereignty over the park, and none over "her" field station.

It certainly was not the case that we hadn't prepared for our mission. We recounted the origins of our work as Peace Corps volunteers in Congo, the change of plans to work in Rwanda, our grant from the Wildlife Conservation Society, and the Chicago meeting at which Dian had said she would take care of everything. When we reached this point in our account, Nsabimana responded with a dismissive laugh and shared a knowing look with an assistant seated nearby. He appeared to find considerable humor in the idea that we would expect Dr. Fossey to make any arrangements with his office. In any event, there was little point in arguing our case further. All we could do was apologize and assure him that we would cooperate fully with ORTPN in the future. Nsabimana gave a brief concluding lecture on Rwandan expectations and, after a two-day wait, we were granted three-month renewable research permits.

The tone of our discussion had actually started to change when we described our research proposal. Nsabimana seemed unappreciative of the research then being conducted at Karisoke, with its primary emphasis on gorilla behavior. In contrast, he saw our proposed work on gorilla ecology and population trends as responsive to some key park management needs. Our parallel focus on human issues addressed even more fundamental Rwandan concerns. We knew that Fossey's research on the social lives and personalities of gorillas had generated global attention, sympathy, and support; now we had an opportunity to demonstrate the relevance of our own more applied research to local and national interests, as well.

We debriefed Dian on our ORTPN discussions the next day at a picnic arranged by the U.S. ambassador, Frank Crigler. We had driven about an hour east of Kigali to meet at an embassy retreat on Lake Muhazi, a marsh-lined body of water surrounded by agricultural and grazing lands. Dian was dismissive of Nsabimana, but otherwise said little about our meeting or anything else. After lunch, the ambassador took us out in a sailboat and told us of his concern for Fossey's emotional state. He said he was her friend, but also felt that her actions, while well intentioned, were not always in the gorillas' best interests. He wanted to be kept informed of events at Karisoke and said that he would do anything he could to help Dian and the gorillas. Not yet forty, he impressed both of us as bright, caring, and exceedingly confident. He certainly was persuasive, as Bill learned when he was talked into taking a spin behind the embassy motorboat on water skis. Only when he got up on the skis did he begin to wonder about the hippos and other creatures in the water. He also noticed a group of small herdboys propped on their wooden staffs, each with one leg raised like so many shorebirds, watching intently as he was whipped across the surface at the end of a motorized tether. Bill could only imagine their thoughts. Amy wisely resisted repeating the surreal experience.

OUR RETURN TO KARISOKE was a welcome relief from the stress of the ORTPN meeting, and the absurdity of Muhazi. Now we could get to work. One priority was to carry out the first gorilla census in more than five years. The last count, conducted over three successive summers in 1971–73, had shown a dramatic decline since Schaller's total of roughly 450 individuals in 1960. Not only had that number crashed to around 260, but the percentage of young had also dropped significantly. Bill's first major job was to assess the current state of the Virunga population through a comprehensive new survey.

Following a few more weeks of familiarization with the gorillas and the forest, Bill set out on his first official census foray on March 4. The target was Mt. Karisimbi, the largest and westernmost of the Rwandan volcanoes, which he intended to census in a systematic progression over several months from west to east. Before leaving, he went up to Dian's cabin to brief her on his plans. She seemed little interested in the census details, asking instead if he wanted to take a gun. When asked why, she simply said, "If you see anyone in the park, shoot them." The suggestion was unequivocal, as was the offer of the pistol, which lay on a table between them. The lanky form of Dian Fossey was folded forward over the table, as she squinted over her reading glasses to gauge Bill's reaction. His manhood was apparently on the line; so were some more important ethical and legal issues about foreigners shooting people for uncertain offenses in their own country. Bill silently recalled the experience of Alan Goodall, who had followed Dian's same blunt advice only three years earlier. Goodall shot and seriously wounded a Rwandan whom he suspected of poaching. Within days, at least five gorillas were herded out of the park in the same area and stoned to death by irate family and friends of the elderly man. Goodall wrote that he saw no connection between the two events, but several others familiar with the killings disagree. Those five dead gorillas from the Cundura region of Karisimbi would not be counted in this census. The gun stayed on the table.

Bill's first contact with a totally wild group of gorillas came on his second day of work on Mt. Karisimbi. Named that day for the river where Bill found them, the Susa group would go on to become one of the great mountain gorilla success stories, eventually accumulating an extended family of more than thirty members. At that time, however, it had the peculiar composition of one adult silverback male, a young silverback, a blackback male, and a young adult of unknown sex. It was an inauspicious start for a family.

That first contact was especially notable for the long-drawn-out "hoot" series from what appeared to be the senior silverback, whom Bill named John Philip because of the nearby Susa River. From behind a partial veil of vines some twenty yards away, he would slowly shake his head, give off as many as fifteen plaintive hoots—owl-like, but increasingly rapid with a rising crescendo—then leap to his feet, give a quick *poketa-poketa* chest beat, and return to his seated position, peering out as if to assess the reaction to his performance. Though the other gorillas didn't flee, this recurrent display seemed to keep them on edge. The young adult stared constantly at the intruder and chest-beat almost as frequently as the silverback. Yet she or he remained in full view barely thirty feet away, feeding almost constantly for more than an hour. This exceptionally long contact also allowed Bill to draw noseprints—the unique set of markings above each gorilla's nostrils—for each of the four individuals, including the distinctive "double seven" marks that would give the young adult its Swahili name of Saba Saba.

Working with Ian Redmond and a Karisoke tracker named André Vatiri, on the next day Bill found no further sign of gorillas, other than old trails and nests of the Susa Group on Karisimbi's eastern or southeastern flank. On March 7, the team planned to move its base camp to the south, but Bill came down with a fever and lay in his tent for most of the day. By nightfall, he was sure that he had malaria, probably contracted during our trip to Kigali to meet with Nsabimana. After three previous bouts in Congo, he knew the symptoms well. He also knew they would quickly get much worse if he didn't hurry back to our cabin and take a curative dose of chloroquine. Early the next morning, Bill set off to find his way back to Karisoke, leaving Ian and Vatiri to continue the Karisimbi census. Growing fatigue made the initial stint hard going, as he had to climb up and over a series of small craters. Once he was on the other side, a thick fog moved in to obscure the view. It occurred to Bill that traveling alone through an unknown forest while seriously ill was not one of his better ideas, but returning to camp proved to be a fairly straightforward process of walking or stumbling downhill toward the Karisimbi–Visoke saddle below. By the end of a four-hour hike, the malarial haze was almost as thick as the mist, and Bill had visions of Cape buffalo behind every bush. It was a surprisingly exhilarating experience, but also a great relief to finally reach our cabin.

A few days of rest, fifteen hundred milligrams of chloroquine, and many cups of hot tea nursed Bill past the worst of the malaria. He wasn't strong enough to go back to census work right away, but he was able to accompany Amy a few times when her study group came close to the field station. Her work with the semihabituated Group 5 was coming along well. As she came to know these gorillas and their personalities, she was beginning to believe that she could actually carry out her ambitious research plan. It was a great opportunity to spend a few hours together with Amy's new extended family.

Field Notes: Leave contact/nest site in V-18 at 11:35. Continue E, NE near Susa, crossing twice before staying on S side. Go over N edge of Campbell Mlima before descending into bamboo. Pass several areas with choice gorilla vegetation but see no other signs of their presence. Arrive Guard Station 13:20. IMR arrives 13:38, and we discuss situation. Decide to return for more nest counts and possible contact. Return via more northerly route; again passing first through bamboo zone where signs of cutting for basket-making purposes can be seen for 30m into the Parc. No further signs of gorillas.

Buooas in V-18 by Susa.

INDIVIDUAL DESCRIPTIONS/NOSE PRINTS:

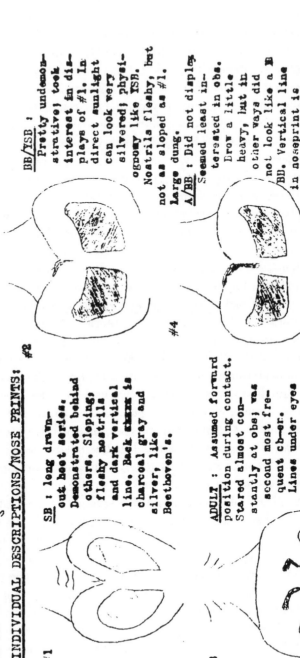

#1

#2

#3

#4

SB : long drawn-out hoot series. Demonstrated behind others. Sloping, fleshy nostrils and dark vertical line. Back chevron is charcoal gray and silver, like Beethoven's.

ADULT : Assumed forward position during contact. Stared almost constantly at obs; was second most frequent cb-er. Lines under eyes quite deep.

BB/YSB : Pretty undemonstrative; took interest in displays of #1. In direct sunlight can look very silvered; physiognomy like YSB. Nostrils fleshy, but not as sloped as #1. Large dung.

A/BB : Did not display. Seemed least interested in obs. Brow a little heavy, but in other ways did not look like a BB. Vertical line in noseprint is striking from all angles.

THE INVOLUNTARY IDYLL couldn't last, and the Fates soon ordered another large dose of reality to speed Bill's recovery along. The Rwandan response to Digit's death and its aftermath had been percolating for three months. Dian and her Rwandan staff had already extracted a certain amount of revenge by capturing and torturing one suspected poacher. Yet while this vigilante action drove one Karisoke researcher to quit and return to the States—a recently arrived anthropologist from Michigan who was appalled by what he saw—it did not seem to greatly concern government authorities. What did concern them was growing pressure from the American embassy, led by Ambassador Crigler, who was well aware of Digit's international stature. This pressure, in turn, was reinforced by thousands of postcards from irate citizens in England and the United States. Few of these cards showed any concern for Rwanda, its people, or their priorities, but they were blunt in their demands for action to protect the gorillas.

In February, the government of Rwanda had named its senior legal representative in the Ruhengeri prefecture, which included most of the park, to coordinate a response to these concerns. In one of his first acts, Paulin Nkubili presided over a summary trial that produced a four-year prison sentence for one of Digit's suspected killers. The swift conviction and exceptional sentence for a poacher were intended to send a strong signal, and they did. Another suspect fled across the border to Congo. Nkubili's next step was also a radical departure from past practice. In late March, he organized a meeting at Karisoke with the *conservateur,* or chief warden, of the Parc National des Volcans, the political leaders, or *bourgmestres,* of the six communities that border the park, Fossey, and himself. Bill was invited to attend as a translator, since Dian spoke almost no French.

It was the first time that any of the political leaders had been to Karisoke, and the first time that the *conservateur* had been invited into Dian's cabin, where the meeting took place. The social and political dynamics were complicated, to say the least, and the meeting was a series of steps forward and back. *Conservateur* Banzubase claimed to be doing his best to protect the park, but felt constrained by a lack of men, equipment, and money. He also felt excluded from the region around Karisoke. This was a jab at Dian's employment of what the Rwandans considered a private mercenary force to patrol roughly 10 percent of the park, or the six-square-mile area that surrounded the research station. Dian countered by saying that park guards were not excluded, only rarely seen. She then went on to make it clear that she felt her people did a better job, and did so without fancy equipment or better pay.

Dian also noted that Karisoke patrols spent much of their time in the Congolese sector of the Virungas: a neighboring national park★ whose international boundary was less than a mile from Karisoke. Although hurt by their own lack of control over Karisoke, none of the Rwandan park or political authorities seemed at all bothered by this violation of Congolese sovereignty. Perhaps they recognized that Karisoke patrols could help to protect Rwandan wildlife beyond their border; perhaps they just didn't care. Left unsaid was the fact that Digit was actually killed in Congo: technically beyond the range—and the responsibility—of Rwandan park patrols blamed for their failure to protect him. Our surprise at learning this from Ian a few weeks earlier was surpassed only by the others' lack of interest in this seemingly important detail. Such was life on a remote frontier, where few could locate the international boundary.

The park patrol discussion came to an end with a final swipe at the *conservateur*. Dian had been casting covetous looks at two nattily dressed and physically impressive Blue Berets—the elite of the Rwandan military—who had accompanied Nkubili to the meeting. She then suddenly suggested that she would welcome patrols by "men like them." The open invitation for military intervention was a shocker, and Bill first asked if Dian was sure before he translated the suggestion for others. Nkubili seemed to grasp the potential for all kinds of trouble in this, and quickly said he would take up the matter with other officials at "the highest level." It was the last we would hear of the idea. Somewhat shaken at the prospect of losing what little turf he controlled, the *conservateur* responded with a pledge to do more with the resources at his command.

The agenda next moved to matters outside the park involving local people and authorities. Dian commended Nkubili on his conviction of the one poacher, but appealed to him to do even more—and to give longer sentences. The *bourgmestre* of Mukingo shook his head at this, gave a derisive snort, and said in French, "You put more value on your animals than on people." Dian glared back at him and stated that Rwanda had "plenty of people—too many." Before Bill could make an editorial decision on translating that comment for the others, she then turned to Nkubili and raised the ante, asking what he would do with politicians who protected poachers. This was translated, but the *bourgmestre* ignored the pointed reference and returned to his earlier theme. *There is not one Rwandan who could value gorillas more than his own*

★ The mountain gorilla habitat in the Virunga volcanoes is spread over contiguous reserves in three different countries: Rwanda's Parc National des Volcans (sixty square miles), Congo's Parc National des Virunga (eighty square miles), and Uganda's Mgahinga Gorilla National Park (twenty-eight square miles).

people . . . we need that land for farms . . . the park gives us nothing, it's just a place for you abazungu [foreigners] *to play with your gorillas . . . when they are all dead you white people will go away . . . remember you are in Rwanda and this is our land.*

The words came out in a torrent, in a tone both menacing and mocking, before Nkubili cut him off. It was his job to end the meeting on the best note possible, so he talked of progress already made, the historic nature of our discussion that day, and the need to follow talk with more action in "our common interest." As his words were translated, Dian stared out the large window to her left. Bill wondered if she was thinking about Digit, who lay buried within her sight less than thirty feet away, or the *bourgmestre*'s challenge. She made no further response except to thank Nkubili and escort the group to the door. Outside the cabin, one of the *bourgmestres* apologized to Bill for the comments of his colleague from Mukingo. They agreed that it would be good to talk again, which several of us did over the coming years. But the group never met again.

Later that week, Dian said that she liked "Uncle Billy," as she had taken to calling Nkubili, and thought he was quite honest.★ She was most excited when discussing the Blue Berets, though: "real men" in her worldview. We knew it was going to take a while to appreciate how Dian formed her judgments of others, but nothing prepared us for her bombshell of a few days later. Almost bursting into our cabin one afternoon, she said that her "spies" had important information about the *bourgmestre* of Mukingo. Asked if he remembered the large, irregular scar on his forehead, Bill said that its jagged shape—along with his combative personality—made him think it came from a broken bottle in a bar fight. But the spies had a different answer: *he had had a lobotomy.* Bill first paused, then started to laugh. But Dian's sad eyes and pursed smile revealed a hurt inside, maybe not because she wasn't believed but because her illusory bubble was so quickly and rudely burst. Meanwhile, as we learned more about the *bourgmestre* of Mukingo's involvement with poaching and illegal clearing of parkland, it was increasingly clear what a powerful and dangerous—and fully capacitated—opponent he truly was.

★ Years later, we learned from Rwandan staff at Karisoke that Nkubili's "honesty" may have been reinforced with cash payments from Dian.

Chapter Three

Mweza

THE TRAP IS BLIND TO ITS VICTIMS. No hunter takes aim with a bow or gun. The trapper selects a likely animal path, then conceals dozens of cable-wire snares along the trail, each attached to a spring mechanism such as bent bamboo. Whatever steps into the trap triggers the powerful spring. The noose is jerked tight and the sharp wire slices through skin and into the bone as the victim pulls and twists to get free. The struggle is generally futile for duiker and bushbuck, at issue only whether they die before or after the poacher returns.

Gorillas are not targeted by trappers. Rwandans will not eat the meat of this or any other primate, and adult gorillas can free themselves by ripping off the wire by brute force. The grisly evidence of their crude removals can be seen in the frequency of maimed and crippled individuals in the Virunga population. Young gorillas lack the strength for such an escape, however, and can be captured alive. In 1978, this fact presented poachers with an opportunity to sell the infants on a cruelly active market, driven entirely by foreign demand. Some of the buyers were sadly misguided seekers of exotic pets. Others represented illicit animal collectors and unregulated zoos. All contributed to gruesome individual suffering and threatened the survival of the entire mountain gorilla population.

In mid-February, we heard of a young gorilla held captive just across the border in Congo. Alain Monfort, a Belgian advisor to ORTPN, the Rwandan national park service, said that the gorilla was seized from poachers as they tried to cross the border and that it was being kept at Congolese park headquarters. Bill reported this to Dian, but she dismissed Monfort as a non-

credible source. A few weeks later, her "spies" reported that the captive animal was a chimpanzee. But when we traveled to Kigali with Dian to renew
our visas in mid-March, Monfort again told Bill that it was a gorilla and that
he and *Conservateur* Banzubaze planned to see it the next day. Based on information that the gorilla may have originated in the Parc des Volcans, their goal
was to retrieve and return the gorilla to Rwanda.

That was enough for Dian—and another blow to the credibility of her
spy network. She whisked us into her Volkswagen van and designated Bill as
driver to race Monfort to the border. Dian had a surprisingly acute case of
acrophobia, which made her reluctant to drive on Rwanda's escarpment
roads and precluded fieldwork in many of the more rugged parts of the
Virungas. We made good time, and the fact that our passports were still at the
immigration office in Kigali posed little problem when Dian paid the appropriate bribes, or *matabiches,* at the twin border posts of Gisenyi, Rwanda, and
Goma, Congo. We were too late to reach the park before sundown, though,
so we spent that night at the home of Noella DeWalc, a friend of Dian's.
Noella, an artist, and her businessman husband, Michel, lived in an old colonial house on the outskirts of Gisenyi. The interior mixed bohemian and
aristocratic tastes, with a spectacular view of Lake Kivu, its myriad islands,
and a surrounding wall of mountains. It was a very peaceful beginning to an
extremely unrestful series of events.

The next morning, we hurried north past the still smoldering volcanoes
of Nyiragongo and Nyamuragira. The rough lava track and a landscape
scarred by multiple flows testified to the volcanoes' prodigious and recurrent
activity. Once around the western base of Mt. Mikeno, we arrived at
Rumangabo, the main headquarters of Congo's Parc National des Virunga.

The *conservateur* walked out to greet us, then led us across a courtyard to
the object of our interest. A guard opened the wooden door of a dank, pu-
trid-smelling shed. He picked up a long pole, then poked at the gorilla until
it hobbled out into the daylight. Surrounded by a handful of park guards
"armed" with sticks and poles and more than a dozen local onlookers, the
young female sat exposed in a bare, muddy clearing streaked with the yellow-
green of her own diarrhea. As she limped around, it was not clear whether
she had lost her foot or simply lost its use. None of us could bear the sight of
this poor, tormented creature and Dian insisted we begin discussions immediately to secure her release. First, however, we asked that a more sanitary
room be made available for the gorilla, with whom Amy had decided to stay.
With much laughter and shaking of heads, the Congolese opened another
room. Amy led the gorilla inside, closed the door, and knelt down a few yards

from her. A soft "belch" vocalization brought the young gorilla immediately into Amy's arms, and tears to Amy's eyes.

The gorilla—probably a four-year-old, though her state of emaciation made her look no more than two—had been caught in a trap at least seven weeks earlier. Yet neither the poachers nor the Congolese park personnel had removed the metal snare from her left leg. The result was a severe and possibly gangrenous infection of a large, pus-filled area around her ankle. Dian argued that removing the wire at that point might cause more harm than good and asked that the gorilla be transferred immediately to her care. The local authorities seemed only too happy to be rid of their charge, realizing that they would be held responsible if it died on their watch. However, the transfer to an *umuzungu*, or foreigner, especially one who would take it to another country, was beyond their power. This required the personal approval of President Mobutu.

For the next thirty-six hours, Bill again served as translator for Dian as she and the senior park warden negotiated for the gorilla's release with top government officials in the distant capital of Kinshasa. The negotiations were surprisingly direct, though hopelessly complicated by the poor quality of two-way communication across the thousand-mile expanse of the Congo Basin. At one key point, the tightly rationed electricity for the nearby town of Goma was turned on so that we would be able to send and receive an exchange of telex messages.

Meanwhile Amy remained with the gorilla, whom we had named Mweza: loosely translated as "can do." During the day, Amy would take her for brief walks in the compound, then go into nearby stands of natural vegetation to collect and bring back preferred gorilla foods. The clinging vine *Galium* was plentiful but required too much strength to chew. Mweza most appreciated the wild celery. The celery also increased her natural fluid intake and roughage as she was taken off her captive regime of diarrhea-inducing powdered milk mixed with unclean water. At night, while Bill and Dian remained in Goma, Amy and Mweza slept curled in each other's arms on a straw mattress in a windowless storage room: the only available accommodation for a mixed species couple. Lacking our own sleeping bags, she appreciated two newly purchased wool blankets that helped to ward off the damp chill of the cold Virunga night.

Late on the second day, we received Mobutu's authorization to take Mweza back to Karisoke, with the promise to release her to the wild when her health improved, or to return her body to the Congolese if she died. We all agreed that carrying her back over the volcanoes on foot would avoid any further problems at the borders—or at least the payment of much larger bribes as we tried to return without our own passports *and* with an undocumented gorilla. So, while Dian hired a chauffeur to drive her back to Rwanda, we spent the first of several nights in our new ménage à trois.

At dawn the next morning, Virunga park personnel drove us several miles to the edge of the forest. From there, we began the long hike back to Karisoke, escorted by eight armed Congolese guards. Before leaving we negotiated with a local woman to buy a cloth *panya,* in which we wrapped the young gorilla on our backs, like an African child. It proved to be an excellent solution for carrying Mweza, although we had problems arranging the sling in a way that didn't cut too deeply into our shoulders. We were sure that an experienced African mother could have straightened us out in no time, but our male escorts professed no knowledge of this apparently gender-specific skill. Fortunately, the weather cooperated and spectacular views helped to distract us as we climbed steadily higher on Mt. Mikeno. To one side, the trail skirted an impressive gorge that cut deeply into the mountain's flank. Dense stands of giant bamboo, with stems six inches in diameter, mixed with forest vegetation that was much more diverse than on the Rwandan side of the range. Above ten thousand feet, the terrain flattened out and we began to pass through alternating blocks of forest and open vegetation. As we emerged into one of the larger clearings, we were surprised by the sudden appearance of a building on our left. Then recognition set in: not because we had ever been there before, but because we knew the remarkable history of this structure, or at least of some of its more storied inhabitants. We were at Kabara.

Kabara's first resident had been the American naturalist Carl Akeley. Akeley came to the Virungas in 1921 to join in a binge of gorilla hunting that was legitimized by its practitioners for the purposes of scientific collection. In the twenty years since Oscar von Beringe's discovery of the mountain gorilla in 1902, forty-three gorillas had been killed or captured by Western-funded expeditions in the Virungas. Five individuals that Akeley shot can be seen to this day in a diorama at the American Museum of Natural History in New York. Moved by subsequent concern for the fate of the remaining mountain gorillas, however, Akeley became a tireless crusader for their complete protection. In 1925, his efforts were rewarded when the Belgian colonial government protected the western half of the Virunga range as the Albert National Park. Later expanded in 1929 to cover all of the volcanoes, Africa's first national park was established "to make the world safe for mountain gorillas." It was a noble and timely gesture, which Akeley hoped to follow up with the first-ever field study of the species. Soon after establishing his base camp in 1926, however, Carl Akeley died of malaria and was buried by his wife, Mary, at the edge of the Kabara meadow.

Mary Akeley went on to write about her husband's work and the natural history of the region. The pioneering long-term study of the mountain gorilla, however, awaited the arrival of George Schaller more than thirty years later. Working from the cabin at Kabara during an eighteen-month period in 1959–60, Schaller was the first to follow and observe wild gorillas on an extended basis. His published observations not only laid the foundation for all future gorilla work, but also set the standard for an entire generation of modern field biologists.

Dian Fossey consumed all of Schuller's accounts and challenged herself to "out-Schaller Schaller." Leaving behind an unsatisfying career as an occupational therapist, she arrived in eastern Congo in 1967 with no formal training, but with an exceptionally strong will to succeed. For the purposes of setting up a permanent site for long-term gorilla research and monitoring, she, too, selected Kabara. But the overall political situation in the region had deteriorated dramatically by that time. Even as Schaller was completing his work, the colonial era was lurching to an unglorious end. The wave of independence that was sweeping across Africa in the early 1960s washed over the Virunga highlands. The Belgian Congo became the Congo Republic in 1960 and the U.N. mandate territory of Ruanda-Urundi separated into the independent nations of Rwanda and Burundi two years later. Both Congo and Rwanda entered subsequent periods of political turmoil, and neither could pay much attention to their now divided responsibility for the Virungas. Fossey's arrival in the region could not have been more poorly timed, as a bloody civil war was reaching a most uncivil climax in eastern Congo. Physically abused and threatened with death, she was fortunate to escape with her life—and the opportunity to start over across the border in the relatively stable environment of Rwanda in September 1967.

Images of Schaller and Fossey loomed large that day at Kabara. And the ghost of Akeley was very much on our minds, too. Vandals had cracked open his burial crypt, the headstone of which lay broken in half across the now water-filled vault below. The inside of the cabin also showed considerable abuse, but it still appeared structurally sound. Nothing could change the beauty of the view outside, however. Shrouded in mid-afternoon fog and draped with mosses and epiphytes, giant *Hagenia* trees emerged as shadowy silhouettes on the forest fringe. The bright yellow flowers of hundreds of much smaller *Hypericum* trees added a rare splash of color. The clearing itself was an unusually large expanse of spongy marsh, dotted with rocky outcroppings and grassy hummocks of drier land.

At that point, aesthetic and historical considerations could have conspired with our own fatigue to keep us at Kabara overnight. This was certainly the preferred option of our Congolese guards. Mweza's health, however, required that we continue on to Karisoke, rather than endure the freezing nighttime temperatures of Kabara without any blankets. We talked the guards out of a threatened strike by asking how President Mobutu might react if the gorilla died, then set off again in the late afternoon. The final leg was unnecessarily long, as our guides claimed to lose their way after nightfall, perhaps to show us that their idea of staying at Kabara was a good one after all. With no alternative, we stumbled on through the heavy underbrush and soggy clearings until the lights of Dian's cabin somehow appeared in the distance. We were completely exhausted when we arrived just before midnight, almost sixteen hours after we had set off. But we were also energized by the day's adventure. So, as Dian examined the gorilla, we talked about what to do next.

Dian kept Mweza the first night, then decided that the gorilla should stay in our cabin and continue to sleep with us. She was frail, with sunken cheeks that accentuated her soulful, imploring eyes. During the day, we took her outdoors when it wasn't raining, and carefully reintroduced her to natural gorilla foods. She walked slowly with a pronounced limp, clearly pushing through considerable pain. We also gave her a sequence of antibiotics as instructed by Dian.

The following week was a roller coaster, as Mweza would seem in quite good spirits one day, only to regress the next. Our shared experiences had bonded us in ways that we could never have imagined, and our own moods mirrored her progress and setbacks. A major breakthrough seemed to come on March 24, Amy's birthday, when Mweza began to feed herself. Blackberries were her special favorite. We cried in joy when she climbed up on our desk and began playfully pulling books and papers onto the floor. But she could not fully recover until something was done about the wire and related infection, which required professional treatment. Here, we ran into a brick wall of incomprehension. French doctors at the Ruhengeri hospital, less than two hours away, had offered to help with gorilla medical problems at any time, yet Dian refused to allow any treatment for Mweza. At first, we thought this was because of a Karisoke policy against any human intervention with gorillas. Yet poachers had caused this situation and we were already intervening heavily. Dian did propose inviting a Belgian nurse, a friend named Lolly, to come up to amputate the infected limb. We argued that this was unacceptable, but were overruled. Lolly was due to arrive on March 30. Only later would we learn of the failed love affair between Dian and the

French director of the Ruhengeri hospital that made her unwilling to ask for help from this obvious source.

Without appropriate care, Mweza's condition worsened, as did our relationship with Dian. On the afternoon of March 28, Dian appeared as Bill was sitting with the resting gorilla outside our cabin. She had clearly been drinking and began to accuse us of not feeding or medicating Mweza. She then asked Bill to get the bottle of liquid antibiotic from our cabin. Holding Mweza's head back, Dian poured the pink liquid down the young gorilla's throat until she gagged violently. Next she tried to stuff some *Galium* into her mouth, but Mweza's teeth were now clenched. As she tottered back up the trail to her cabin, Dian mumbled, "She's dying."

A few hours later, Mweza stopped breathing as the three of us lay huddled together on the floor of our cabin. Without thinking, Bill began to give her mouth-to-mouth resuscitation while Amy pushed rhythmically on her chest. To our amazement she began to breathe. We sent a message to Dian to tell her what was happening, but there was no response. Less than an hour later, Mweza again stopped breathing. Resuscitation worked again, though this time Mweza was so startled when she came to—not surprisingly, given her state and the white face looming over her—that she bit Bill's lower lip as he finished a breath. With visions of flapping flesh where his lip had once been, he pried himself loose before the terrified youngster could do any real harm. The next time resuscitation was needed, we discovered that a mouth-to-nose technique works much better, given the flat, disklike shape of the gorilla's nostrils.

As the night went on, newly arrived researcher David Watts, whom we had met in Chicago with Dian, joined in our vigil. We repeated the resuscitation effort two or three more times. In the end, nothing could help. Not the ridiculous ice packs and hot water bottle that Dian sent down after several requests for help; certainly not her own drunken, incoherent appearance at our cabin around midnight, when she again accused us of neglect before Bill removed her from the cabin. Mweza died before dawn.

MWEZA WOULD NOT BE the last gorilla that we would see die, many under more horrible circumstances. But she inspired the most grief—and anger. Anger at our own helplessness, anger at Karisoke's lack of any system to provide proper care, anger at the deeply disturbed mental state of Dian Fossey, and anger at poachers and a world that could put a bloody price on such intelligent, feeling, and beautiful beings.

Over our first hundred days in Rwanda we were exposed to the roles of poachers, politicians, and individual personalities in the life-and-death struggle of the gorillas. Events ran together in a dizzying blur. But it was clear that we weren't going to be doing research in an ivory tower. The real world didn't beckon, it beat down our door.

Chapter Four

Close Encounters

DESPITE MANY DIFFICULTIES, our first few months in the Virungas were also a time of great wonder and beauty. The lives of gorillas and the forest world in which they lived stood in sharply favorable contrast to the problems of the human world.

By 1978 much was already known about gorillas. George Schaller had published *The Mountain Gorilla* fifteen years earlier, followed in 1964 by his widely read popular account, *The Year of the Gorilla*. Schaller's books were the gospel as we knew it and they retain much of their vitality today. It would be another five years before Dian Fossey published *Gorillas in the Mist,* but the world was already very familiar with her work through two photo-studded magazine articles and a full-length television documentary by the National Geographic Society. Appearing at a time in the early 1970s when the emerging environmental and women's movements were hungry for heroes, these media productions made Dian a household name and her gorillas a cause célèbre. In the process, they destroyed several lingering myths about the species and turned King Kong into the Gentle Giant of the Virungas.

We had digested all of this information and more before our arrival in Rwanda. But nothing could have truly prepared us for our time in the presence of such exceptional creatures.

AMY SPENT FAR MORE time with the gorillas than Bill. Her research on feeding ecology and habitat use required a minimum of five to six hours of direct observation each day, and she ultimately logged more than two thousand hours among them. The initial months were a time of gradual introduction, though: a two-way process of learning about each other that created the bonds of trust and understanding that are essential for any effort to reach across the gap between species. This is perhaps especially true for the relationship between gorillas and humans, where the gap can appear so small as to create the dangerous illusion that it does not exist at all.

The exceptional gorillas with whom Amy spent most of her time were known by the unexceptional name of Group 5. As a research group, it was secondary to Group 4, which had been Dian's primary focus, and not all of its members had even been identified. Still, Group 5 was partially habituated to the presence of observers, thanks to the efforts of previous researchers Sandy Harcourt and Kelly Stewart. Their work, however, like Fossey's, had focused on gorilla behavior and consisted almost entirely of observations of social interactions during the gorillas' midday rest period. They had not tried following the gorillas of Group 5 closely while they moved. Harcourt encouraged Amy's research, but advised that the gorillas might not tolerate it. Following at close quarters was exactly what was required, though, if we were to advance our understanding of gorilla food requirements and preferences—and, ultimately, their survival needs.

Members of Group 5 immediately accepted Amy's presence in their midst, but resisted the idea of being closely trailed. The white ape was apparently welcome to join them at rest, but not while they moved and ate. Their displeasure was generally limited to cough grunts—gruff-sounding coughs directed at the offending party and intended as a warning to stop whatever she was doing. On two occasions the dominant silverback, Beethoven, grabbed her forearm, which seemed to disappear within the grasp of the silverback's immense hand as he gently, but forcefully, squeezed. Amy took the warnings to heart and worked at making herself less intrusive by trying a variety of tactics combining distance, posture, and appropriate vocalizations. Within weeks, a mix of both known and novel techniques was yielding the desired result. Paradoxically, the key was moving closer to the gorillas. In this way she was always in sight, rather than staying farther away and repeatedly coming in and out of view as individuals moved through the dense understory of Virunga vegetation.

Some research was possible even in the early stages of this habituation process, but it was primarily a time of observation and learning about Group 5 and its members. Beethoven was the undisputed patriarch of this extended family. Darkly silvered across much of his four-hundred-pound body, he was both an imposing and calming presence. From his mighty haunches and massive chest to his regal sagittal crest—the bony ridge down the middle of his head that anchored his powerful chewing muscles—Beethoven looked the part of the leader. A second male, Icarus, had the more sculpted body of a younger silverback and was even more physically impressive. Yet his brighter and less extensive silvering, uncertain demeanor, and tendency to remain on the outer fringe of the group underscored his subordinate status. Beethoven was presumed to be the sole mate of the four adult females, and father of the seven subadults—immature gorillas less than eight years old—who completed his full family. It is possible that he was also the father of Icarus.

Gorillas are overwhelmingly social beings. This is never more obvious than during their late morning or midday rest period. For some, this is serious siesta time, but even in repose gorillas appear to like nothing better than to form a simian daisy chain, with each member in direct physical contact with as many others as possible. Amy was deeply moved one day when Ziz, a blackback male, rolled over during a siesta and laid his hand against her arm, linking her to part of the family chain. For those not sleeping, grooming is another way to stay in contact and reinforce personal bonds through the ritualized removal of debris from each other's rich coat of hair. The heavy Virunga rainfall can be of some help, but the cleanliness of the mountain gorilla's coat is largely a function of constant grooming of the four- to six-inch-length of thick, coal black hair over much of their bodies. The end result is a healthy, glistening sheen day in and day out, regardless of conditions around them—and a tightly knit family.

Group 5's composition also allowed a clear view of the social bonding that is central to the mother-infant relationship. Gorilla infants spend their first three years sleeping with and nursing from their mothers. During their first six months they are almost never out of direct contact, held in their mothers' arms during feeding, lying on their stomachs at rest, and clinging to their chests or backs as they move through thick vegetation. These young infants are also a constant focus of grooming. From six to eighteen months, infants are allowed some limited freedom of movement, but rarely much beyond their mothers' reach or sight. Even at rest they are often constrained by a discreet, yet forceful, maternal footlock around their ankle or wrist. Roaming distance increases steadily through their third year, and interactions with other infants and older juveniles multiply dramatically during this time. But individual differences also start to appear at that age.

The young infants of two females in Group 5 clearly illustrated the role of personality in gorilla development. Pantsy was roughly twelve years old and a first-time mother of Muraha, a sixteen-month-old female. By curious coincidence, Pantsy's mother, Marchessa, also had a sixteen-month-old infant named Shinda: a smaller male with grizzled brown tufts of hair who was technically Muraha's uncle, though they were born only four days apart. From Amy's earliest observations, there were obvious differences between the two infants. Muraha was far more outgoing and physical, whereas Shinda appeared shy and retiring. Muraha always beat her age-twin in wrestling matches. Wrestling was an uncommon experience for Shinda, though, because Marchessa was very protective of her son and only rarely allowed him to have any contact with others. Pantsy, on the other hand, seemed almost indifferent to the whereabouts of Muraha at that age, letting her move quite freely among the rest of the group and play with older siblings. Perhaps Pantsy figured that Marchessa would help to watch over her granddaughter; perhaps Pantsy was more inexperienced than indifferent. Most likely Shinda and Muraha and their mothers simply had different personalities, as would dozens of other gorillas we would come to know in the years ahead.

Another aspect of gorilla behavior became obvious as Amy began to follow individuals while they fed. Gorillas in groups are supremely social. On the move, plowing narrow trails through thick ground cover in search of food, they operate in a very different context. They are competing for food resources that, while generally abundant, can be quite limited in terms of both quantity and quality at any one place or time. So when two gorillas covet the same juicy stalk of wild celery, immediate hostilities can erupt. This is unlikely to happen between a young gorilla and an adult, since a cough grunt from the latter is almost certain to settle the matter. Competition between adults, however, is another story. Nothing in the literature had prepared Amy for the sudden outbreaks of intense, almost maniacal screaming that irregularly punctuated the otherwise placid feeding bouts of Group 5. Generally no more than ten seconds in duration, these interactions could make seconds seem like minutes and, in some cases, ended only with a sharp bite from the victor. Beethoven even intervened on rare occasions to settle matters with a bite of his own. More commonly, disputes ended when one individual gave in to the superior vocal or physical display of the other. Mostly, these encounters revealed a generally nonviolent, if spectacularly aggressive, mechanism for dealing with competition.

～へ～へ～

I F THE GORILLAS OF GROUP 5 fell short of pacifist perfection, their lives were nevertheless marked by an exceptional degree of tranquillity. And for the young, there was plenty of time for play. Gorilla games are similar to those common to most human cultures around the world. Tag, wrestling, and king of the mountain all have their Virunga variations. Play was most common during group rest periods, when two or more young gorillas would start to chase each other through the surrounding underbrush and then engage in a freestyle wrestling match. A fallen *Hagenia* tree in their midst made the perfect prop for a rough-and-tumble version of king of the mountain, with the added complexity of a slippery trunk and dangling vines as alternate attack routes to the top. And the intertwined limbs of bent and broken *Vernonia* thickets formed a remarkable imitation of a playground jungle gym— or is the jungle gym a fair copy of the *Vernonia* clump? Whatever one's perspective, young gorillas play as much as one third of their waking hours, sometimes even enticing their elders to join in. Gorilla youngsters may not laugh and scream, but they do emit a stuttering "chuckle" and certainly enjoy themselves as much as any human children at play.

Members of the youth brigade of Group 5 made a concerted effort to entice Amy to join in their play. Her arrival in the group was frequently greeted by Pablo, a nearly four-year-old, as an opportunity for cross-species play. His reputation as a gorilla juvenile delinquent had preceded him, however, as had the unfortunate history of an earlier Karisoke field assistant who apparently spent much of his short-lived time with Group 5 playing with Pablo. Great as the temptation to play might be—and it was very strong—Amy was determined not to give in. To leave her role as observer to become a member of the group would alter the gorillas' behavior and undermine her own research in the process. Far more important, entering into their social lives would diminish their independence and the integrity of their way of life.

Resisting Pablo proved difficult, however, because of his persistence and his preferred tactic of ambushing researchers from behind. This could take the form of an aerial assault from his perch on bent bamboo, or simply grabbing our backpacks and yanking us to the ground if we tried to walk past without paying him proper attention. If our crash landing attracted the unwanted attention of Beethoven, Pablo would look away like an innocent bystander. He would have whistled in mock distraction if he were able. Ziz, an eight-year-old male, took a novel approach one day when we were together observing Group 5. Bill was already on all fours when he felt an arm drape across his back. In an instant, Ziz swung his entire weight onto Bill's back and lay there, facedown and gently rocking. Though Bill's first thoughts were of how to spurn the blackback's amorous advance, he soon realized that Ziz just wanted to play. Ignoring the two-hundred-pound gorilla on his back was not an option, so Bill tried to rock him off, but this only caused Ziz to chuckle and rock more in return. The solution came when Bill dropped his front end suddenly, sending the gorilla to an unceremonious headfirst landing. Ziz picked himself up, pushed off pointedly on Bill's face with his left rear foot, and walked away without a backward glance.

Even Amy's policy of passive resistance didn't guarantee avoiding all play. One day, as she sat in a thicket taking notes on other family members, Muraha and Poppy—the two-year-old daughter of Effie—tumbled down from above and began chasing each other around her. Soon they added a new element to the game by grabbing some of the many dangling vines and pulling them as they ran. After ten minutes of this "circling the maypole," they had exhausted the vine supply and moved on to other amusements, leaving Amy so thoroughly and tightly wrapped that she had to pull out her Swiss Army knife to cut herself free.

Our clothing and other paraphernalia were also keen objects of attention. Boot laces were always there to be untied—without any interest in learning how to retie them. Backpacks were also fun to spirit away and usually attracted a crowd for close inspection, though no gorilla had yet mastered the mystery of the zipper. The gentleness with which they treated all objects was truly amazing—especially in comparison with the known destructive tendencies of their close cousins, chimpanzees. This extended even to our Zeiss binoculars, which at $600 a pair we were not inclined to leave lying around. But on the few occasions when a gorilla did get hold of them, they were handled with a soft touch worthy of fine crystal, then carefully placed back on the ground. The interest of the younger gorillas also extended to our bodies. The freckles on our arms were particular objects of grooming attention for Tuck, a bold six-year-old, who tried on several occasions to remove them with his fingernails and lips. Tuck's curiosity extended further one day, when he reached in the top of Amy's loose-fitting sweatshirt to touch a breast. It was a time when the gorilla's characteristic gentleness was definitely appreciated.

Gorilla play didn't always require others. Dangling from bent bamboo and slowly spinning from vines were favorite solitary distractions. So was just sitting and watching individuals and events around them—watching with all-too-human eyes that reflected thought and allowed the easy illusion of understanding. We would endlessly imagine what their thoughts might be, but the bridge that linked our species spanned a deep chasm of incomprehension.

A favorite time in those early months was the end of the day, when we would return to our cabin, change into dry clothes, and talk about the day's events. Hot tea took the edge off the rude Virunga climate and fresh peanuts roasted in a spicy oil made up for a lack of lunch in the field. As we prepared our daily one-pot meal of some variation on beans and rice, Amy recounted her day-in-the-life with Group 5. Dropping her role as scientist she would comment on the personalities of different individuals, psychoanalyze Pablo, or talk about the comic antics of "the kids." If Bill had been out with another group that day, he would respond in kind. We had already shared many experiences in our eight years together and we would later have two boys of our own to observe, enjoy, and endlessly discuss, but those evening talks in our first few months, relating our encounters with such remarkable creatures and their parallel world, were exceptional. After dinner it was back to work. Amy would pull out her pack of waterproof notes from the day to transcribe her data. Both of us would type up more social observations for Dian and group movement notes for the long-term Karisoke records. There might be time for a letter home, or maybe even to read. But by nine o'clock, as the ground temperature approached freezing and the humidity hovered near 100 percent, it was often too cold to write or turn pages. Then we would dive under our two layers of sleeping bags to warm up and sleep deeply until just before dawn.

THE COMFORTS OF A SHARED BED and time to talk became a luxury as Bill's work increasingly took him away from Karisoke. But when in camp he would help monitor Group 4 or unhabituated groups that ranged beyond the research periphery. Group 4 was especially interesting. It was the most habituated of the research groups, the primary focus of Dian's work for almost five years. After completing her dissertation in 1974, though, Dian began to spend much less time with the gorillas. Kelly Stewart and Sandy Harcourt and a succession of others then monitored the group's activities. Both before and after the death of Digit, this was Ian Redmond's principal responsibility, but Bill would fill in when Ian was not available.

Group 4 was notably different from Group 5. Not only was Uncle Bert the sole silverback, but he was much younger than Beethoven. His broad saddle was a brighter silver, and he seemed to carry his head and shoulders in a more upright manner—though Beethoven was no less imposing for his mild slouch and modest paunch. The strong, musty odor that is characteristic of all silverbacks also seemed more pungent in Uncle Bert. There was no secondary silverback: that had been Digit's role until he was killed. But two young males were positioned as heirs apparent: eleven-year-old Beetsme and ten-year-old Tiger. As in Group 5, there were several adult females in Group 4. The ancient and weary-looking Flossie was mother to almost-seven-year-old Cleo and three-year-old Titus; Macho had seven-year-old Augustus and two-year-old Kweli; and ten-year-old Simba constantly carried Mwelu, whose still pink face and plastered hair gave away her newborn status. Yet despite the almost equal number of young gorillas, there was much less play within Group 4 than in Group 5, and even the subadults showed less interest in human observers.

Overall, the impression was of a much more subdued gorilla family. Was this due to Digit's recent murder? Was it their family personality? Perhaps Group 4 was just more accustomed to the near constant presence of human observers. Whatever the reason, Group 4 offered an opportunity to know an array of new individuals and their various family relationships, and to speculate on the reasons behind their behavior.

One day, a torrential downpour suddenly swept in from the Congolese side of the range. Despite the intensity of the storm, none of the gorillas sought shelter, so Bill decided to hunker down and join them. For more than two hours they sat on the exposed western slope of Mt. Visoke: the gorillas stoically hunched over, females shielding infants as the rain poured off their long hair, while Bill shivered as he watched pools of water collect in the folds of his full-body rain gear and wondered what he was doing there. The reward, if not the reason, came quickly in the wake of the storm's departure, as a wave of sunlight poured over us. Group 5 would have begun moving and feeding as soon as the storm ended. Group 4 not only stayed in place, but the entire family lay back, its members stretching out their arms to expose hairless chests and armpits while basking in the exceptional solar warmth. With little behavior to note, and his own body to warm, Bill stripped off his cold clammy rain jacket and flannel shirt and joined them.

If sunbathing at ten thousand feet weren't reward enough, the retreating storm stripped the usual surface cloud cover from nearby Mt. Mikeno, five miles to the west. Bared to view and bathed in light, Mikeno's sheer rock face looked like some up-thrusted granite spike with no relation to the other, less defiant volcanoes. The idea that George Schaller had climbed to its 14,557 foot summit without technical equipment seemed preposterous, even for a giant of field research. But Schaller's view from the top could not have surpassed that day's spectacle, as no fewer than eight waterfalls were visible at once. Most were ephemerals: free-falling cascades that danced in the light for ten or fifteen minutes until the concentrated runoff ran out and the spigot closed shut. Three larger falls were still visible after nearly a half-hour. At that time, Bill realized that the gorillas were stirring and Uncle Bert himself stood in full strut above him. Any notion that he, too, was enjoying the view faded with Bill's realization that Uncle Bert was impatient to move downslope and find some of the forty to fifty pounds of food that he needed to consume each day. Bill was blocking the way, so he squeezed to the side, leaving what he thought was enough room to pass. Uncle Bert saw matters differently and gave a whack to Bill's kneecap with his knuckles as he ran by. The bruise lasted a long time, serving as a reminder of how much was left to learn about proper gorilla etiquette.

Group 4 also occupied very different habitats from Group 5. They spent much of their time on the gently sloping western flank of Mt. Visoke, where extensive stands of nettles provided the major food resource. These were stinging nettles that grew to heights of six feet and there was no way to get through a day among them without experiencing one or two sharply burning sensations wherever they touched your skin, even through thick pants and long-sleeved shirts. Gorillas seemed impervious to direct contact and devoured the nettles in large quantities, though only after careful removal and folding of the stinging leaves in a way that made swallowing easier and less risky. Recounting this dietary preference to older Belgian friends one night over dinner, we also learned that young nettles—carefully peeled—were a common ingredient in soups during the harsh days of the German occupation in World War II. Whether because of its taste or the memory, the Belgians never offered to prepare the soup from our ample larder.

Despite their bountiful presence, nettles did not meet all of Group 4's nutritional needs. After days of feeding on the lower slopes the group would move suddenly, often higher up on the mountain. One time they made a beeline out of the saddle and didn't stop for over an hour as they climbed three thousand feet straight up, almost to the summit. Moving first under broad *Hagenia* trees, Bill followed as they quickly passed the treeline and moved onto ever narrower ridges covered with rugged heath. Climbing higher, they entered the subalpine zone, where giant forms of cold-adapted shrubs predominated over an open landscape of grasses, mosses, and lichen-covered rocks. One of those shrubs, called *Lobelia,* was the primary attraction for the gorillas, as most of the adults fed on the white pith of its shiny green six- to eight-foot-long stems. The plants' characteristic "headdresses" of bayonet-shaped leaves had been broken off and left nearby in scattered clumps. After taking some notes on feeding, Bill left the group to explore Visoke's 12,172 foot summit. There, a thick blanket of tussock grass and moss completely covered the crater rim and continued down several hundred meters to the almost perfectly circular lake within. It was a fantastic place where wind and clouds rushed past to constantly alter the view. Sundown comes fast at the Equator, though, and Bill realized that the gorillas had already left. After two false starts down ridges that came to terminal cliffs, he found a trail leading down to the saddle and back to camp just after dark. The gorillas had stopped to make their night nests about five hundred feet above the saddle, but climbed right back to the top of Visoke again the next day. They then repeated this cycle for a third day. It was a great way for Bill to stay fit, and the Visoke summit was a magical place. But there was no obvious reason why the gorillas of Group 4 made such a rigorous migration up and down that mountain three days in a row, nor why they then moved just as abruptly back into the saddle to stay. Group 5 behaved similarly several times later in the year. These were questions for Amy to ponder and address, if possible, through her research.

W ITHIN SIX WEEKS, the gorillas were tolerating Amy's decision to fol- low them more closely, and she was gaining more information. Patience and a modified approach had rewarded her with an exceptional research opportunity. That is, until the day in July when Liza suddenly left. Liza was an established member of Group 5. Her earlier offspring, Nikki, had transferred out of the group before Pablo was born in August of 1974. Although precocious in his ability to cause mischief and quite independent, Pablo was not yet four at that point and still counted on Liza for some breast milk each day and a warm bedmate at night.

So it was a shock when Amy learned from David Watts that Liza had abandoned Pablo and the rest of her family. David usually worked with Group 4, but was watching Group 5 for two days while Amy caught up on her field notes. On the first day, David said that he heard another gorilla vocalizing not far away, but didn't realize what had happened until he noted Liza's absence on the second day. At that point, David returned to camp to tell Amy and Dian. Dian was in a bad mood and initially refused to send a team to search for the missing gorilla. The next day, however, Bill left with one of the best trackers, Vatiri, to find Liza. They backtracked to the day of her departure, finding a solitary exit path that led directly to the trail of another band of gorillas. After two days of pursuit across more than a dozen ravines, up into the alpine zone, and down the far side of Visoke, they finally caught up to the other group. At Bill's approach, a young-looking silverback made a series of short but impressive bluff charges. Later visual identification confirmed that Liza had transferred to Group 6 and its volatile silverback, Brutus. She appeared to be very comfortable with the situation, even if Brutus did not.

It is normal for gorilla females to transfer to other families around their eighth year, when they approach sexual maturity. This provides an effective mechanism to avoid inbreeding. Yet within Group 5, Liza was one of several females believed to have bred with their father or half-brother. This might even explain Pablo's mildly crossed eyes and two webbed fingers, which are common indications of inbreeding. What was exceptional in Liza's case was her moving on and leaving a juvenile son behind.

The complications this would bring to Amy's life were clear on her first day back with Group 5 after Liza's departure. As a heavy rain began to fall, Pablo strode right up and huddled by Amy's side. Soon, the motherless child nudged his head under her arm and the two sat entwined for an extended period. Pablo seemed content for this arrangement to go on indefinitely, but Amy's emotions were a battleground. She knew that she couldn't, and shouldn't, act as Pablo's surrogate mother. Her role as researcher was one thing, but there was a much more fundamental issue of trust in the gorillas themselves. They aren't our wards; they are independent beings with social systems evolved over eons that had kept the species alive. With trust that Liza knew what she was doing, that Pablo was sufficiently independent to get by on his own, and that other gorillas would help out if needed, there was no need for Amy to treat Pablo like an orphan. He simply had to get out from Amy's shelter and get on with his life as a gorilla.

So said the head. Amy's heart would lag, but if maternal emotions couldn't be altered, behavior could. With a farewell hug and a gentle nudge from his all too briefly adopted mom, Pablo was sent back to his own species. He would return several times over the next few days, only to find Amy's arms tightly closed in resistance to his approach. No gorilla females showed any adoptive tendencies toward Pablo, either. Beethoven, however, almost immediately showed a degree of paternal care that was unknown before. When it rained, there was refuge under Beethoven's massive frame. During rest periods Pablo was always welcomed at his side. And at night, Pablo shared the warmth and security of his father's nest. It was a side of adult gorilla males that was rarely seen and a treat to watch over the next several months. As painful as it was, Amy was convinced she was right to resist Pablo's entreaties.

As they went about their lives, the gorillas of Group 5 introduced Amy to their world. During her first few months, they moved through most of what she would come to document as a five-square-mile home range. It was a richly varied world. To the south a descending line of adjacent craters testified to a distant past of volcanic venting from Karisimbi's eastern flank. Now dormant, the Five Hills were most notable for their rich mix of herbaceous plants and rare concentrations of blackberries, adored by gorillas and researchers alike. It was also a preferred habitat for elephants, where we once counted twenty-seven in a herd as they thundered past our position on First Hill.

Moving north, Group 5 would usually cross the Karisimbi–Visoke saddle somewhere below the Karisoke station and move quickly onto the slopes of Visoke. This was their zone of most concentrated use, where the dried bowls of former night-nesting sites dotted the landscape like relics of some mythical ground-dwelling giant bird. It was also an area divided by numerous streams and ravines, each with its own special character. *Bonde ya Mifupa,* or Bone Ravine, was a steep and narrow cut where the gorillas rarely lingered. *Bonde ya Maji* was named for its permanent source of water, but it was also a magical place where dangling vines and saturated mosses draped the giant *Hagenia* trees perched precariously on its abrupt flanks. Here the Ruwenzori turaco might reveal its presence with a distinctive metallic gargle as it hopped from hidden branch to hidden branch, then flash its crimson underwings in brief spectacular flight across the ravine. To the east, *Bonde ya Kurudi* was large but easily crossed, and covered with good gorilla foods. Yet Return Ravine was named for the simple fact that upon reaching this point, Group 5 always turned around or headed south back into the saddle.

As it neared the lower park boundary at an elevation of around 8,700 feet, the *Hagenia-Hypericum* forest mixed with and ultimately gave way to dense stands of bamboo. Twice each year, for several months at a time, Group 5 would sleep in the forest and move down each day to feed intensively on bamboo shoots and leaves. Aesthetically, this was one of the least interesting areas in the park, but it was critical for both research and conservation. Bamboo offered the most desirable and nutritious food for gorillas when the tender young shoots were available. Yet it was the bamboo zone that Rwandan farmers had converted to farmland most intensively in recent years, and widespread illegal cutting in the park reflected bamboo's value as a building material. Live bamboo also made an excellent spring mechanism for snare traps, which were most common near the border of the park. Even more worrisome were traps of suspended logs intended to break the backs of Cape buffalo when triggered by their passing below—but which were equally lethal for gorillas and researchers. Any time spent along Karisimbi's bamboo apron made it clear that this was a primary zone of human–gorilla contact and conflict. Amy always felt some relief when Group 5 would reach the Susa River and turn north toward the Five Hills or back to Visoke. The Susa per se was generally not a barrier to the gorillas, any more than Return Ravine. It stopped flowing altogether during the late summer dry season and it had numerous crossing points. On one May day in 1978, though, the Susa became a raging barrier.

Although the Virungas have a well-deserved reputation for their wetness, rainfall is not constant. It is highly seasonal, linked to the monsoonal patterns of the Indian Ocean to the east. July and August are almost totally dry, with rainfall building steadily in the following months through early December. Rain then drops off sharply, but continues intermittently through mid-February, when the long rainy season officially begins. By April and May the rains are especially heavy and downpours can last all day and night, sometimes totaling several inches in twenty-four hours. The volcanic soils quickly turn to boot-grabbing mud at this time and, unable to absorb any more rainfall, send the excess runoff into the myriad watercourses that cut the mountains. As most of these have steep, rocky banks, water levels can rise dramatically and dangerously in a very short period of time.

On May 10, Amy was following the trail of Group 5 as it moved from the slopes of Mt. Visoke southwest toward the distant Susa. Nearly constant rains over several weeks had turned the broad saddle region into a supersaturated swamp. The gorillas showed no interest in stopping under such conditions and, as the deluge continued, they worked their way toward the Susa. Amy had no difficulty following their trail of pug marks in the mud, but did have trouble keeping up as the earth itself seemed to grasp at her every footstep. Finally, she heard a strange roaring sound that rose above the steady din of the rain. As she moved closer, the source became clear: it was the Susa—three feet higher and many times wider than Amy had ever seen it. Wisely, the gorillas had not tried to cross at that point, but instead moved downslope along a rocky ridge that still contained the engorged river. After several hundred yards, a parallel ridge on the other side forced the Susa into a narrow channel no more than twenty feet wide. It was here that trail sign showed that the gorillas had attempted to cross.

Amy scouted the situation. The distance was too great to leap across; the gorillas must have walked. But how? Stepping into the swollen watercourse, she was soaked up to her hips before the powerful current forced her to retreat. When her successive forays failed to find a crossing, she moved farther downstream, where she quickly came to a twenty-foot waterfall. Her eyes were drawn to the pool below, where she searched in fear for drowned bodies. Finding none, she returned to the original crossing. Did they all make it? She could picture the silverbacks and even adult females having sufficient bulk to resist the current and succeed in crossing. What about the younger ones like Pablo and Tuck? Did other gorillas lend a hand? Form a chain? Were Muraha and Shinda on their mothers' backs, or in their arms? Did these mothers walk upright on two feet to keep their youngsters above the water? It must have been an extraordinary sight to witness, but all that really mattered was whether they survived. Amy returned to Karisoke, where she spent a restless night worrying about the gorillas. The next morning she returned with Emmanuel Rwelekana, one of Karisoke's best trackers, to help with a search. The Susa was already much lower, and they crossed easily. Within an hour they located the group and proceeded to confirm a full count of healthy gorillas going about their normal lives. How the gorillas managed to cross the raging Susa was a secret they kept to themselves.

M<small>OUNTAIN GORILLAS ARE BEAUTIFUL</small>, much more so than their lowland cousins. Their long black hair rounds off the edges of angular jaws, bony brows, and sharp crests; it softens their otherwise powerful frames. Their faces and foreheads are also less sloped than those of their almost equally long-haired cousins, the Grauer's gorillas of eastern Congo. Then there are the eyes, the deep brown reflecting pools in which we sometimes see ourselves. Finally, there is the dignity with which they carry themselves and which gives them a certain aura.

Quince, with her perfectly proportioned body, shiny black hair, and dazzling eyes, was perhaps the most beautiful mountain gorilla we ever encountered. Nearly eight years old, she was a female entering maturity. While still a willing playmate for Ziz, Tuck, or Puck, she might soon become the mate of Beethoven or Icarus. More likely, she would leave to join another family. Her life in late 1978 was full of all the possibilities for a gorilla of that time and place, and she seemed to carry herself accordingly.

Yet in October, Quince began to slow down. It was a time when many of the gorillas had been sick with coughs that indicated bronchial problems due to especially cold and wet conditions. But Quince had no cough or runny nose. At first there were few signs beyond a tendency to stay at rest a bit longer. Then her head began to hang as she walked and her stops became more frequent. The other gorillas altered their movements accordingly. Distances covered each day dropped dramatically, while the number and duration of rest periods increased. Food resources in the lower saddle, just above the bamboo zone, seemed ample for even an extended stay. But Quince's appetite declined, too.

Within a week she had visibly worsened. She was spending at least half of each day in a series of poorly made day nests, staring down when she wasn't sleeping. Dian had no advice or available means of intervention and assistance, so Amy's daily routine became an increasingly distressing vigil. By the end of the second week, Quince was barely moving. Untold pain had dimmed her once lustrous eyes to a dull regard of surroundings that seemed to hold little interest. The other gorillas had not lost interest in her, however, and were now feeding in shortened loops that took them out to untouched food patches, then back to Quince. They continued to make new nests each night, often within one hundred yards of their previous night's site. Unlike the others, who constructed sturdy bowls of interlaced plants and branches that provided insulation from the cold damp forest floor, Quince could summon only enough energy to tuck a few unconnected branches beneath her.

Amy found Quince in such a nest in the early morning of October 20. Lying on her stomach with her face buried in her arms, she might have been blocking out the pain. But she didn't stir and the last warmth was fading from her body when Amy moved close to touch her. The other gorillas seemed to acknowledge her death and had already moved away. Amy hurried back up to camp to tell Dian, then returned to Group 5 with the Rwandan camp staff, who would carry Quince's body away. Some tissue samples were removed for later analysis before Quince was buried alongside Digit and a growing list of others behind Dian's cabin. We never heard any results from tests that might have been done, and to this day the cause of death remains a mystery.

The loss of another gorilla was hard to accept, especially for Amy, who had watched helplessly as the process unfolded in seeming slow motion. But whereas humans were responsible for every step of Mweza's suffering and death, Quince at least died of natural causes—prematurely to be sure, but a gorilla's death in a gorilla's own world, surrounded by her own kind. It was an all-too-rare event for that time in the Virungas.

THE DAY OF QUINCE'S DEATH, Group 5 set off on a long trek reaching north along the park boundary. Before Amy caught up to them the next morning, they made a 180-degree turn and headed rapidly in a direct line toward where they had last seen Quince. Amy quickened her pace and caught up along their flank, just in time to be rewarded with an exceptional sight. First Icarus, then Puck, went straight to her nest and placed their faces on the exact spot where Quince had breathed her last. Each then sat back and stared off into space. The two sat side by side as others passed near the nest site. Then the entire family moved off silently into the surrounding forest.

The return of Group 5 to the site of Quince's death was an unparalleled event. What would they have done if the body had been left there? Would they have returned again and again, as some elephants do, as she turned to bones? There was no record of a precedent among wild gorillas at that time, though several members of Group 4 later reacted to a death in a similar manner.

Quince's death, Liza's transfer, and Pablo's attempt to adopt Amy as a surrogate mother moved us deeply. Each incident offered profound insights to another world; each exposed serious limits to our understanding and ultimately raised more questions. More revealing was the experience of the gorillas' day-to-day routine: their familiarity and harmony with their environment, their complex social relations, their individual personalities, and their interactions with the human apes among them. It was a perspective born of continuous contact that revealed both the common and the exceptional in an unbroken flow of unforeseeable events. We knew from our earliest days that we were among the luckiest people on earth to be sharing that experience.

Chapter Five

A Swamp Runs Through It

AFRICA'S TWO GREATEST RIVERS originate in Africa's two greatest lakes. Lake Victoria gives rise to the Nile, while Lake Tanganyika drains into the Congo. This geographical conceit, however, ignores the fact that both great lakes are nourished by thousands of smaller waterways that flow down from the surrounding mountains and highlands. Many of these have their source in Rwanda.

An extensive marsh occupies the saddle area between Mts. Karisimbi, Mikeno, and Visoke. About twenty minutes west of the Karisoke field station this wetlands-meadow complex comes to a gentle rise, which separates the waters of the Congo and the Nile river systems. On one side of this unassuming hummock, an eighteen-inch-wide grassy channel collects the clear Virunga waters and sends them on their way toward Lakes Kivu and Tanganyika, out through the Lualaba River, and on to the mighty Congo. Twenty feet away, a similar conduit forms a stream that gains in size and power as it flows in an easterly direction through camp. Joining forces with the Susa, Mukungwa, Nyabarongo, and Akagera rivers along the way, these waters enter Lake Victoria before the final four-thousand-mile push down the Nile and into the Mediterranean.

Besides being a geographical curiosity, Camp Stream was the lifeline of our existence. Fed by waters from both Karisimbi and Visoke, it provided a steady source of water that was rare in the Virungas. Even constant rainfall didn't guarantee any surface water over large areas of the park, where porous soils and lava rock formations quickly drained the water underground. Standing water without a flow was suspect for health reasons in any part of the park. But this was rarely a problem with Camp Stream, whose clear waters we drank untreated in all but the peak of the dry season as it flowed thirty feet from our cabin.

In the rainy season, Camp Stream's character could change in less than an hour, as runoff from the latest downpour turned it into a torrent of crashing, swirling white water. The thought of riding a tube at high water down its rollicking cascades, constrained within a rocky channel no more than ten feet wide, was a recurrent fantasy. But we didn't have a tube—and the water temperature was barely above freezing.

At one level, temperature and rainfall information were irrelevant in the Virungas. It was always cold and wet. Yet a great deal of variability hid in those two adjectives. A 5-degree-Fahrenheit drop on a cloudless evening at ten thousand feet sent the temperature below freezing, while a 10-degree rise on a rare sunny day brought on a quick striptease of outer clothing layers as the thermometer approached a balmy 70 degrees. On our arrival, we were surprised to learn that Karisoke had no reliable means to keep climatological data other than a tiny plastic rain gauge under some trees and a department store thermometer attached to a window on Dian's cabin. When Bill requested better equipment from Rwanda's Bureau Météorologique, we received a professional high-volume rain gauge and a min-max thermometer that recorded daily highs and lows. With their installation in appropriate sites, Rwanda gained its only high-elevation rain forest monitoring station, Karisoke entered the realm of modern meteorology, and we were made precisely aware each day of just how cold and wet it really was.

DESPITE ITS DRAMATIC RISES, Camp Stream never overflowed its steep banks to flood the main residential area while we were at Karisoke. It didn't have to: we lived in a perpetual swamp. From the spot where the Congo and Nile waters parted ways less than a mile west of camp, the entire saddle area was a broad, sloping wetland. Some of its waters took the fast track via Camp Stream, but most moved more slowly as a sheet of steadily flowing, nearly freezing water. Situated on the eastern edge of the saddle, Karisoke occupied slightly higher ground and escaped the main flow from the upper marsh. But this location only placed us closer to the slopes of Visoke, from which a tributary sheet of unchanneled water seeped almost constantly across the station as it sought to join Camp Stream.

One of the great attractions of Karisoke was the fact that so much of it was left in a natural state. No effort had been made to drain or channel large areas of marsh, which retained their cover of grasses, sedges, and the occa-

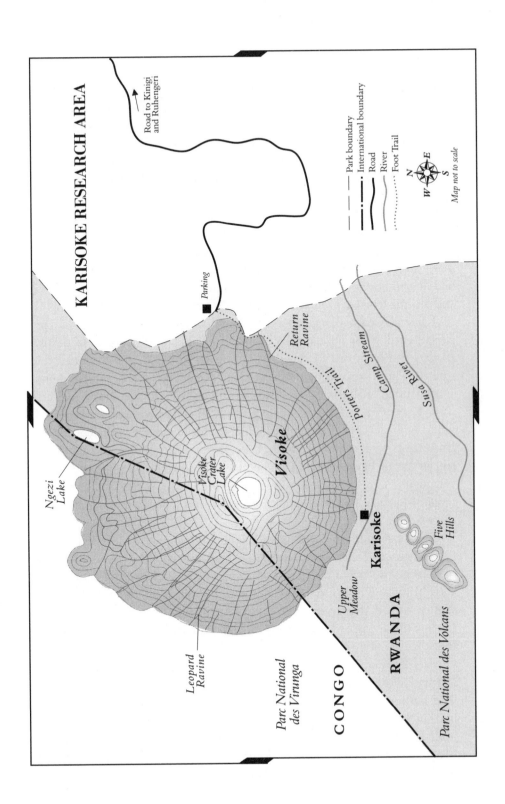

KARISOKE RESEARCH AREA

Road to Kinigi
and Ruhengeri

Parking

Return Ravine

Porter's Trail

Camp Stream

Susa River

Visoke

Visoke Crater Lake

Ngezi Lake

Leopard Ravine

Upper Meadow

Karisoke

Five Hills

Parc National des Virunga

CONGO

RWANDA

Parc National des Volcans

- - - Park boundary
- · - · International boundary
——— Road
River
······· Foot Trail

N
W E
S

Map not to scale

sional alpine *Lobelia,* which washed down as seeds or seedlings from the surrounding slopes. On higher ground, *Senecio* shrubs mixed with purple-crowned thistle and thousands of yellow-flowered *Hypericum* saplings—a relative of St. John's-wort that grows to tree size in the Virungas—to form dense thickets. These were separated by a network of grassy clearings and natural seeps. Towering over everything were the giant forms of *Hagenia abyssinica.* Almost one hundred feet tall and six feet in diameter at the base, these trees reached their greatest size in the area immediately around Karisoke. They are improbable life forms, seemingly designed by Dr. Seuss. Their massive trunks give rise to irregular, exuberant crowns, seasonally festooned with grapelike clusters of fruit. Their limbs jut out at impossible angles—angles made all the more remarkable by the tons of saturated mosses and lichens borne by those limbs during most of the year. More frightening still was the fact that so many *Hagenia* limbs hung over our frail tin cabins.

Karisoke's natural vegetation attracted a variety of animal species. The *Lobelia* outside our cabin produced a three- to five-foot-tall spike of purple flowers and buds that bloomed from bottom to top over the course of a month. These attracted a steady stream of northern double-collared and scarlet-tufted malachite sunbirds—Africa's answer to the hummingbird, though sunbirds lack the ability to hover. Cinnamon bracken warblers and white-browed robin chats skulked under the ground cover of thickets, while Ruwenzori turacos hopped though the canopy above and emerald doves ripped the air like low-flying jets. White-naped ravens came on occasional scavenging forays, usually in pairs making loud raucous calls. Far more irritating than their vocalizations, however, was the scraping of their claws on our tin roof.

The most endearing of the local mammals was the black-fronted duiker. Standing no more than knee-high, this small antelope boasts a lustrous coat of rich chestnut over most of its body, with a blaze of black on its snout. Delicate black and white markings decorate the inside of its ears. A pair of slightly curved, dagger-sharp three- to four-inch horns provide its only defense, while elongated hooves afford support in marshes. Elsewhere in the forest, duiker were extremely shy because of their status as a prime target of hunters. Around Karisoke, however, the protective presence of researchers and staff allowed them to tiptoe quietly about their lives with little concern. The bushbuck, a deer-sized antelope with foot-long spiraled horns, also lived in the surrounding forest but was rarely seen. Most commonly, it made itself known through its disconcerting habit of barking exactly like a dog. More than once in our first few months, we were startled into poacher alert mode by the sharp barks of a bushbuck, imagining instead the hunter's dog.

Occasionally, we were rewarded with the sight of its white spots and lines against a coat of reddish brown, with black and white markings on long elegant legs.

We rarely spotted a Cape buffalo around camp, but its abundant dung piles reminded us that it was a common nocturnal visitor. What this half-ton savanna-dweller was doing at ten thousand feet in a rain forest was never clear to us, but the presence of Africa's most lethal creature was to be respected. And its tracks along the forty-yard path to our outhouse meant that only the most necessary trips were taken there at night. Other animals tended to stay away from camp, although elephants passed within fifty feet of our cabin one night and gorillas came within two hundred yards on our first Christmas Day. A leopard was seen by others less than a mile away.

Those animals that did wander into the Karisoke complex walked where they wished, free of fences or other barriers. People generally walked on stone trails. The imperfect solution to life in a swamp was a minimal network of slightly raised trails, reinforced with rocks of various sizes. Around Dian's house, built on the highest ground, these worked quite well. For those of us at the lower end of camp, it was futile to try to stay dry, especially at night when the risk of a sprained ankle from rock-hopping in the dark far outweighed the discomfort of soaking feet. The solution was to abandon any concern about wetness. Equipped with army jungle boots that allowed water to flow freely in and out, and good wool socks, it was Bill's standard practice each morning to stomp into the first pool of standing water or mud and proceed with the day. Amy took a more gradual approach to the same soggy result. Once dry again at night, though, there was a strong urge to stay that way. Amy eventually joined Dian and several others in wearing rubber boots around camp, but local markets didn't sell Bill's size thirteen.

THERE WERE ACTUALLY few destinations within the bounds of the research station. Six structures spread over an area barely larger than a soccer field. With the vegetation, the condition of the trail, and the state of our relationship, though, the one hundred some yards from our cabin to Dian's could seem much longer.

The buildings at Karisoke differed in size, but all shared the same style. Three-inch-diameter poles of either *Hypericum* from the park, or *Eucalyptus* purchased from outside, formed the rectangular frame. Additional support came with the construction of the floor: a platform of rough-hewn planks raised about three feet above the bare ground. Smaller *Hypericum* saplings, laced together with vines or wires, provided support for the roof. Thin sheets of *mabati,* or corrugated tin, were then nailed onto this skeleton to complete the external structure for both roof and walls. Interior designs were also quite similar across buildings, with a layer of stiff bamboo mats providing a false ceiling for false insulation, while papyrus mats draped the walls and covered the floor. Our cabin measured roughly thirty by twelve feet with a partial partition erected between our sleeping quarters and a combination kitchen-office. We also had three windows and two ill-fitting, but functional, doors. Two very noisy metal spring bed frames, one desk, three chairs, a counter, and a stool completed our furnishings. It offered all that we needed and provided suitable habitat for the seven species of rodents that shared our cozy home.

The Karisoke buildings did little to keep out the elements. Direct rainfall was generally warded off, but at the risk of severe hearing impairment during intense storms. And while the rain might not get in from above, the humidity from below was a constant companion. Kerosene lamps helped if you sat close, but mostly we just dressed warmly. When Amy was alone in camp, she commonly slept in sweat pants, wool socks, a turtleneck shirt, vest, and even a knit cap.

Fire would have helped, and it was an option. We had a small wood-burning stove that held about an hour's worth of wood. Unfortunately, the wood was almost always wet. It was also chopped from dead *Hagenia* in the park. This had always been the accepted practice at Karisoke, and there did appear to be an ample supply: Dian alone burned six to nine wicker baskets of wood per day in her two stoves and open fireplace, versus our one basket. But we wondered what the ecological impact of removing so much fallen wood might be. And it was disturbing to freely consume a resource that local people could be arrested for using. In a modified moral bailout, we reduced our consumption to a minimal level, making sure that we also dried clothes or plant samples whenever we had a fire going. Dian eventually rescued us from this ethical compromise by completely cutting off our wood supply in a pique over some unperceived sin on our part.

A FIRE ALWAYS BURNED in the open pit that constituted the true center of life at Karisoke. The pit was partially covered with two sections of corrugated tin to protect the large fire used to boil water and cook food. It was a place to find warmth of many kinds. In the rare moments when they didn't have other work to do, it was the gathering place for the Rwandan camp staff. There they could escape the chill air, dry their rolled-up tobacco leaves on the hot stones, and share stories of the day, all the while stirring the ever-present pot of beans that provided their breakfast, lunch, and dinner. Someone who had just come up from the valley might bring news of another's family, or how the crops were doing. All of the men were farmers and none worked more than half-time at Karisoke. The fire pit was also the place for Rwandan staff and foreign researchers to meet and talk.

There were three steady positions at the research station. The housekeeper, or *mutu ya nyumba,* was occupied mostly with work in Dian's cabin, but was also responsible for weekly laundry for others in camp. Two men split this position in alternating twenty-day shifts; each was known by a single surname in keeping with Karisoke practice. Kanyarogano was the senior "boy," as Rwandans had learned to call the job from the Belgians, who in turn must have picked it up from the British in East Africa. He also carried the unofficial title of Dian's principal in-camp informant. He played the role to comic effect with constantly shifting eyes and sideward glances, as well as a tendency to lurk, rather than walk, around the station grounds. It didn't take long to realize that he was really a decent fellow with an unfortunate set of behavioral tics and a reputation to match. The alternate housekeeper, Basira, had a more direct approach and more personable character.

The woodcutter, or *mutu ya kuni,* had the lowest status of the three main camp jobs. Two men generally split this position on ten-day shifts, with no apparent hierarchy between them. Nshogoza had worked at Karisoke longer, but also had been dismissed at least once before. Perhaps for this reason his public character tended toward the dour; but in person, he was a very likable man who saw himself as more than a woodcutter and clearly wanted to better himself. His partner, Rukera, couldn't have been more different. Constantly smiling, he played the buffoon to the hilt and was the butt of constant jokes when he was in camp. Still, he was a serious worker who seemed impervious to cold and pain. He spent his days barefoot, provided with neither boots nor rain gear, stalking the cold, wet saddle in a never-ending search for fallen *Hagenia* trees. These he would chop up, load into baskets, and carry back to camp on his head at an average rate of twelve to fifteen loads per day. Often he balanced an entire log on his head so that he could carry it back to chop closer to the fire. It was on one such occasion that his axe-head glanced off a log and sliced deeply into his bare right foot. As Rukera hobbled into the Pit and sat on a wooden bench, Basira called us to advise on treatment. By the time we arrived, Ian Redmond had sprung into action, so we joined the other camp workers watching as Ian threaded a very large needle he used to repair his boots. Rukera barely reacted as the needle pierced his thick skin and Ian closed the three-inch gash with a series of tight stitches. Afterward, he joined the others in laughing at his own misfortune. The next morning, Rukera left camp alone to walk the several hours down off the mountain and back to his house, with a borrowed rubber boot on the injured foot.

Gorilla trackers enjoyed the highest status among the various Karisoke workers. They owned their own rain gear and boots, they always carried a machete or curved *umuhoro,* and they walked with the swagger and self-assurance of fighter pilots. Their work also took them away from camp and the watchful eyes and strange ways of *Mademoiselli,* as they referred to Dian. Yet while their standing was primarily due to their work with gorillas, Karisoke's raison d'être, they had very little contact with the gorillas themselves. This was a camp rule: trackers were to lead researchers to the gorillas, but remain out of sight. Dian did not want the gorillas to feel comfortable around any black people, only white researchers, ostensibly to limit the threat of poaching. However, this rule was necessarily violated on a regular basis at the critical point when a tracker would come upon the gorillas—just as poachers might come upon the group. Still, the Karisoke trackers would then back away and either return to camp or stay for several hours under cover until the researcher was finished.

Especially in our early months, when the trackers' skills and familiarity with the terrain were most needed, the time on the trail was a great opportunity for discussion. Swahili was the working language in the Virungas, since none of the staff spoke French and none of us had yet mastered the intricacies of Kinyarwanda—one of the most complex of the Bantu family of languages. Swahili, though, was a regional lingua franca, and one that we were fortunate to have learned during our two years in neighboring Congo. We covered a range of topics in our walking talks. Some were about the gorillas: the creatures they only knew from trail sign, swaying bushes, vocalizations, and all-too-brief sightings. Many discussions had to do with life in America. Over time, we could ask more questions about Rwanda. Political subjects were generally off-limits, even though all the men were Hutu, the ethnic group that made up President Habyarimana's power base in northwestern Rwanda. Political decisions were often referred to as *maneno ya Mungu,* or "God's will," reflecting the men's apparent acceptance of matters beyond their control.

Several trackers worked at Karisoke while we were there, their number at any one time depending on need. Generally there was one full-time tracker's position, as for housekeepers, split into two twenty-day shifts. Little Nemeye always took one of those shifts. He was a short, thin young man of about twenty-five, with a friendly nature and a winning smile. He was clearly Dian's favorite at that time, and an excellent tracker. Able to focus on the most subtle of signs, Nemeye rarely followed a dead-end trail. When it came to the gorillas themselves, though, he seemed less interested than some others in the details of their lives. It was a job, and he did it very well.

Vatiri was a frequent partner of Little Nemeye's. Named for the rare passage of an automobile, or *voiture,* at the time of his birth, Vatiri was the trackers' tracker. Acknowledged by all for his skills, his successful location of a researcher lost at night became a Karisoke legend. So, too, was his recovery of a car key dropped along a three-hour trail through thick brush. Yet as good as he was with gorillas and lost scientists, Vatiri's true passion was tracking poachers. Called up from the valley whenever suspicious signs or sounds were detected, he seemed to relish the chase. While captures were rare, he often returned with a load of wire traps, machetes, and other contraband seized from poachers' camps. Short missions of this sort—with a potential for performance bonuses—also suited Vatiri's interest in spending more time on his farm than in the forest.

Rwelekana, who helped Amy pursue Group 5 across the flooded Susa, was another exceptional tracker who preferred to be at home. Although wages were never high at Karisoke—roughly one dollar per day in 1978— this was still more than camp staff could earn in the agricultural sector. Rwelekana converted his meager earnings into a series of land purchases. This strategic acquisition of farmland was extremely shrewd in the face of the growing Rwandan land shortage, and a reflection of Rwelekana's wisdom and industriousness. He was also the tracker most curious about the gorillas, about forests and wildlife in other places, and especially about life in other African countries, in Europe, and in the United States. We would come to know him much better while doing census work, but he was always an enjoyable companion at Karisoke.

Big Nemeye, so named for his precedence over Little Nemeye as well as his solid physical stature, was a former tracker who was *fukuza,* or fired, by Dian. His banishment could be lifted under certain conditions, but his temperament and more limited skills made him a less attractive regular employee than the others. Big Nemeye, too, would later help with census work, where Bill would learn that his skills were greatly augmented by a willingness to work hard and that his character was greatly tempered by abstinence from alcohol.

Whichever trackers were in camp were most likely to be found around the main fire as the day came to a close. Fieldwork finished, the smokers would dry their tobacco before rolling it in whatever paper they could find. Often these were the lined and ink-stained pages from their own children's used school notebooks. When Bill sometimes offered the men factory-wrapped Impala cigarettes, he would joke that he was improving their health. Besides, far more smoke entered their lungs each day from the smoldering wet wood of the Pit than from any number of cigarettes.

Smoke or no, the fire was warm as the cold night air rolled down off the mountain, and the Pit was the center of Rwandan staff life. Discussions were not that different from those at other workplaces around the world, including the right to complain. With time, the topic would invariably return to the central focus of all life at Karisoke: the increasingly strange and reclusive Dian Fossey. Rarely seen outside her cabin, Dian almost never visited the gorillas anymore. Regardless, her presence was felt in many ways. A single name shouted across the compound would send the housekeeper or woodcutter running, and even the most hardened trackers had learned to heed the call. Failure to respond could result in summary suspensions, usually rescinded within the month. More serious infractions might cause someone's hard-earned *mushahara,* or salary of $10 to $20 worth of Rwandan francs, to be waved in his face and then tossed in the fire. The victim rarely responded until he returned to the Pit, where the others would console him with the reminder that *Mademoiselli ana kichwa sawa toto.* All would agree that the strange white woman was, indeed, childlike. They also knew that with little education and no French, they had few if any employment alternatives outside of her camp.

THE DENSE TREE COVER, frequent low clouds, and Dian's dark moods could make Karisoke a foreboding place. Cold and wet were constant companions. Shades of gray sky, the height of clouds, and the intensity of rainfall distinguished days from each other. The equatorial day length hardly varied, and sunrise and sunset were blocked by massive volcanoes to the east and west. Yet Karisoke was also a magical spot that could stir the imaginations and passions of those of us fortunate enough to live and work there.

Chapter Six

Gorillas by the Numbers

M WEZA'S DEATH HAD lowered the Virunga population by one. How many more gorillas had fallen to poachers? How many had been born since the last census five years earlier? How many still survived? These were some of the very basic—and critically important—questions that Bill's census was supposed to answer.

The February census of Mt. Karisimbi—interrupted by malaria—found only fifteen gorillas in two groups. This was disappointing given Karisimbi's massive size and ample habitat. The next target was the gorilla-rich core area of Mt. Visoke. The composition of the two main research groups was well known, but before moving on to the completely unhabituated population we needed to confirm the numbers in three peripheral groups closer to Karisoke. These families were not followed on a regular basis, but occasionally interacted with Groups 4 and 5. If these interactions involved the transfer of a female, as in the case of Liza, Bill would usually set off with a tracker to follow and identify the group, and confirm the presence or absence of the female in question. For the census, though, we needed more rigorous attention to group numbers and composition.

Nunkie was one of the great gorilla success stories. He first appeared on the northern slopes of Visoke as a lone silverback in 1972. Over the next five years, he settled into a small area of rugged terrain almost directly above the research camp, squarely between the larger home ranges of Groups 4 and 5. Whatever Nunkie may have lacked in preferred habitat, he seemed to make up for in personal magnetism. As of 1978, he had attracted at least four females—two from Group 4—and his highly vocal copulatory binges were a matter of record. So were three young infants. Nunkie seemed to have secured his place on the mountain by occupying a less desirable range at the upper limit of the forest zone. He rarely ventured below ten thousand feet and thus had no access to the choice bamboo stands at lower elevations. His home range of steep slopes and ravines provided protection, however, as well as ample food supplies of celery, thistles, nettles, and thick clumps of preferred *Vernonia* shrubs.

Approaching Nunkie's group in 1978 was always something of an adventure. Clambering over some ridge or into a rocky cleft, Bill never knew who he would meet first: a calm female like Papoose or Petula—transfers from Group 4 who would tolerate his close presence—or a minimally habituated transfer from a completely wild group. If the latter, her screams were certain to bring a swift response from Nunkie. Crashing through the undergrowth, he would soon appear in massive full strut, pursing his lips and swatting at any nearby vegetation as he displaced his gaze first to one side, then the other. Nunkie had never made his peace with the presence of white apes.

For the census, Bill was fortunate to first encounter Papoose and her two-year-old, N'Gee, as well as a nearby Group 6 transfer, Pandora. In their company, he sat in relative calm and counted a total of eight individuals, including an only slightly affronted Nunkie. Completing the census work, however, required leaving the family to do a series of at least three successive nest counts. Only in this way could we get the most accurate count possible, due to several helpful aspects of mountain gorilla behavior and biology.

As gorillas move through their range, they spend each night in a different location. At each of these sites, every gorilla above the age of three to three and a half constructs his or her own "nest" by loosely weaving a blend of plant stems, vines, and leaves into a bowl-shaped sleeping platform. Infants sleep in their mother's nest up to the age of three, or until the next infant is born. Even more conveniently, mountain gorillas usually defecate in the bottom of their nests during the night or early morning. It is believed that the dry, warm dung provides added insulation from the nighttime chill. The result is a store of information for census workers. Silverbacks are readily identified by their significantly larger dung, as well as the abundant presence in the nest of white and gray hairs. Subadults from six months to eight years old can be quite accurately aged according to relative dung size. Further distinctions can be made among adults, since the presence of both infant and adult dung in a nest indicates that the adult is a female rather than an immature male. In the case of Nunkie's group, this information provided unambiguous, repeated counts of eight individuals: one silverback, four females, and three infants. This count stood as two return visits to the group failed to provide visual evidence of a newborn infant who might not yet produce solid dung. Nunkie had done well indeed.

The next census target was another peripheral group that occupied an equally rugged range on the eastern slopes of Visoke. Whereas Nunkie had a reputation as *nguvu*—powerful—Group 6's silverback, Brutus, was simply known as *kali*—nasty. Bill had encountered an agitated Brutus once before, when Pablo's mother, Liza, transferred to his band from Group 5. Bill would learn firsthand just how nasty he could be at a later date, but his contact of May 27, as recorded in his field notes, was enough to confirm conventional wisdom.

> *Coming over a small rise at 11:45, I find Brutus 5m away. He screams twice and retreats through a bush tunnel. Members of the group are heard descending into the ravine beyond. Brutus wraagh at 11:50. . . . At 11:54 I believe Brutus has followed others and I enter the tunnel; while inside, Brutus screams and charges to 4m. This is repeated at 11:57 to within 2.5m, terminated with a sweeping vege-swat. I retreat and wait for evidence that Brutus is no longer guarding the other end of the tunnel. . . . At 12:12 I crawl through without mishap.*

Tunnels were to be a recurring theme in Bill's relationship with Brutus over the next two years. On that day, though, the young silverback moved his group across the ravine and sat calmly in full view on the other side for more than an hour. Bill counted six individuals during this visual contact, then confirmed a total of eight from a series of nest counts. This was a decrease of three adults, at least two of them females, since the last count four months earlier by Ian Redmond. Such a loss from natural transfers might at least partly account for Brutus's agitated state. It also raised concerns of poaching deaths in a home range that was very close to human settlements outside the park.

The day after finishing the Group 6 nest counts, Bill moved on to the last peripheral contact. Peanuts had been known by Karisoke researchers for many years: first as a blackback in the since disbanded Rafiki group, then as leader of his own group with multiple females. In late 1977, Peanuts was badly beaten in a fight with another silverback, who took his females and left him so severely wounded that Dian thought he would die. He survived, though, and continued his solitary patrols on Visoke's lower slopes, no doubt hoping to again attract some females. On May 28, Bill approached Peanuts to see if he had succeeded, perhaps with some of the missing Group 6 members.

At 11:03, Peanuts parts vegetation to peer at obs 6m below. He stares for 3 minutes, then sits and scratches. A tear-shaped drainage is still visible from his right eye wound. After a soft slur/hoot at 11:08, he moves to within 4m. He stands at 11:15, moves to within 2m and sits at 11:17. . . . At 11:22 he gives a brief hoot/chestbeat before moving to one meter away: there he knuckle-stands for 2 minutes before placing his left hand 6 inches from obs.

At that point, Peanuts's downtrodden physical appearance, his lack of any family or companions, and his apparent interest in social contact across species lines led Bill to place his hand over Peanuts's and give it a few pats. He couldn't know what that gesture of intended consolation meant to a gorilla, but the silverback sat down beside him. For the next ten minutes, Peanuts remained within arm's reach, distractedly grooming himself and casting occasional glances toward Bill. With any extended eye contact, he would turn shyly away, then look up and around as he scratched his massive chin, producing a sound like a fork scraping thick boot leather. Whatever Peanuts might have been thinking, a flash of brilliant crimson from the underwing of a passing Ruwenzori turaco jolted him into action. Moving several meters away, he inserted himself into a thick clump of vegetation and spent most of the next hour busily consuming vast quantities of lush thistle and celery.

As Peanuts finally ambled off through the undergrowth, Bill felt both privileged and saddened: privileged to have felt the powerful bond of close contact initiated by Peanuts, yet sad that he was the sole such contact that Peanuts might have for months or even years. We had already seen enough of gorillas to appreciate that they were supremely social creatures. The solitary male might be an evolutionary necessity in a polygamous society, but Peanuts was a tragic figure nonetheless.

ADDING THE RESEARCH and peripheral groups to the earlier count from Karisimbi, Bill reached a total of fifty-eight gorillas. To learn any more he had to move beyond the Karisoke contact zone, establish a series of base camps, and conduct multiday censuses of the remaining wild gorillas.

Bill and Rwelekana shouldered their heavy packs and followed the trail north from Karisoke through an open forest dominated by the contorted shapes of giant *Hagenia* trees. Uprooted trunks littered the grassy glades, highlighting the risks of growing at such precarious angles with heavy loads of mosses and epiphytes. In death, many of these trees had felt the axe of the Karisoke woodcutters. The trail skirted the saturated edge of the upper meadow before leaving the broad saddle to rise through dense stands of five-foot-tall stinging nettles in the heart of Group 4's range. Climbing higher into more rugged terrain, Bill and Rwelekana left the nettles behind. Mt. Visoke's steep slopes are cut every few hundred meters by sharply eroded ravines. Ranging in size from mere clefts to minor canyons, they were lumped as *bondes* in the simplified Swahili spoken at Karisoke. Most were given names. *Muti kufa* (Dead Tree), *Ndege* (Bird), and *Kulala* (Sleep) ravines were all within the range of Group 4. Next came the more evocatively named *Bonde ya Chui* (Leopard), *Bonde ya Kujiua* (Suicide), *Bonde Mkubwa* (Grand Canyon), and the world of unhabituated gorillas beyond.

Rwelekana knew this area from his eight years of experience at Karisoke and led Bill unerringly to the best crossing site for each ravine. Beyond the steep gorge of Kujiua, we came to our intended bivouac site: an abandoned leopard cave at 10,270 feet on the edge of Bonde Mkubwa. The Grand Canyon was named for its boxlike shape and sheer walls with few transit points. Leaving our packs, we climbed a rocky ridge along the canyon's western flank. The footing was slippery and the gnarled giant heath required that Bill maintain his six-foot-two frame in a constantly stooped posture. Still, the ridge provided an excellent view into the canyon bottom a few hundred meters below, where Rwelekana's expert eyes searched for signs of any recent gorilla passage.

The basic technique for a gorilla census is to walk up and down the ridges and ravines of each volcano in a systematic fashion looking for trail sign. As gorillas move through the thick herbaceous vegetation in which they find most of their preferred food species, even lone silverbacks leave a path of bent and broken plants. Larger groups may not flatten a wider trail, but the passage of many individuals leaves a deeper and longer lasting mark. The easiest way to find these trails is to walk along a ridge and look for telltale cuts through the herbaceous mat that colonizes the exposed sides and bottoms of the ravines below.

Peering into the Grand Canyon, Rwelekana did not take long to find what he was searching for: *Iko njia ya ingagi*. A gorilla trail. Even Bill's less experienced eyes could detect the dark line of flattened plants that carved through the undergrowth perhaps eighty meters below. But he was not prepared for the silverback who soon emerged at the head of that line. We not only had the great good fortune to come across fresh trail on our first day out, but the gorillas themselves were immediately visible.

At 12:30, a silverback crosses the bottom of the canyon followed by 2 adults and a juvenile/young adult. . . . SB [silverback] chestbeats as I descend. . . . While these 4 climb far side, at least 4 others remain on S side and are climbing back up toward me. . . . An adult female with pursed lips plays rear guard, climbing out on limb of a large Hagenia; from there she cbs [chestbeats] and slaps trunk frequently . . . at 12:52 a YA with prominent wart joins "Purser" in tree, followed by a juvenile at 12:55. Both cb and slap trunk . . . a second SB hics out of sight on near side. . . . Purser is on her way down at 13:03 as SB crashes through group and moves into canyon; all follow. I remain above and climb out on a limb to get a better view and count as they cross to the far slope. Between 13:08–13:12 I count 8 more individuals. Added to the 4 already across this makes a group composition of 12: 2 silverbacks, 7 adults, 1 young adult, and 2 juveniles . . . they move out of sight about 80m up the N slope of the canyon and continue to cb infrequently.

To observe an unhabituated group for forty-two minutes was one thing; to make visual contact with what would prove to be the entire family was completely unexpected—and almost never repeated in the rest of the census. Bill was able to see eight of the individuals well enough through high-powered binoculars to draw their noseprints. Most animals display some individual identifications, such as the variations in a zebra's stripes or the marks on an elephant's ears. The gorillas of Groups 4 and 5 were known to us at a single glance by their body type and posture. But to identify unhabituated and less-well-known gorillas, we depended on the unique patterns of deep lines that creased the skin above their noses. Sketches of these noseprints were kept at Karisoke for the purpose of monitoring individuals who were not members of long-term study groups. The value of this system was evident when, upon our return to camp several days later, we were able to match four of the prints with those of three females and an adolescent from Peanuts's former group.

Viewing gorillas is informative and personally satisfying, but the heart of census work is the nest count. So with less than three hours of light remaining, Rwelekana and Bill turned away from the gorillas and climbed back out of the canyon to look for the group's morning trail. Finding it less than a hundred meters up the ridge, we backtracked southwest across a broad herbaceous slope and into the Kujiua gorge. After just over an hour, we reached the nest site of the night before on the far rim. Twelve nests were counted, but none contained infant dung: a disturbing finding for a group that probably had four to five breeding-age females among its seven adult members.

Following a rapid return to the Grand Canyon, Bill and Rwelekana retrieved their packs and settled into the evening's accommodations. The cave proved to be more of a recessed ledge in the canyon wall, perhaps six feet deep and no more than twenty feet long. The only sign of a leopard was some very old scat in one corner; but the single steep entryway to one side and the sheer drop-off before us made for interesting thoughts about a possible return of the *chui,* as the leopard is known in Swahili. After sharing a pot of beans and rice, however, we stretched our sleeping bags out on the rock ledge and fell quickly into a deep sleep.

The piercing screams that next ripped the nighttime stillness were of a kind and intensity that Bill had never imagined possible. After taking stock of where he was—and the limited options for escape—he turned on a flashlight and saw that Rwelekana, too, was awake. The screams returned, building to a terrifying crescendo. It was easy to imagine some poor creature being ripped to pieces and eaten alive, sinew by sinew, somewhere in the dark and brutal world beyond our cave. Was the leopard the killer? Would it drag the remains back here? Seeing Bill's widening eyes, Rwelekana smiled and whispered, *iko imperere.* Bill recognized the Kinyarwanda word for the tree hyrax, but he couldn't accept that the shrieks could emanate from a mammal the size and appearance of a hedgehog. Some scientists classify the hyrax as the closest living relative of the elephant. Others report that its bansheelike vocalizations proclaim the hyrax's territory and its readiness to mate. Before Bill could rationalize either of these facts with his own perceptions, he fell back to sleep.

The next several days were less eventful, but nonetheless full. Additional nest counts confirmed the total of twelve individuals and no infants in the first group. We found two more families totaling six and eight individuals on the northeast slopes of the mountain. We dismantled an active poacher hut on the appropriately named Anger Hill and destroyed a patch of marijuana near another poacher cave. Unfortunately, we ran out of food and had to restock before we could finish all of Mt. Visoke. To save time, if not effort, we took the most direct course back to Karisoke. This meant crossing a succession of ravines midway up the mountain until we reached one of the research access trails. There we crossed a very fresh gorilla trail that we had to follow in case it revealed another unknown group. Within minutes, we were face-to-face with a screaming, ground-slapping silverback. Given the style and the setting—another claustrophobic tunnel of vegetation—Brutus seemed the likely candidate. We confirmed this identification as he repeated his ambush act four more times while we tried to move down the ridge toward Camp Trail. There we parted ways, Rwelekana hiking down to his home off the mountain, while Bill climbed the final half-hour to camp. From high above, Brutus continued to vent his displeasure.

B ILL'S PLAN TO return to finish the Visoke survey was delayed by another curious decision by Dian. She was quite keen to learn of Bill's findings and initially seemed encouraged that he thought there were more gorillas to be found on Visoke. It was disappointing not to find more young, but the overall numbers were good. So Bill was shocked to receive a note from Dian the next day stating that she would no longer be able to afford the cost of the census effort. At a maximum of $10 a day for two Rwandans and all the beans they could eat, expense was an improbable explanation. She was more likely concerned that the somewhat positive numbers would conflict with her public statements that the mountain gorilla population had continued to decline and was now below two hundred individuals. It was a dark view that increasingly matched her mood, for which she desired neither proof nor refutation.

The problem was resolved after a trip to Kigali, when ORTPN advisor Alain Monfort contributed $300 toward the census from a small Belgian research fund that he controlled. Within a month, a new team was back in the field. Big Nemeye took over as tracker, since the revised schedule no longer fit with Rwelekana's commitments to farmwork. An occasional Karisoke porter and tracker-in-training named Antoine Banyangandura was hired to guard our camp and keep the daily bean pot cooking. A Peace Corps volunteer named Roger Palm even contributed a few weeks of his vacation time to the effort.

On July 21, the reconstituted census team established a base camp at Ngezi, named for the small lake-filled crater that jutted out from Visoke's northeastern flank. Our first task was to clean the filth and litter left by poachers in an abandoned cabin at the end of the lake. This tin structure dated from an earlier short-lived effort by Dian and a European friend to provide a camp for tourists far away from Karisoke. The cleanup completed, we took advantage of the waning light to relax at the lake's edge before sundown. A pair of yellow-billed ducks paddled across a stretch of open water, breaking Visoke's reflection on its surface. From the thickets around the lake came the bell-like *goong-goong* of the reclusive buff-spotted crake, a small bird whose long feet enable it to walk on floating vegetation. A broad bowl of herbaceous vegetation just beyond the lake was marked with trails that indicated possible gorilla use. Far above, a soft glow briefly flourished, then faded, as the sun set behind Visoke's summit.

It would be the last relaxing moment on this trip. On the first day out, we discovered two new groups whose composition clearly showed that we had not counted them before. The first group appeared on a ridge as Bill and Nemeye were showing Roger how to count nests. When the gorilla group crossed a nearby ravine, we had a perfect view of a magnificent family of twelve individuals: two silverbacks, four females with four infants, and two other subadults. As we watched, the gorillas settled down to feed and rest in remarkable calm. Then came voices and a silverback scream. Then more voices, human voices, from below. As the gorillas fled, five European tourists emerged from the bushes, shocked to find other humans instead of their quarry. They were in the company of a poorly dressed park guard, who looked as though he desperately wanted to be somewhere else. These were some of the few hardy tourists who paid a $5 fee, or an equivalent bribe, for the chance to see wild gorillas in 1978. Though sanctioned by ORTPN, such forays were led by inexperienced guides who knew nothing about gorillas or how to behave in their presence. Few groups succeeded in actually observing gorillas, but these people had at least made vocal contact. Whether their guide wanted to push the experience any further was not at all clear.

Our team went on to confirm a count of twelve individuals in this group, which we recorded as Group 11 in the final census results. We added a second group of six gorillas over the following days. This completed the survey of Visoke proper. We next moved out onto a broad plain bounded by a series of hills to the north of Ngezi. Located entirely in Congo, the area was covered with a thick scrub vegetation and had little to recommend it as likely gorilla habitat. Yet it had to be searched. We covered a large area in two days of forced ten-hour marches. Surrounded by a sea of *kibaba nzovu*—elephant nettles that packed a wicked twenty-four-hour sting—and exposed to the rare dry season sun with no surface water, the Virunga Wonderland was transformed into Green Hell National Park. But these conditions didn't keep other humans away. Fresh poacher trails crisscrossed the plain and we cut dozens of traps. At one spot hyena prints overlay fresh duiker tracks, both of which were followed by the small barefoot prints of a pair of poachers, probably pygmies. We added our lug-soled boot prints to the mix and imagined the intriguing possibility of a group encounter. But the sight of smoke pulled us off the trail toward a distant hill. Fresh duiker remains greeted our arrival at the smoldering fire. Inside one of two small huts, Nemeye discovered a plastic sack of a white powder used to poison elephants at watering holes. We also recovered wire traps and machetes before we destroyed the huts. Climbing the steep flank of Ngezi late on July 24, Nemeye found and cut three more fresh snare traps within five hundred yards of camp.

Arriving at the edge of Ngezi around 5:00 P.M. that same day, we were exhausted but relieved to know we had finished the difficult plains section of our census. We would move on to the bamboo hills to the east of Ngezi with renewed hopes of seeing gorillas. Antoine was waiting as we approached the cabin. He bowed his head as he held out a note for Bill in both hands. The watchman's posture said to expect the worst.

Uncle Bert has been killed. We don't know where the other gorillas are or if they are alive. Please come back to camp as soon as you can. Dian.

Chapter Seven

Massacre

THE TRAIL FROM NGEZI wound uncertainly as it sought the path of least resistance down the crater's rugged eastern flank. Dense stands of *Vernonia* and bamboo provided a low ceiling of almost constant cover, while a gauntlet of rocks and exposed roots demanded attention to the trail below. Along the way, rocky outcroppings afforded release from the cover and the opportunity for Bill to look up and out at the world beyond his tired feet. To the northeast, dozens of low-lying craters gave way to the jagged peaks of Sabyinyo, the notched flat top of Gahinga, and Muhavura's classic cone. To the east and south lay the rich farmlands that covered the Virunga piedmont like a patchwork quilt. Looking at their neat rows of crops, stone boundary walls, and permanent households, Bill couldn't imagine that these settlements had been carved from the park's lower reaches less than ten years earlier.

Adding to the air of normalcy were the sounds that rose to reach Bill as he descended the path. They were the sounds of another African day coming to an end: greetings called out as people returned from fields and markets; shouted invitations for men to gather for drink and talk; calls from women to collect their children to help prepare for supper—all punctuated by the bleated resistance of goats being led to evening lockup on the tethers of little boys. It was hard to believe that anyone from this bucolic setting could be associated with the cold-blooded killings that had caused Bill's forced march back to Karisoke. The fatigue of the full census day and the shock of the news from Dian had put his mind on autopilot as he passed quickly along the edge of the park to join Camp Trail at its base. A final shot of adrenaline shortened the climb up to camp.

A MY'S NOTE ON THE DOOR of our darkened cabin said simply *at Dian's*. A few minutes later, we held each other in a wordless embrace at the door of Dian's cabin. Inside, Bill was surprised to see Betty Crigler there. The U.S. ambassador's wife was a close friend of Dian's and a frequent visitor. Dian emerged from another room looking even more stooped and haggard than usual. She briefed Bill on the situation as a gallery of gorilla photos gazed down from the wall behind her. Among the faces half illuminated by the gas lanterns was that of Uncle Bert. Dian confirmed that Uncle Bert was found dead and decapitated that morning, and stated her conviction that others were dead or captured, too. David Watts had actually followed Group 4's trail up onto Visoke's slopes after the killings and was amazed at their relative calm. But he could not confirm the presence of all the other gorillas before darkness settled in. With that report, Dian retreated to her room.

Betty Crigler expressed concern about Dian's state of mind: *She's blaming herself for the deaths; she thinks it's revenge for how she's dealt with poachers.* Given that suspected poachers had been subjected to a variety of tortures and other abuses carried out under Dian's direction, the prospect of retaliation was not unthinkable. But there was nothing to support this view either. Our ability to think clearly about anything at that point was seriously limited, in any event, and we left Betty without a response to her concerns.

Dian reemerged to join us at the dinner table, where she barely touched her food and avoided all eye contact. Her tortured face spoke volumes about the depth of the hurt inside. David, too, showed the extreme pain of the day's events. He had discovered Uncle Bert's still warm but headless body just before eight o'clock that morning. It would have been a brutal experience for anyone, still more so for someone as gentle and sensitive as David: a very quiet and serious researcher whose main emotional outlet was playing the violin. He would show a much tougher side in dealing with the latest tragedy to befall Group 4.

We returned silently to our cabin sometime before midnight. Physically and emotionally drained by the day's events and exertions we collapsed into an all-too-brief sleep.

A T DAWN ON THE MORNING of July 25, we set off in a large group for the site where Uncle Bert was killed. Amy and David followed Group 4's trail with Little Nemeye and Basira, while Vatiri and Kana—an occasional Karisoke worker—set off on the day-old trail of the poachers. Bill stayed behind with Rwelekana and Rukera to reconstruct the attack and to search the site for more bodies. Our starting point was the large spot where Uncle Bert's blood had poured out and dried to form an incongruous dark red mat on the crushed celery and other greenery below. The vegetation was flattened in all directions, reflecting the chaos of the attack. It was difficult to separate gorilla flight trails from those of the poachers and of the Karisoke staff who had briefly searched the area the day before.

After twenty minutes of searching, Bill followed what looked like the trail of a solitary individual. There he discovered a large black form, facedown in an area of thick sedgelike plants. He called to Rukera, who helped turn the body over to reveal the gruesome death mask of Macho. Her face was locked in a contorted grimace as she fell from a single rifle shot in the back. With Macho beyond help our concerns turned to Kweli. There was no sign of Macho's two-year-old son anywhere near her, though what could have been an infant's narrow path did split off from her trail to join the general maze beyond.

Nearly two hours of intensive searches turned up no further sign of Kweli and no other bodies. The official death toll stood at two. Additional camp staff were then called to help carry Macho's body back to Karisoke. Bill stared at her tortured face as the young mother was loaded on a crude litter: her namesake eyes now dimmed forever. She and Kweli were the first gorillas to greet us six months earlier, on our first day in the Virungas. As the procession headed slowly back to camp on a hazy dry season day, Bill wondered why the poachers didn't take her head, too. She hadn't run more than thirty yards before she dropped. Certainly her killer could have found her and hacked off her head as they had with Uncle Bert. Did they leave her to pursue Kweli, a more valuable live infant? But there was no sign of poachers following the infant's trail. Did David arrive to interrupt the poachers in the act? Did they just kill for the fun of it? Was it revenge?

Vatiri returned late that afternoon with hard evidence that this had not been a chance encounter between the gorillas and their killers. The poachers had camped for at least two nights on a hill just west of the attack site, from where they had apparently monitored the group's movements. They had certainly watched David's daily coming and going and must have guessed that he would return to the group on the morning of their attack. This investment of time showed clear premeditation. Yet the willingness to take a risk also indicated an expectation of reward. So, again, why didn't they take another head? Why didn't they pursue Kweli? Or did they?

The return of Amy and David's group provided the last bits of information to ponder for that day. They had followed Group 4 well into Congo, where the remaining gorillas finally slowed down to eat and rest. This allowed David and Amy to approach close enough to see that Kweli was indeed alive and with the others. But the young gorilla had been shot, too. This seemed to confirm that the killers were after the silverback's head and a live infant. These were the two most valuable commodities in the cruel international market that was driving local poachers to pursue mountain gorillas. Whatever the reason for the latest killings, it still didn't explain why the poachers abandoned their grisly task in such apparent haste. If David had been close enough to interrupt them, he would have surely heard the shots.

The next day, Vatiri returned to track the poachers to where they exited the park in the commune of Mukingo—home of the *bourgmestre* with the attitude and the scar. Meanwhile, Dian made a rare trip down the mountain to enlist the local prosecutor Paulin Nkubili's support for the arrest of previously known suspects from that district. We could only help David follow Group 4 and hope for the best.

It was surprising to see that Group 4 had fled far into Congo, almost to the base of Mt. Mikeno. It was also a shock to recognize that we had been in the area before. Five months earlier, Dian had sent us with Ian and several Rwandan camp staff to frighten the gorillas away from what she believed to be a major poacher area. Our instructions were to align ourselves on the side opposite the direction we wanted the group to travel, then make as much noise as possible until they fled. Ian fired a pistol in the air on at least two occasions to achieve the desired effect. After two days of such ambushes Group 4 was back on the slopes of Visoke, apparently unaware of our involvement in the charade. Or at least they didn't hold it against us when we contacted them in our role as researchers the next day. It was uncomfortable to frighten them as we did, though, even under orders and with the best of intentions. Now it was much more discomforting to wonder if we had done the right thing. *Might they have been better off on Mikeno? No, Mikeno is a poachers' paradise. But look what happened . . .* There is a lot of time to think when walking through the forest. Sometimes too much.

As we watched Group 4 with David, Kweli was everyone's preoccupation. He was clearly in pain, and seemed reluctant to move his left arm. Yet Beetsme and Tiger both groomed his wound, and he often slept with his older brother, Tiger. Perhaps he had a chance.

Meanwhile, Beetsme began what could only be described as pathetic attempts at silverback displays and vocalizations. Beetsme was believed to be eleven or twelve years old in the summer of 1978. He had migrated into Group 4 several years earlier: the only recorded instance of a solitary male transfer between groups. When asked her thoughts on his origins, Dian responded with a surprised, "Beats me!" The name stuck, along with her peculiar spelling. Beetsme was too young to take up the leader's mantle under normal conditions; but this was an encouraging sign in an exceptional situation. If he could assume the role and responsibilities of a silverback, he might hold the group together and avoid the possibility of further deaths. At greatest risk were one-month-old Frito, three-month-old Mwelu, and perhaps three-and-one-half-year-old Titus. Under normal circumstances, an established silverback is likely to live into his forties. By that point, he should have produced a successor from within his own bloodline. This was the situation with Beethoven and Icarus in Group 5. In Group 4, though, Uncle Bert had been killed in his prime; his only possible successor, Digit, had been killed six months earlier. These deaths created a situation that only rarely occurs in nature: a group of females and young infants with no mature males. This void would almost certainly be filled by a new silverback, with or without his own group, taking control. In the process, he would likely kill the youngest infants in the group.

Infanticide is driven by the ruthless logic of genetics. Gorilla males, like those from most other species, seek to breed as soon as possible with new females. This assures that their genes will be passed on to future generations. Yet gorilla females cannot become pregnant during the approximately three years that they nurse their young. The brutal means to end a mother's lactation—and quickly restart her reproductive cycle—is to kill her infant. This is unlikely to happen when the successor is related to the deceased silverback, as would have been the case with Digit. But when an unrelated silverback takes over a group, the youngest gorillas are at great risk.

The death of Uncle Bert provoked a change in Beetsme. Within a week his vocalizations gained resonance and his displays seemed more confident. He began to harass Flossie, charging, hitting, and biting the mother of newborn Frito. Twenty-two days after Uncle Bert's death, Beetsme gave a killing blow to Frito. Flossie carried the dead infant for two days while Beetsme continued to harass her. Once she dropped the body, David saw Beetsme copulate with Flossie. But for all his aggression, Beetsme could not keep control of Group 4. Within days of the infanticide, Flossie, Cleo, and Augustus all transferred to Nunkie's Group. Two weeks later, Flossie and Cleo would transfer again, this time to the Susa group.

The remaining adult female, Simba, did not transfer with the others, and Beetsme made no attempt to harm her infant, Mwelu. This may have reflected the possibility that Beetsme was Mwelu's father, since earlier records had noted that Simba and Beetsme had copulated, even while Uncle Bert and Digit were alive. Beetsme also continued to help Tiger groom Kweli. By

September, Kweli seemed to be on the road to recovery from his bullet wound. David noted that he had no difficulty in keeping up with the others and that he was feeding normally. He stopped playing, but David felt that this could be attributed to depression after the loss of his mother and father. Then in October, Kweli quickly deteriorated. He stopped eating and whimpered frequently. The next morning, he didn't seem able to move from his nest. Beetsme tried to prop him up, but Kweli slumped back into his bed. After an hour the other gorillas moved off to feed nearby. Then David took advantage of a powerful rainstorm to lift Kweli and carry him back to Karisoke. He was not breathing when he arrived at camp. David and Dian could not resuscitate him.

The day after Kweli's death, Tiger, Beetsme, Titus, and Simba circled back to the exact spot where they had last seen Kweli. After a brief inspection of his nest, they moved off to feed. In December, Simba followed the other females and transferred to Nunkie. Within days, Nunkie had killed Mwelu and was copulating with Simba. Barely one month later, Beetsme, Tiger, and Titus joined forces with Peanuts. Five gorillas were now dead from this one poacher attack; six, counting Digit. Flossie, Simba, Cleo, and Augustus all transferred to other groups. Peanuts, Beetsme, Tiger, and Titus were on their own in a group with no females. The future of an entire lineage—the primary focus of Dian Fossey's work and the best-known family of gorillas in the world—was irreversibly altered by the July massacre.

WE DID NOT SEE the deaths of Frito and Kweli, but we experienced them through David's eyes and emotions. For us the most brutalizing aspects of the 1978 killings were the autopsies of Uncle Bert and Macho. Before going to Ruhengeri to press for the arrests of suspected poachers, Dian left brief instructions to cut the gorillas open, recover any bullets, and remove parts of certain organs before burying them. Yet there were no proper implements for the grisly task. Instead, we cut through the gorillas' incredibly thick skin with a Swiss Army knife and opened Uncle Bert's rib cage only with multiple blows from a machete. Once inside his massive chest cavity, we were stunned to see how even his great size and strength were no defense against the ravages of a single bullet, probably fired from a high-powered military rifle. Entering through a tiny one-quarter-inch hole in his chest, the projectile had shredded soft tissues and shattered his spine before exiting through a much larger hole in his back. Macho was the same. We never recovered a single bullet from any of the gorillas shot during that terrible time.

We had come to help save gorillas. Now we were surrounded by dead and decapitated bodies and were up to our elbows in their blood. It was impossible to make any sense of events, but this was the reality of gorilla conservation in 1978. Our research might help us understand the African poachers' world, the harsh economics of local poverty, and the power of foreign markets. But nothing could ever justify the killing of such innocent and peaceful creatures. Nor could we begin to comprehend how anyone could be so sick as to pay for a gorilla's disembodied head or its brutally orphaned infant.

Section Two

Pieces in the Puzzle

Chapter Eight

Final Count

*S*ABYINYO. The teeth of the old one. Time and erosion have removed any trace of its once massive cauldron, leaving only five jagged incisors with wide gaps between. Below its peaks, a dark carpet of bamboo softens the sharp edges of deep canyons and crevasses gouged from its flanks.

At the beginning of the twentieth century, the colonial empires of Britain, Belgium, and Germany intersected on Sabyinyo's highest tooth. In their rush to carve up Africa, the European powers did not feel constrained by their profound geographical ignorance of the vast continent. Instead they sat around a table at the Berlin Conference of 1885 and drew lines along known rivers, presumed watersheds, and reported landmarks. Yet explorers from Baker and Burton to Speke and Stanley had avoided the mountain kingdom of Rwanda, steered away by guides who knew only of the Rwandans' fierce reputation as fighters.* The result was a complete lack of information about the region, including the failure to note the existence of either Lake Kivu or the Virunga chain of volcanoes. Stanley did record the name of a single volcano called *Mufumbiro,* and it was on that dubious spot that the Berlin conferees marked the shared boundary point of the British, Belgian, and German colonies. Years later, the first Europeans to actually visit the area would discover that Mufumbiro did not exist. In its place was a long chain of partially active volcanoes. Ultimately, a tripartite commission would allocate

---

* Stanley overruled his guides' advice and attempted to enter Rwanda across the Akagera River. He was repulsed by a thick fusillade of spears and arrows, then circumvented the region as he moved on to explore the Congo.

93

the Virungas' southern flank to the German colony of Ruanda-Urundi, its northwest sector to the Belgian Congo, and a smaller northeast sector to British East Africa.

Oscar von Beringe was a young military officer assigned to oversee Germany's investment in Rwanda. In 1899 he was the first white man to explore Mt. Visoke. Three years later, Captain von Beringe returned with a patrol to display the German presence near the trinational border. Climbing a narrow ridge at about 10,400 feet on the east side of Sabyinyo, he spotted a group of "black, large apes." He proceeded to shoot two of the animals, "which fell with much noise into a canyon. After five hours of hard work," one body was recovered. The West had discovered the mountain gorilla and von Beringe's name would be forever linked with that of his prey, *Gorilla gorilla beringei.*

FOLLOWING THE JULY MASSACRE of Group 4, life at Karisoke slowly returned to a semblance of normalcy. Amy and David visited the gorillas every day. Dian returned to the darkness of her cabin. One suspected poacher was held in the Ruhengeri prison; at least one other fled to Congo. But there was no solace from the loss of another five gorillas from the group Dian had studied for years, no way of replacing five more gorillas lost to the population.

On August 5 Bill left with Big Nemeye to resume the gorilla census in the eastern half of the Virungas. The first camp was situated in the saddle between the Muside crater and Sabyinyo's western flank. Nestled in a thick blanket of bamboo, it was an excellent location with permanent water. Sabyinyo's multiple peaks were backlit by the first morning light. One drawback to the site was made clear on the first morning when our breakfast of tea and honey was interrupted by the blare of loud music. The offending radio was held by the lead man in a string of thirteen smugglers, each carrying newly purchased goods from Rwanda. A few minutes later, fifteen men approached from the opposite direction, each bearing a fifty-kilo sack of coffee beans on his head. In this way we learned that the Muside–Sabyinyo trail was a major smuggling route openly condoned and encouraged by Rwandan authorities who benefited from both ends of the trade. The coffee was purchased from sources in Congo and Uganda, where local currencies were nearly worthless at that time, and resold by Rwandan entrepreneurs for a handsome profit in Kigali. Meanwhile, the smugglers used their Rwandan francs to purchase commercial goods in Ruhengeri—radios, bicycles, corrugated tin roofing—that were not available across the border in their own countries. A startling image of Rwanda as a regional economic powerhouse began to emerge over morning tea, as groups of up to seventy-six smugglers would pass in a steady stream over the following days.

Thoughts of the Smugglers' Highway disappeared within minutes of our leaving camp under the watchful eyes of two local *zamus,* or guards. Climbing west, we found fresh dung almost immediately. The gorillas couldn't be far away. Nemeye smiled and announced that this large, multilobed bolus was left by an even rarer Virunga resident: an *ngurube mukubwa,* or giant forest hog. Fresh furrows indicated where it had been digging up roots just before we arrived. Perhaps a solitary old boar, its souvenir pile of steaming dung was as close as Bill would ever get to that elusive five-hundred-pound tusker in all his years in Africa. The dark cover of bamboo and *Vernonia* gave way to more open habitat as Bill and Big Nemeye climbed higher. Slight *Hypericum* trees lined the Muside crater rim, their willowlike branches and yellow flowers lifting and falling with currents of wind that flowed over the summit. Farther west, dozens of smaller craters and an equal number of deep gashes along parallel fault lines recalled to a more violent volcanic past. Over the next several days, this narrow six-mile stretch of low-lying terrain yielded thirteen more gorillas in two small bands.

After watching Sabyinyo at sunrise and sunset from camp, we next turned east to tackle that most difficult of all the mountains. Twenty-six days of extreme effort later, we had counted another twenty-seven gorillas in four widely dispersed groups. The gorillas on Sabyinyo spent most of their time feeding on the herbaceous vegetation that lined its deep canyons and ravines. On many days we could cover no more than a single canyon, as sheer walls blocked passage to the next and forced us to retrace our steps to the point of entry. Often, we could see where the gorillas had used their natural four-wheel-drive to scramble over a precarious ledge to reach the next canyon, leaving their upright cousins behind.

Vegetation, too, was a problem on Sabyinyo, especially the bamboo thickets that covered most of the mountain below 9,500 feet. Never easy to traverse under the best of conditions, a natural die-off had turned much of the lower bamboo zone on western Sabyinyo into an impassable tangle of dead stalks and vines. The lush greens of the Virunga forest gave way to dried yellows and browns. Our boots kicked up clouds of dust. Heavy poaching in the area had also caused most of the elephants to flee, removing nature's best bulldozer and trail maker. Nemeye and Bill took turns hacking their way to higher ground with machetes. One day, Bill's impatience with slow progress through this clutter drove him to crawl up and over the interlaced bamboo thicket. This natural trellis was a precarious perch for someone his size, but he gave a self-satisfied smile to Nemeye as he passed him overhead. The smile lasted for another ten yards until the fragile canopy broke and Bill crashed about fifteen feet to the ground, where he impaled his neck on a broken shaft of bamboo. Nemeye rushed forward, his dark eyes a mirror of concern as Bill carefully removed the three-inch shard. When only a small amount of blood leaked out of the wound, Nemeye's eyes softened and a smile appeared. Kneeling on the forest floor, we shared the African ritual of laughing at disaster narrowly averted: laughter that is even stronger when the disaster is brought on by one's own foolishness.

Strong bonds form during a census. For seven to ten days at a time the tracker and researcher are each other's rescue team, co-workers on counts, social support units, commiserators in the rain, dinner companions, and tent mates. On Sabyinyo's long census days we would also entertain each other with songs. Big Nemeye's tastes tended toward the religious and patriotic, while Bill favored an eclectic mix of Western folk, rock, and show tunes. One day, Bill was working his way through the score of *Oh What a Lovely War!*, a musical of mostly antiwar parodies sung by soldiers in World War I. As he sang "When This Lousy War Is Over," Nemeye recognized the tune from "What a Friend We Have in Jesus." He seemed pleased that there might be some hope for Bill's salvation, and especially happy to find a song they could sing together. Nemeye added his religious lyrics in Kinyarwanda to Bill's secular version, blissfully unaware of the incongruent content of the new duet. The tune became a daily favorite for the rest of the census.

Another time in the bamboo zone, Bill grew weary of the endless folding of his tall frame under angled stalks. A random thought led his mind into a parody of the Oscar Mayer theme song—a cultural relic from graduate school in Madison, Wisconsin, home of the Oscar Mayer wiener.

> Oh, I wish I were a little tiny Mutwa,
> That is what I'd really like to be.
> For if I were a little tiny Mutwa,
> Then my back would not be killing me.

Nemeye listened, then asked why Bill was singing about a *Mutwa,* or pygmy. Bill explained that he would feel a lot better if he were only five feet tall, like a pygmy, and didn't have to bend over constantly. Surprisingly, Nemeye grew angry at this response. He stated with considerable indignation that *no white man wants to be an African, so he certainly doesn't want to be a pygmy.* This led to an interesting discussion of human origins as we worked our way up another deep gorge with no sign of gorillas. In Nemeye's theory, white people must have arrived first on earth because they owned most of the money and material goods. The Tutsi arrived next, which explained why they were Rwanda's traditional overlords and continued to enjoy comparative wealth and status. Hutu farmers, like Nemeye's ancestors, came later and found only the land to claim and work. Last to arrive were the poor Twa, pygmies who were left to forage in the forest. It was an enlightening insight that turned the widely accepted sequence of ethnic arrival in the region—first Twa, then Hutu, Tutsi, and European—on its head. Yet it also had its own clear logic from Nemeye's perspective.

If Big Nemeye held ethnic prejudices, he kept them to himself. He was a member of the large Hutu majority that now dominated Rwandan politics after centuries of Tutsi rule. Virtually all of his neighbors across the fertile Virunga lava plain were Hutu. He purchased goods from Tutsi merchants in the town of Ruhengeri and pursued Twa hunters in the park. Still, one of Nemeye's favorite songs had the recurrent refrain: *I'm a Hutu, you're a Tutsi, he's a Twa—we're all Rwandans.* Bill grew tired of the tune, yet appreciated its ecumenical message of ethnic harmony.

Rwandans do not make a habit of revealing much about themselves and are generally much more taciturn than their outgoing neighbors from Congo and Uganda. They're not shy about asking foreigners questions, though, so evenings around the campfire often became "tell us about America" time for Nemeye and the camp guards. Political issues were not a major interest, but American women—their pants and personalities—were a source of bafflement and intrigue. So was the practice of buying land: a foreign concept that was spreading rapidly into Rwandan life, especially in the fertile lava zone around the Virungas. Mechanized farming, supermarkets, and superhighways were popular subjects, as was the Apollo space program. Yet somehow moon landings and moonwalks were more easily accepted than coinoperated vending machines. The idea that you could drop coins into a metal box and that food—sandwiches, hamburgers, fruit, hot and cold drinks— would drop into your hands evoked the most bafflement, head shaking, and laughter at the ways of the *abazungu*. Explaining all this in Swahili was a challenge for Bill, but it was an entertaining way to end hard days of fieldwork before heading off to sleep.

O N SABYINYO'S EASTERN SLOPE, we found our fourth and final group of gorillas on that mountain: a small band of six individuals. We picked up their trail on an exposed lava fin near what we assumed to be the unmarked Rwanda–Uganda border. It was eerie to think that von Beringe might have taken aim at gorillas on that very same ridge in 1902. We followed the trail, doing nest counts along the way and catching one brief look at the group as they moved steadily northwest around the mountain. The ragged line produced by illegal farmlands encroaching on the Ugandan side was clearly visible below. Earlier, we had spent considerable time counting gorillas within the Congolese sector of the Virungas, but working without authorization in Uganda was different. Idi Amin had recently failed in an effort to take over the northwest corner of Tanzania and now Tanzanian forces were massing to counterattack. Radio Uganda was reporting the need for vigilance along the entire southwest border.* We didn't expect to find Ugandan troops at ten thousand feet, but neither did we want to become guests of President Amin, either. We returned one more day to complete our nest counts, then ended our work in the Ugandan sector.

At our base camp, a surprisingly large and active hyena population chortled and hunted around us through the night. This fantastic auditory experience was totally unexpected at high altitude in a rain forest. Less positive was a tale we had recently been told of a tourist in Kenya who slept in an open tent wearing his boots. A hyena, attracted to the scent of leather, pulled him out by one boot at night and mangled his foot before it was driven off. Even if Bill didn't sleep with his boots on, it was not a reassuring thought. Added to our concerns about military movements, we slept little and were happy to move on in the morning.

Our last camp was in the Gahinga–Muhavura saddle. There, a contingent of Rwandan soldiers stood guard inside a small tin hut. They were nervous, though it wasn't clear whether this reflected the military situation across the border with Uganda, two hundred yards away, or their fear of wildlife in the park. They were not an impressive group, and several seemed happy to have company.

We were now in an area of the park that had never supported many gorillas, even in Schaller's day. Yet we still needed to cover a large amount of terrain. Fortunately, Gahinga had very few ravines of any significance and its tall stands of bamboo formed cathedral-like arches that were much easier to walk through than the bamboo thickets to the west. At lower elevation on Gahinga, though, we did encounter strange *Mimulopsis* trees, whose raised stilt roots were adept at tripping tired legs. We found some old gorilla nests at several sites, but no sign of passage within the past six months. Our own presence, however, triggered volleys of sharp *Piao!* alarm cries from resident

* Tanzanian troops eventually did attack through Rwandan territory, but passed farther east through the more accessible Akagera Park. Amin's troops were quickly routed and the brutal dictator fled to Saudi Arabia.

troops of golden monkeys (*Cercopithecus mitis kandti*), a medium-sized monkey endemic to the region whose vivid golden mantle is as striking as its cry.

Muhavura would be our last challenge. Its 13,540-foot cone made it an almost perfect bookend to Karisimbi in the west. Fortunately, most of the Ugandan sector was covered with a recent lava flow that could not support gorillas, so we did not have to risk an encounter of the military kind. During our brief forays into the Ugandan sector we saw no sign of wildlife, though we did hear hunters using dogs with bells. Twice gunshots echoed from below. Hiking on Muhavura was not too difficult: its slopes were steep, but without any significant physical or vegetative barriers. Much of the mountain's mass was above the treeline, where an open steppelike environment predominated. Northern double-collared sunbirds flashed their metallic green and purple plumage while brilliant red and green long-tailed malachite sunbirds darted manically about this alpine moorland, their long, thin beaks adapted like those of hummingbirds to drinking nectar. Bushbuck browsed here, too, but quickly barked and bolted on sight.

Finding no gorillas one day, we decided to explore the summit. Near the top we were forced to crawl through a dense thicket of gnarled, moss-covered alpine *Senecio,* a woody plant that grows over several centuries to a height of ten to twelve feet. This entangling complex extended for hundreds of yards in what seemed like the world's largest natural jungle gym. Our reward for passage was a perfectly round crater lake perched on Muhavura's summit. The lake was no more than twenty yards across, its surface reaching to within eighteen inches of the rim. Bill wondered if it ever overflowed, then giggled as he knelt to drink from the cold dark source at the top of the world. Clouds came and went in seconds. During clear moments, we could see lakes far to the north in Uganda and the crowded fields of Rwanda to the south. With the return of clouds, a chill crept under our skins. We reluctantly left our watering hole with a view as the day drew to a close.

Our sole gorilla contact on Muhavura was also the last of the census. Late one afternoon we encountered fresh gorilla sign. We backtracked to make two nest counts, then returned to contact the group and complete our count the next morning. We confirmed the presence of seven individuals: one silverback, two adult females, another unsexed adult, and three young gorillas. Leading downslope, the fresh trail of compacted herbs wound past the yellowed skeleton of a jackal, its foreleg still held in the death grip of a poacher's wire trap. Nearby, we could hear the gorillas. Bill moved closer until he saw a hand reach into a bush barely twenty feet away. He then climbed a tree, revealing the face of an older female, sitting in a day nest eating *Galium.*

With a severe stare and a reddish brow . . . she stares back for nearly a minute, moving her head back and forth, giving time for a noseprint. She then flees silently with others moving ahead of her. . . . By 13:10, fear dung is prevalent.

Silent flight and diarrhea were common reactions of wild gorillas when confronted by humans. They were also signals for us to leave them alone and complete our final nest counts.

For four more days we moved methodically counterclockwise around Muhavura, until we reached the open lava field near the Rwanda–Uganda border along its eastern slope. We crossed and explored sixteen ravines in that time, several of which contained prime gorilla habitat. Yet we never saw another gorilla, nor any sign of other animals besides birds. We had seen the same sort of unoccupied habitat at the extreme western end of the park on Mt. Karisimbi. This meant that there was still room for the gorilla population to expand. But it also meant that these sectors far from the central park headquarters were killing fields where poachers ruled and where gorillas had been exterminated.

On our last afternoon we walked out of the park and down to an old colonial estate at Gasiza, near the base of Mt. Muhavura. The house needed paint and the foundation was cracked, but its size, stone arches, and panoramic views testified to a more glorious past. Now it was the home of Drs. Alain and Nicole Monfort. Alain and Nicole were Belgian ecologists with a deep love for the Akagera National Park in eastern Rwanda: a savanna-wetland complex with a warm, dry climate. Alain disliked the rude Virunga environment, but had worked to improve park management and security during his brief tenure in the north. As hosts, Alain and Nicole received the first unofficial census results and other news from the forest. But mostly we enjoyed good food, good wine, and good company. After weeks of functional Swahili, Bill was happy to use his much richer French vocabulary.

SOLVING CONSERVATION PROBLEMS is like putting together a jigsaw puzzle for which there is no picture on the box, many parts are missing, and there is too little time to examine all the remaining pieces. Our challenge was to fill in the most complete picture possible through a combination of biological and socioeconomic research. The census provided the first pieces to the very complex puzzle of how to better understand and protect the mountain gorillas.

The hike home from Gasiza to Karisoke covered about twenty-five miles and took just over eight hours. It was ample time for Bill to reflect on what he had learned and what it might mean for the gorillas. George Schaller had carried out the first Virunga census in 1959–60. A simmering civil war in Rwanda prevented him from surveying certain areas, but his combination of direct counts and estimates based on habitat characteristics★ produced a baseline total of four hundred to five hundred individuals. The next census was conducted by a team of Karisoke researchers over a three-year period between 1971 and 1973. Their findings showed a dramatic decline to 250 to 275 gorillas, with a parallel decrease in average group size and percentage of young. As he walked through farm fields along the park edge, comparisons with the current numbers churned in Bill's head.

The good news was that the population appeared to be stabilizing after its free fall in the 1960s. Our minimum count was 252 individuals, which would later be extrapolated to an estimated population of 260. Better yet, we had found forty-two infants under three years old, versus only thirty-three in the same age class only five years earlier. Average group size had also increased, but only because the number of groups had declined from thirty-one to twenty-eight. The bad news was that gorillas were avoiding the eastern and western extremes of the park and concentrating in the center, especially around Mt. Visoke. On Mt. Mikeno, where Schaller had counted more than two hundred gorillas, there were only eighty-one. Mts. Sabyinyo, Gahinga, and Muhavura supported only thirty-four gorillas in five groups. Objectively, it was a mixed message at best. Emotionally, it was a great relief not to find further losses, especially after the bloody events of the past year.

The bright red and yellow flower displayed on the outer fence of a household compound, or *rugo,* signaled that fresh sorghum beer had been brewed inside. Big Nemeye, coming off an involuntary stint of sobriety for the census, homed in on the signal like a bee foraging for nectar. Soon he had negotiated several large bowls of *umusururu* for himself and each of the porters. The light brown liquid flowed down their throats as easily as the beer's Kinyarwanda name flowed from their lips. Bill declined a proffered bowl because of his distaste for the surghum chaff that came with the beer. With team spirits higher we continued west toward Karisoke.

Near the base of Sabyinyo, the typically rich soils of the Virunga piedmont gave way to a hard lava pan. Hollow thumping sounds revealed the presence of extensive natural tunnels below our boots. Nemeye commented without elaboration that many local people hid, and sometimes died, in these caves during past times of troubles. Aboveground, the shallow soils sup-

---

★ Schaller's estimates for inaccessible areas of the Parc des Volcans would be borne out by subsequent censuses of those same areas.

ported patches of grass between rocky outcroppings, providing a rare opportunity for herdboys in this region to graze their cattle. Bill assumed the same cattle would move illegally into the park once we had passed.

Beyond Sabyinyo the soils deepened and the land returned to its natural richness. Lush rows of potatoes and beans extended to the horizon, juxtaposed with large square patches of white pyrethrum flowers. It was an attractive landscape, set against the backdrop of towering volcanoes. Only the occasional *Markhamia* or *Hagenia* tree stood in mute testimony to the fact that all of this land had been torn and cleared from the park just ten years earlier.

WHEN RACHEL CARSON published *Silent Spring* in 1962, the Western world was forced to confront the devastating effects of DDT on wildlife and human health. DDT emerged as a potent insecticide during the post–World War II chemical revolution. It allowed American and European farmers to greatly expand their control over damaging insects and increase their yields of fruits and vegetables. But it also killed pollinators and other beneficial insects. It moved insidiously up the food chain to poison frogs, fish, birds, and people. When politicians finally listened to the scientific evidence, DDT was banned and the race was on to find less toxic alternatives. Kenya was already growing pyrethrum, a daisylike flower that produced a natural insecticide called pyrethrin. With a new world market to satisfy, exports of pyrethrin skyrocketed. Kenya's efforts to expand production, however, ran up against its limited amount of high-elevation land required to grow pyrethrum. Rwanda already grew some pyrethrum in the Ruhengeri region and there was more highland habitat in the park.

With money from the European Development Fund—and no environmental impact assessment—the Rwandan government rushed forward with a program to spur pyrethrum production. Twenty-five thousand acres of lower elevation forest habitat were cleared from the Parc des Volcans in 1968 and 1969. Five thousand families were awarded five-acre plots, called *paysannats,* on which they were supposed to maintain 40 percent of their land in pyrethrum. In a country beset by chronic land shortages, it was a popular program among northern Rwandans—even if large blocks of the cleared land were illegally claimed by senior political figures from the region.

By the late 1970s worldwide demand for pyrethrin was already in decline. Western laboratories had succeeded in synthesizing several less toxic alternatives to DDT, with shorter and less problematic supply lines than those stretching all the way to landlocked Rwanda. Local farmers adapted by ignoring pyrethrum production quotas and growing more marketable—and edible—white potatoes. Gorillas, too, adapted to the loss of habitat by moving higher on the mountain. But with the lower park limit now at almost nine thousand feet, the gorillas were exposed to near freezing temperatures every night. Pneumonia, already their number one cause of mortality, would kill even more of the very young, old, and sick. Though no one ever recorded how the gorillas had formerly used the area cleared for *paysannats,* it was certain that they had lost a part of the forest rich in many of their preferred foods, including large areas of bamboo.

It was impossible to stand at the foot of Visoke at the end of that day and feel optimistic, despite the positive census findings. The surrounding landscape looked as if people had been living there for centuries. Yet 40 percent of the Parc des Volcans—22 percent of the entire Virunga forest—had been cleared in the past ten years. Habitat that had supported mountain gorillas for millennia had disappeared in a relative instant. Whatever the census numbers said, it seemed unlikely that there was enough habitat left for them to survive within the retreating park boundaries.

Chapter Nine

Life in a Salad Bowl

PUCK AND TUCK sat shoulder to shoulder, looking out over the fields and thatched *rugos* that stretched to the south as far as the eye could see. When Amy found a better vantage point, she could see that they were also eating thistle. But why had they walked out of the park and ventured almost thirty yards across open farmland to clamber up on that rocky knoll? Did they know that gorilla food was hidden in the midst of the unappealing potatoes and pyrethrum? Puck was barely nine, four years older than his brother Tuck. Both were too young to remember the area before it was cleared from the park; but both were still young enough to let curiosity lead them on a reconnaissance run into new territory.

Amy had never seen any gorillas beyond the park boundary before. Now all she could do was imagine their motives—perhaps the view on a rare clear day?—and wait with some concern for them to rejoin the group. Within ten minutes, Puck and Tuck had finished their snack, ambled up the gully between two rows of potatoes, and slipped back into the forest. They would not visit the fields again while we were there.

APPLIED CONSERVATION RESEARCH was extremely rare in the 1970s. The few people committed to long-term fieldwork in African rain forest environments were almost all behavioral scientists. Jane Goodall and Dian Fossey were leaders in this field and their work fed an insatiable global appetite for information about the lives of chimpanzees and gorillas. As the plight of these species became increasingly clear, however, so did the need for sound information about how to save them. Amy's study of mountain gorilla feeding ecology was designed to inform those responsible for gorilla conservation about these endangered creatures' most essential requirements for survival.

Every species has basic needs: food, water, security, social support, opportunities for reproduction. Food and water come first—and mountain gorillas derive both from the rich plant life of the Virungas. The recent loss of 40 percent of the forest habitat of the Parc des Volcans raised deadly serious questions about the long-term viability of the remaining 260 Virunga gorillas. Recovery to the population levels of four hundred to five hundred gorillas recorded by George Schaller seemed out of the question.

Almost twenty years later, Schaller's 1960 study of the mountain gorilla remained the only significant source of information on gorilla feeding ecology. In addition to his comprehensive analysis of their behavior and social organization, Schaller documented the gorillas' use of diverse habitats on Mt. Mikeno in the Congolese sector of the Virungas. He also introduced the use of direct observation to record the foods consumed by gorillas. This was a great improvement on the indirect methods of dung and stomach content analysis that were considered the only feasible way to obtain information about wild animal diets at that time. Yet despite Schaller's Promethean efforts, there was much more that we needed to know about the gorillas' food and habitat requirements. Was Mikeno representative of the rest of the park? How much did the gorillas actually consume of the different foods available to them? Did they have preferences? If so, were choices based on quality or availability? Ultimately, could the now diminished forest meet their long-term needs?

ONE DAY IN JUNE, as our first long rainy season approached its end, Amy was watching Beethoven feeding about ten feet in front of her. Hearing a noise, she looked away to see who was moving through the thick undergrowth nearby. She had not yet seen all members of Group 5 and wanted to complete her daily count. Reassured that Pantsy and Muraha were accounted for, she turned to find that Beethoven had moved silently behind her, where he was reaching for a plastic bag she had stuffed with *Galium*. Amy reached as well, but lost the race. Beethoven grabbed the bag, strutted a few feet away, and sat down with a confident look. As Amy watched in bemused shock, he reached into the bag and began to cram folded "wedges" of

Galium into his maw, chewing steadily until he finished his meal. Dropping the bag by his side, he then nimbly removed a few strands of the clinging vine from his hairy forearm and ambled off to nap.

For Beethoven, this was a déjà dinner. For Amy it was lost data. As part of her research, she would monitor one group member for five hours almost every day. During that time she recorded everything the individual ate, what was within reach but wasn't eaten, and what other gorillas within view were eating. She had devised a simple method to determine quantities of food consumed: replicating each individual's feeding by sitting nearby and gathering equal quantities of whatever was being consumed. When the feeding was over, Amy's duplicate meal was bagged, brought back to our cabin to be weighed, then dried for later nutritional analysis. If she had any qualms about choosing foods of similar quality to those eaten by the gorillas, Beethoven certainly dispelled them on the day of the great *Galium* heist.

Within a few months, the gorillas of Group 5 completely tolerated Amy's almost constant presence. Every other day, she would leave the cabin at the break of dawn and hurry off alone to catch the group before it began feeding. On rainy days, she would often find the entire group still in bed. Most other mornings she might find several adults still lounging sleepy-eyed in their night nests, provoking nothing more than a few belch vocalizations. Amy returned the two-tone sound, like a deep clearing of the throat, to acknowledge her presence. The younger gorillas were usually already up and active, waiting for their parents like eager children on Christmas morning. After the last adult rolled out of bed, a decision would be made in silence and the group would move off behind either Beethoven or an older female like Effie. At this point, Amy would move close to her focal animal and stay near that individual throughout the morning. On alternate days, when she did five-hour afternoon focals, as these intensive observation periods were called, she would catch up with the group in late morning and then stay until their last feeding session ended and they began to make their night nests.

Amy's discovery that the gorillas were more at ease when she stayed close to them opened a new window on the world of gorilla feeding ecology and behavior. From three to six feet away, every thorn on a thistle, every stinging hair on a nettle, is clearly visible. These were important foods for the gorillas, though ones they treated with respect. The long leaf of a thistle plant provided a natural sheath into which the gorillas would carefully fold the spikes along the leaf's edge. The gorillas then strategically placed this packet in the side of the mouth and ground it under the crushing power of their molars. Nettles received similar treatment, though in this case the gorillas would slide the undersides of two fingers up the stalk, collecting all the leaves in a bundle with the irritating surfaces pressed against each other and away from their skin. Then they twisted off this tuft and consumed it with the stinging hairs aligned away from the point of entry to the mouth. Gorillas apparently tolerated the natural defenses of these two plants because of their combined nutritional and high water content. Only wild celery contained more liquid, which would sometimes run down the gorillas' chins as they chewed noisily on the hearty stalks. Celery was also the most palatable of these plants to our tastes, though we had to be terribly thirsty to tolerate the bitter aftertaste of its younger stems. Amy sampled nearly all of the gorillas' foods and came to especially appreciate mature celery as a midday drink substitute.

Bamboo was the gorillas' most desired food. During the five months of the year when young shoots were in season, Group 5 was never far from the lower elevation zone near the park border, where bamboo flourished. Here the gorillas spread out in a fan formation to increase their odds of finding the randomly located sprouts. As individuals discovered the three- to four-foot pointed shoots, they first looked over their shoulders for competitors, then sat to enjoy the delicacy in private. After pulling the full shoot from the ground, they carefully unwrapped the paperlike sheaths and noisily devoured the succulent core. As they moved on, neat circular piles of spiral wrappings were left in their wake, helping Amy to track the group across the generally bare terrain beneath the full cover of bamboo. This canopy was broken in some places and bamboo mixed with other plants. One plant that thrived under these conditions was *Droguetia,* a vine covered with one- to two-inch-long leaves. The gorillas had a fondness for *Droguetia,* especially in combination with bamboo. Later analyses would reveal that this was not just a taste preference but a powerful form of nutritional complementarity.

Bamboo shoots were the gorillas' most concentrated form of protein. They provided abundant liquid, too, and made up more than 14 percent of Group 5's annual diet. Schaller had also noted the importance of bamboo. Yet between his study area on Mt. Mikeno in Congo and Group 5's range on the border of Rwanda's Parc des Volcans, there were gorilla families that almost never ate bamboo. One of those was Group 4: the ill-fated clan of Uncle Bert and Digit, the long-term study group of Dian and now David. Occupying the higher elevations in the Visoke saddle, Group 4 rarely descended low enough on either side to find bamboo stands. Perhaps it was a cultural matter and they really didn't like bamboo, or perhaps they preferred their concentrated stands of nettles. Maybe they didn't like the competition from other gorillas. Left forever unknown is the question of whether or not Group 4 might have had easier access to bamboo if extensive lower elevation stands had not been converted to farm fields.

One unexpected discovery was the gorillas' interest in ants. Amy had already seen that the gorillas would actually fight over *Vernonia* galls infested with insect larvae. Claims of vegetarian purity were dealt another blow as she collected evidence of their predilection for ants. At first she saw gorillas popping driver ants into their mouths as they groomed themselves. But Amy thought this might just be the easiest way to kill and rid their fur of a hard-biting pest that had somehow made its way onto their bodies. Then she discovered whole driver ant bivouacs ripped apart by the gorillas, after which prodigious amounts of undigested ant body parts could be found in their dung. This was far from the carnivorous and even cannibalistic tendencies discovered in chimps, but it indicated a more complex diet than had generally been presumed.

The gorillas were certainly selective in their food choices. Bamboo shoots and galls were sources of competition, bruised feelings, and occasional physical violence. So, too, were black raspberries and stalks of juicy celery during the dry season. Shelf fungus offered a rare treat whose taste and high protein content were apparently worth fighting for. If younger gorillas discovered the fungus, they would usually nibble on it until displaced by an elder. Older gorillas would simply break off a large chunk and then try to slip discreetly away from the rest of the family. Gorillas rarely carried food of any kind, but shelf fungus and bamboo shoots were notable exceptions.

Some other resources were equally rare, but shared in a much more communal fashion. Every couple months, Beethoven led his band to a particular site on the flank of Bonde ya Kurudi, or Return Ravine. There they would sit like patients outside a doctor's office—adults sitting, youngsters playing—waiting their turns as one gorilla at a time entered a small cavelike opening. Once finished, that gorilla would exit with rouged lips and a belly covered with reddish clay. Later analysis confirmed the clay to be high in iron, as Amy suspected, but it remained unclear why the gorillas needed this mineral supplement. Nor has this habit been noted in any other Virunga gorilla group. Another Group 5 favorite was the root of the giant *Lobelia* plant. Here, too, a purposeful expedition was required to reach one of several subalpine locales where the plant thrived on exposed high mountain slopes. At the site, the adult gorillas would spread out to find choice plants, while the youngsters waited nearby for access to the products of their digging. Once the roots were exposed, the gorillas selected strands that they pulled between their teeth, efficiently stripping and eating the prized epidermis. This paperlike root bark was the one food source that Amy did not collect. In addition to finding no substitute technique for removing the bark, she didn't want to add any burdens to the already stressful, high-elevation existence of the *Lobelia*.

Its importance to the gorillas thus remains a mystery. What is certain is that gorillas exert considerable physical and cooperative effort to obtain this rare resource.

One indelicate feeding habit of mountain gorillas seemed to hold particular appeal for a single Group 5 lineage. Effie and her offspring, Puck and Tuck, were all partial to eating the dung of other gorillas. It was an irregular habit, most common on rainy days, leading one former researcher to quip that it was "a hot meal on a cold day." In more prosaic terms, the behavior likely signaled a dietary deficiency, or a need to replace depleted intestinal fauna, that was met by obtaining the desired supplement direct from another gorilla source. Whatever the rational explanation might be, it was not attractive to watch four-year-old Tuck munching on a steaming lobe of Beethoven's dung. But then the gorillas are not here to meet our expectations.

A MY SPENT MORE THAN two thousand hours observing Group 5 to complete her feeding ecology study. Five-hour focals formed the heart of her work, augmented by a single twelve-hour, dawn-to-dusk focal on one individual in each of the six age and sex classes. The mentally and physically demanding work required an extremely high level of concentration and attention. While the demands of recording the details of gorilla feeding for hours at a time numbed the mind, the rude Virunga climate numbed the body. Rain gear could keep most of the body dry, but Amy's feet were a lost cause. Rubber boots kept the water out only to replace it with sweat from the inside. And no leather boot in those days could earn the Virunga "waterproof" label. Cold was even worse than wet, and the combination of the two could easily cause hypothermia if one didn't stay active. Moving around wasn't an option, however, during extended focals. Amy learned to read the feel of the air—low clouds were ever-present and not too helpful—to predict the arrival of rain and quickly throw on a thick oiled-wool sweater under her rain jacket. The broad leaves of certain plants, when available, provided welcome cover for her boots. But nothing could protect her hands. If the gorillas continued feeding in the rain, Amy would write inside a plastic bag. This kept her paper dry, but rubbing against the condensation of the bag's interior produced chronic deep cracks in her fingers, which didn't heal for weeks at a time during the long rainy season. Worse, the exposure and tight writing style required by small notebooks caused constant cramps. Sometimes her fingers just stopped working.

Cape buffalo encounters offered a great way to restart the blood flow.

Amy's schedule of alternating dawn departures and dusk returns assured daily exposure to peak buffalo activity periods. Moving quickly through the thick Virunga understory, one can easily miss the creature's bulky black form in dim crepuscular light. Suddenly, a six-foot-tall, half-ton frame looms in the path only twenty feet away. Three-foot horns curl out from a massive base on the head to form dagger-sharp tips. Dark liquid eyes reveal no apparent thought process inside. The combination of lethal force, limited analytical powers, and a nasty temper make the Cape buffalo the most deadly of Africa's vast array of dangerous species. This was not reassuring when Amy was armed with only a walking stick or machete and good climbing trees were rare. Fortunately, despite a score of close encounters with buffalo, she was never injured. On the few occasions where she was actually charged, she dove into nearby bushes and held her breath as the buffalo rumbled past. Twice she was able to climb a *Hagenia* tree and watch as the buffalo stood for an interminably long time staring in her direction before finally slipping away. Others have not been so fortunate, and Karisoke guides and researchers have been injured by irate buffalo. The rich wet meadows that occupy all saddle areas may help explain why this savanna denizen is so plentiful in the high-elevation Virunga rain forest. What is certain is that the presence of Cape buffalo adds an element of excitement to any walk through the woods.

One morning around seven, Amy returned almost breathless to camp. She had just run all the way back from First Hill, where she had left Group 5 late on the previous day. Instead of picking up a fresh gorilla trail, however, she entered a vast tangle of uprooted and twisted vegetation that covered the entire eastern flank of the small crater. Elephants! Looking around, Amy soon detected a trunk waving like an entranced cobra over some small, nearby trees. Next a solitary eye appeared. Then a swishing tail. After months of seeing only their sign, Amy was surrounded by live elephants. She ran to share the news with Bill, who had stayed behind to work in camp. Together we ran back up the hill. We saw no elephants but young trees cracked around us. Finally, another trunk appeared above some tall shrubs and we realized that we were now upwind of this olfactory periscope. The forest turned silent for a few minutes and we decided to seek higher ground. Just as we arrived on a slight knoll, all hell broke loose thirty yards below as the entire herd charged past. We counted twenty-seven individuals and even remembered to snap a dozen pictures. The photos showed a line of smallish heads, ears, and tusks that resembled those of the lowland forest elephant. Later analyses would indicate that the Virunga elephant was possibly an intermediary between the forest and savanna races. At the time, we were rewarded with a kaleidoscope of fast-moving body parts and what felt like an earthquake as the herd rumbled past. And an indelible memory.

Despite the widespread presence of Cape buffalo, not to mention elephants and leopards, our only injuries in the Virungas came from falls in the difficult habitat and—on two exceptional occasions—from the gorillas themselves. The first of these injuries occurred when Amy found herself in the wrong place at the wrong time. On the south side of First Hill, late one afternoon, Beethoven cough grunted and rushed to displace Effie at a juicy blackberry bush. Amy had been sitting by Effie taking notes and was shoved aside by the onrushing Beethoven. The push itself was harmless, but her foot was wedged against the base of a tree stump and her knee twisted as she fell awkwardly to the side. The contest was typically brief, with a few screams and perhaps a bite before Effie fled the scene. Amy rose unsteadily to her feet, then felt a sharp pain when she placed any weight on her right leg. So she cut a young *Hypericum* to make a staff and hobbled slowly back to camp. As night fell, Bill set out to search for her, only to find her as she approached Camp Stream. Fortunately the knee healed quickly and Amy was back with the group in a few days.

A MY'S TIME WITH GROUP 5 also provided greater insights into gorilla behavior and group dynamics, especially the importance of learning. Gorilla mothers nurse their young for at least three years. During the first six months, the infants depend almost entirely on the natural richness of maternal milk. But even before they eat solid food, Amy noted, infants receive their first maternal lessons in what to eat from the Virunga smorgasbord. Clinging to their mother's hair, gorilla babies not only see what their mother is eating but also are covered with a steady rain of edible remains. Mouthing these morsels without swallowing gives a first taste of what lies beyond their mother's breasts. By the age of six months, most babies begin to experiment with solid foods. Mothers still provide their primary instruction in what is edible and—even more important in a world with its share of poisonous plants—what is not, by confiscating inappropriate items. Still, social influence from others is strong at an early age. If Muraha saw Pablo or Puck chewing on a piece of decomposing wood, the same type of item was likely to end up in her mouth, too. Amy's research showed that young infants were the most likely to imitate the eating habits of those around them.

Learning is the foundation for the mountain gorillas' intimate knowledge of their complex habitat. The home range of Group 5 covered five square miles with an altitudinal range of more than four thousand feet. Sectors of two major volcanoes, several smaller cones, dozens of ravines, two large streams, a broad saddle region, and an extended edge of contact with human settlements made up this range. Each of its distinctive characteristics had to be learned. Beethoven's beelines to the clay cave and subalpine food sources were almost certainly part of a mental map, committed to memory as he followed his father through the forest many years before. So, too, were his pinpoint approaches to *Bonde ya Daraja,* or Bridge Ravine, where the group would cross in stately procession across a fallen *Hagenia* tree (while we usually took the low road about twenty feet below). Beethoven generally led on longer treks between two points, yet adult females seemed to have equally detailed knowledge of where to locate food once they were at a particular site. Ziz, Puck, Tuck, Pablo, Poppy, Shinda, and Muraha were certainly storing all this information for the benefit of the next generation as they otherwise ate and played their way through their youthful years.

B Y MAY OF 1979, Amy had finished several months more than her goal of a full year's cycle of data collection. It was time to begin an extended period of vegetation sampling. Knowledge of what the gorillas selected and rejected from the forest's bounty would make sense only in comparison with an assessment of overall food resource distribution and abundance. This information would come from thirty sample plots—250 meter by 250 meter quadrangles—that Amy had evenly located across Group 5's home range. Working together, we could complete an assessment of one quad in five or six hours, plus walking time to and from the site. The almost thirty days we spent on vegetation sampling provided a rare opportunity to spend time together in the field after fifteen months of demanding parallel work schedules. It allowed us to revisit many places and rekindle memories from earlier times with the gorillas. It was also work that neither of us has ever felt any desire to do again.

The basis and bane of vegetation sampling is the random numbers table. This clever invention assures that within the mathematically selected primary sample area, the researcher does not somehow bring biased elements into the study. So within each quad, we systematically laid out five straight transect lines with string every thirty meters. How far we walked along that string and how far we turned either left or right to find our nested 100-, 10-, and 1-square-meter sample plots, however, were subject to the dictates of the random numbers table. And it *was* a dictator. No sooner had we returned from a precarious descent into a ravine than the next magic number would require that we go back down over the same cliff. This perverse sequence also seemed to guarantee return passes through especially nasty patches of elephant nettles. At times like these, only Amy's strong commitment to good science overwhelmed Bill's strong desire to sample someplace more accommodating.

We completed vegetation sampling in mid-June. The heavy rainy season was ending with a vengeance, and we had experienced several days in a row of powerful afternoon storms. On our last day, we were determined to start early and finish by noon if at all possible. We were on schedule when a rare flash of lightning signaled the start of an hour-long blitz of hail. The marble-sized iceballs stung our exposed hands and heads as they shredded much of the surrounding vegetation. By the time we finished our last samples we were standing in ice water in a daze, but were still aware enough to realize that we were both in the early stages of hypothermia. Our solution to the problem—to pack our bags and race back up the mountain to camp—made sense given our poor mental state and excellent physical condition. Our half-hour descent turned into a ten-minute return sprint uphill that warmed our bodies, leaving us almost delirious with laughter as we pulled up about fifty yards below our cabin. There we held hands and continued walking until a strange, deep ripping sound brought us to a halt. The sound turned into a roar as a giant *Hagenia* tore its roots from the earth and lurched toward our cabin. The tin *mabati* was no match for the tons of wood and water-soaked moss that crashed through our roof, cutting our home almost in half. Had we arrived five minutes earlier, we would have been peeling off our wet clothes exactly where the colossus came to rest by the side of our bed. Instead we spent the remainder of the day cutting up the tree, replacing a few key pieces of the wooden framework, and patching new panels of corrugated tin in place. Among the few advantages of living in a tin shack is its ease of repair.

O UR SAMPLING REVEALED one disturbing fact about the Virunga ecosystem. In more than 1,500 plots, we found not one *Hagenia* sapling. Mature *Hagenia* were quite common from the park boundary at 8,800 feet up to the treeline at 10,600 feet. But there was little sign of any regeneration of this Virunga giant: the largest life form and one of only two major tree species in the entire park. Gorillas didn't eat any part of the tree, but they did scale its sloping trunk to feed on its abundant lichens and ferns. Birds, squirrels, hyrax, and many other creatures certainly depended on the tree for food and shelter. Outside of our sample areas we did see some signs of regeneration, but always as shoots from the fallen trunks of dead trees. We also knew of one patch of young saplings in a disturbed area near the park boundary. Our finding was not definitive, but it did raise serious questions about Karisoke's continuing use of *Hagenia* as its primary source of firewood—a practice that ended a few years later when Sandy Harcourt became director of the station.

⁓⁓⁓

A MY'S SEVENTEEN MONTHS of research revealed that the Virunga gorillas made very selective choices from among more than one hundred distinct food types. They preferred more nutritious and higher quality foods whenever possible and actively sought diversity in their diet. Bamboo shoots made up the majority of their diet during one third of the year. Furthermore, many of their foods were relatively widespread and abundant, with the notable exception of bamboo. Some areas with high food values appeared in our sampling, yet were rarely visited by Group 5. Apparently, they didn't need to visit them. The gorillas lived in a giant salad bowl that still offered an ample supply of diverse, nutritious foods.

This conclusion was contrary to our expectations—and to widely held perceptions. Citing their low numbers and loss of habitat, a growing number of writers and conservationists were already consigning the mountain gorillas to the dustbin of evolutionary history, or to a constrained life of captive breeding. Fossey herself spoke openly of their imminent extinction. Ian Redmond had told us that he believed it was our task "to record all that we could about the gorillas' lives and behavior before they disappear." Ian and Dian were as committed to the gorillas as anyone could possibly be, but we believed that our overriding challenge was to save the mountain gorilla, not document its demise. Our research was demonstrating that the gorillas had stabilized their population, with sufficient land and food to maintain their numbers and perhaps grow. In a later demographic analysis we would show that the population even had the reproductive potential to return to Schaller's level of four hundred to five hundred gorillas. We were determined to use this new information to help make a positive case for mountain gorilla conservation in the wild.

Yet many problems persisted in the salad bowl paradise. Poaching continued throughout our tenure at Karisoke. The long rainy season of 1979 was punctuated by the constant coughing and sniffling of gorillas who no longer had a lower elevation refuge from periods of extreme rain and cold. And the rising tide of human population lapped steadily higher on their forested shore. Without a strong dike of political support, the gorillas would be swamped.

Chapter Ten

Sex Changes and Songfests

PUCK AND TUCK WERE BROTHERS. But that was before Puck gave birth.

From the time Dian Fossey began her work at Karisoke in 1969, Group 4 was the primary focus for research. Other nearby groups were monitored irregularly to track their movements, membership, and interactions. But only in the mid-1970s did any serious research begin on Group 5, when Sandy Harcourt began working at Karisoke and used data from both Groups 4 and 5 for his behavioral research. When Sandy left Karisoke to complete his doctorate at Cambridge University, Group 5 was again left to infrequent monitoring until our arrival in 1978. At that time, Dian gave us a family composition list for Group 5, but there was some confusion over certain of the twelve listed individuals. Amy quickly determined that there were actually fourteen in the family: two silverback males—Beethoven and Icarus; four adult females—Effie, Marchessa, Pantsy, and Liza; two blackback (eight- to ten-year-old) males—Ziz and Puck; three younger males—Tuck, Pablo, and Shinda; and three younger females—Quince, Poppy, and Muraha.

This structure remained quite stable, except for Quince's death and Liza's transfer, throughout our time at Karisoke—until November 14, 1978, when things changed dramatically. Amy had taken two days to catch up on transcribing her field notes and was sitting at our desk when David Watts knocked and walked in. David was about five feet ten, with a sturdy build, long black hair parted down the middle, and round wire-rimmed glasses. He continued to monitor what was left of Group 4, but had taken advantage of Amy's absence that day to take some comparative notes on Group 5. Once inside our cabin, he tried to maintain his usual dry demeanor, but a smile quickly took over his entire face. "Puck had a baby." No response. "I didn't see it happen, but I think she must have had it while I was there because she was still licking off birth fluid."

She. Baby. The words finally kicked in and we sat there stunned. It was great news, unbelievable news. But it would be dishonest to say that Amy didn't feel some disappointment, too. Day after day, month after month she stayed with the group, rarely missing a day in the field. Yet first she had missed Liza's transfer—and now a birth! Gorilla females give birth only every four years, so it is a rare event. No one had ever witnessed a birth in the wild, although Kelly Stewart believed that one occurred while she was with Group 4. Does the mother go off alone? Or do others help? Does she squat or lie down? What does she do with the umbilical cord? Does she vocalize in any way? How amazing it would be to watch a new gorilla enter the world. The thoughts came in rapid succession, until the most fundamental question took over: How did Puck have a baby?

Gorilla males have disproportionately small genitalia for their size. At the same time, mountain gorillas are very well endowed with long body hair that covers all but parts of their hands, feet, chests, and faces. So, while it is easy to see the penis and scrotum on very young gorillas, they virtually disappear from sight after the age of two. Somehow, Puck's lack of a penis had eluded detection by Dian in "his" early years, and her true gender had subsequently gone unnoticed by Sandy and others. We wondered if there were other such cases.

Over the next few weeks, Amy and David took advantage of their acceptance by the gorillas to investigate who had what below the bulge of their bellies. With Puck now out of the closet, only three other older subadults needed to be checked. Rest periods provided the best opportunity to sit next to the selected individuals and engage in mock grooming—something neither Amy nor David would do except under such unusual circumstances. Pablo was first in Group 5 and he took the grooming and checkup in stride, even lying on his back for better access. If he expected a full body rub, though, he was disappointed when Amy confirmed his maleness and moved on. Tuck provided an opportunity the next week when he came to inspect Amy's red bandanna. While the five-year-old sat to examine the bandanna, Amy methodically parted the hairs between Tuck's legs. There, a small black protuberance was clearly visible, but it didn't match the expected size or shape. After further examination, she determined that Tuck in fact had a sizable clitoris. Tuck seemed unbothered by the inspection, but Amy came back to the cabin in disbelief. Puck and Tuck were sisters! A few weeks later, David would report that four-year-old Augustus from Group 4 should henceforth be known as Augusta.

Three of the four older subadults in the two main study groups had been incorrectly sexed. Amy sent a quick note to Sandy Harcourt and Kelly Stewart at Cambridge to let them know that they might want to review any gender analyses in their nearly completed doctoral theses. Amy's own work would need some modification, too, since Puck's feeding patterns would now be analyzed as those of a pregnant female, not a growing young male. Puck would also be added to the list of females who did not transfer out of their birth families, but instead stayed to breed with either her father, Beethoven, or half-brother, Icarus. The good news, from a conservation perspective, was that the Virunga gorilla population now had three more females who would likely produce a total of ten to fifteen surviving offspring for the next generation.

IN THE DAYS AFTER the Great Virunga Sex Change, it was interesting to consider how presumed gender differences might have influenced our own perceptions of individual behavior. Puck and Tuck were both very active and confident gorillas. Both liked to play and seemed to have equal, if not superior, standing in comparison with the other young males in their group. We had actually discussed whether Puck or Ziz was more likely someday to emerge as the dominant silverback in Group 5 if Icarus remained on the periphery or went off as a solitary male. Tuck was the most curious of all the gorillas, always approaching to inspect new patches on our clothes or boots, examine freckles, or tug on bootlaces. Puck once took Bill's binoculars and lifted them to look through the larger end. As she moved her hand slowly back and forth beneath the binoculars, Bill wondered what she thought of the "little" fingers she saw through the smaller lenses. Exceptional as much of this behavior might be for gorillas, it seemed to us to be stereotypically "male": outgoing, assertive, exploring. Now we could shed our preconceptions and see that it was normal female gorilla behavior, too.

For an alternative explanation of Puck's and Tuck's standing in Group 5, one had to look no further than their mother. Effie was the highest ranking of the clan's females. In conflicts over food, Effie's posture, cough grunts, and apparent willingness to fight if necessary almost always carried the day. She was one of the first to enter the clay cave and had a priority seat at the feast of the roots. If reinforcements were needed, she had at least three offspring in the group; four, if the young silverback, Icarus, was hers, too. Even

Beethoven yielded to Effie at times when she decided to follow a different trail or seek a better nest site.

One day, a powerful storm swept down from the saddle and caught us all off guard. At the first clap of thunder, the females bolted for a nearby fallen *Hagenia*. Effie arrived first, followed by Pantsy, Amy, and Tuck. Bill arrived a few seconds later and found only enough space to cuddle up next to Tuck, leaving his lower body out in the soaking rain. At first, Beethoven sat stoically in the gorillas' classic rainy weather position: arms folded, head tipped down, with water running off his long hairs. This "wet Buddha" position was usually an effective strategy for at least an hour. After five minutes of exceptionally heavy downpour, though, Beethoven rose up and strode over to where the five of us crowded into our crude *Hagenia* shelter. He stood in full strut with his chin out, facing toward Effie but not looking at her, pursing his lips. It was a tense moment, at least for us human observers. Finally, Effie ended the standoff with a series of sharp cough grunts. Beethoven did not respond at first, then moved silently on to seek another respite from the rain. How would Beethoven have reacted if we alone had been in his preferred spot? Would he have tried to displace us? Without Effie to take charge, we certainly wouldn't have resisted. Yet there were also signs by that time that Amy had attained a certain standing within the family hierarchy. Sometimes it was nothing more than Marchessa or Pantsy ceding right-of-way to her on a trail. Sometimes it was a prime seat next to Beethoven during a rest period. These were subtle signs of an intermediate status. But whatever Amy's standing might be, it was always behind Beethoven and Effie.

Group 5 also had lineage hierarchies, in which the mother-daughter duo of Marchessa and Pantsy was clearly secondary to the dominant matrilineage of Effie, Puck, Tuck, and Poppy. Vocal hostilities between these alliances could be especially intense, particularly over access to choice foods. If Pantsy were challenged by Effie, Marchessa might be drawn into the fray. Yet despite much impressive screaming and posturing, the results predictably favored Effie and rarely produced any casualties. In two or three instances where the level of aggression rose especially high, Beethoven intervened to stop the fight with a quick bite to the neck or shoulder of one of the females. Nonviolence, however, was the general rule and family life continued as before.

INTERACTIONS WITH OTHER gorilla groups were more intense and unpredictable than internal squabbles. With Group 5, these interactions occurred every few months, often when the group was near the edge of its home range. Once, a pair of silverbacks without any other family and known as Batman and Robin appeared out of nowhere and set Group 5 to flight for several days, before the apparently dynamic duo returned to their unknown home near Mt. Mikeno. More commonly, Group 5 encountered Group 4 or 6, or Nunkie's group along the borders of their vaguely defined and overlapping territories. Occasionally it met the Susa group, Bill's first census discovery, along its namesake river. Contacts with Nunkie produced some of the most impressive displays. He was a massive silverback with a successful track record of attracting females—and a willingness to use force, as he had shown in killing Simba's infant, Mwelu. Yet interactions between Nunkie and Group 5 never reached this point of physical aggression. Nunkie would typically run back and forth on a ridge or other elevated piece of turf, ripping up bushes and small trees, which he would then swat on the ground. Beethoven and Icarus would respond in kind with chestbeats, roars, and loud slaps that shook the ground. In the end, Nunkie would usually defer to the double silverback display, leaving an area of flattened vegetation behind as he climbed silently higher on Visoke's slopes. But interactions with Group 6 seemed to elicit a higher degree of tension. Liza had already transferred to this family, and Brutus was a young, aggressive, and often agitated silverback who seemed intent on adding to his harem. Physical violence loomed as a distinct possibility when these two groups met, though Beethoven often led his clan on a preemptive retreat while Icarus covered their rear.

We rarely observed two gorilla groups interacting. One day we walked no more than thirty minutes downhill from our cabin to meet Group 5 at Return Ravine. This was one of their clear home range boundaries: a line from which they systematically turned back toward the core of their range. That day, most of the gorillas were deep in the ravine and we stayed on the edge to watch. Soon, we realized that there were too many gorillas for Group 5 alone. Eventually we counted twenty-five. Group 5 spread along the southern flank of the ravine, while Group 6 was equally dispersed across the opposite bank. Several gorillas from each group were feeding; others were sitting together watching the strangers on the other side. Beethoven and Icarus were nowhere near the front line of contact along the stream. Brutus was uncharacteristically calm. Suddenly we realized that Ziz, Group 5's blackback male, had crossed the divide and was walking among his would-be antagonists. Our pulses quickened as he approached Brutus, then sat down near another Group 6 member closer to his size. Nothing happened. Our attention then switched to Liza. Was she there? Would she seek out Pablo, or would he cross over to spend some time with the mother who had abandoned him less than a year before? Liza was present, but made no move to approach anyone in Group 5. Pablo sat in plain view with the rest of his extended family, but showed no interest in a reunion, either.

A few members of each group eventually crossed the stream and calmly intermingled, lightly feeding and resting. After about forty-five minutes, Beethoven rose to climb out of Bonde ya Kurudi and back toward the heart of his home range. Most of the family followed. Brutus and company moved off in the opposite direction, leaving Ziz to sit alone for a while before he turned to rejoin Group 5. In their wake, there were no broken trees or flattened vegetation. No screams echoed off the mountain. We hadn't heard a single cough grunt. Over the years we would see other interactions, mostly nonviolent but highly dramatic. This was the only completely peaceful encounter that either of us would witness.

IF GORILLAS RARELY encountered their own kind, interactions with other species were even less common. The bright flash of a Ruwenzori turaco's underwing might attract an individual's gaze for a few seconds. A ground thrush's nest once caught the attention of Puck, Tuck, and Pablo. They sniffed and gently touched, but didn't harm any of the eggs inside and left the nest where they found it. Another time the ground began to shake and everyone in Group 5 moved close together, as Bill imagined an earthquake or a new round of volcanic activity. Then Icarus stood up on two feet and cough grunted loudly at the five or six Cape buffalo thundering by, as if to chastise them for disturbing the Virunga peace.

Twice, Amy observed direct physical contact between gorillas and another animal. A chameleon kept Group 5's youth brigade amused for at least five minutes one day. Tuck, Pablo, Poppy, and Muraha gathered for an up-close inspection as the adults went about their business. The green and yellow reptile remained motionless on a plant except for the steady, independent rotation of its turretlike eyes. Finally Tuck poked at its tail, provoking a few slow steps with its wondrously cleaved feet. One by one, the gorillas then moved away, leaving their Virunga cohabitant in peace. On the second occasion, the group came across an infant black-fronted duiker lying under a tangle of vegetation. Its mother had probably left the tawny bundle in this hiding spot as she moved off to feed. Again the adult gorillas acknowledged but showed little interest in the small creature. But the younger gorillas quickly gathered around the new plaything for closer inspection. The duiker seemed fearful, and Amy wondered if she should intervene to set it free. But the gorillas showed only curiosity as four of them drew closer, sniffing and staring at their quarry for almost two hours. Periodically, one would pull timidly at the stems covering the duiker, then jump back to watch its response. Finally, Pablo could not resist poking its hindquarters, causing the delicate creature to emit a soulful bleat that scattered the young gorillas. Moving on, they left the still immobile duiker to wait for its mother in peace.

M OUNTAIN GORILLAS are not very vocal. Their most common sound is the flatulent by-product of their vegetarian diet. Actions and body language play a large role in gorilla communication, reinforced by perhaps fifteen distinct vocalizations. Most of these vocalizations are quite brief and appear to serve clearly understood functions. Belch vocalizations—contrary to their name—are soft rumbling sounds used by gorillas (and human observers) to acknowledge each other's presence. Cough grunts are staccato barks from deep within the chest that warn of displeasure. A *wraagh* is a prolonged loud vocalization, more like a roar, that signals serious anger. Singing, however, is entirely different.

Every few months, more frequently during dry spells, Group 5 would break into song. Usually in an area where the entire family was feeding on abundant high-quality food, one individual would start a low rumbling sound, breathing loudly in and out in a modulated tone. This might remain a solo performance and last no more than a minute. Often, however, others would join in, adding gender- and age-specific basses, baritones, tenors, and sopranos to the mix. The result was a chorus of intertwined melodies, rising and falling in a natural rhythm that might continue for several minutes: a gorilla Gregorian chant in the Virunga cathedral. These impressive performances doubtless served some other function, but we chose to enjoy them as unrestrained expressions of individual happiness and group harmony.

Chapter Eleven

Island Refuge in a Rising Tide

THERE IS NO PRIVACY in Rwanda. Wherever we walked out of the park, there were women working their fields or children tending goats. A drive in any direction passed a steady stream of pedestrian traffic. If we sought relief behind some bushes, the eyes of curious children would appear on the other side. People were spread everywhere across the country, not concentrated in the cities. In 1978 the combined populations of Kigali, Ruhengeri, Gisenyi, Butare, and Cyangugu totaled less than 200,000. The other 4.6 million Rwandans lived in a uniquely dispersed settlement pattern in which isolated households dotted the landscape. The lives of these people depended almost entirely on what they could grow or graze on increasingly small and fragmented parcels. In a poor nation the size of Vermont, land hunger was intense and the country's wildlands were at risk.

AT LEAST TWO THOUSAND YEARS AGO, the ancestors of today's Hutu farmers arrived in the highlands among the Great Lakes of east-central Africa. There they found a favorable climate and a mosaic of natural wetlands, grasslands, woodlands, and forests that covered the region's rolling hills and rugged mountains. As they had throughout their long migration from the other side of the Congo Basin, these Bantu farmers probably first settled along the forest fringe in what is now eastern Rwanda. Using both fire and the Iron Age tools that they brought from West Africa, they began to clear the moist montane forest to expose its rich soils. Perhaps as important as their iron axes and hoes, the immigrants brought the domesticated banana: a highly nutritious perennial food source. The natural forest retreated, replaced by a blanket of banana stands broken by open fields of finger millet and peas.

The proto-Hutu were not alone in Rwanda. Twa pygmies predated them by millennia, though their low population density and hunter-gatherer practices left little mark upon the land. They, too, preferred to live along the forest–savanna edge: a transition zone between two key habitat types that was rich in the plants and animals that supported their way of life. With the arrival of Bantu farmers, the Twa likely entered into barter arrangements with their new neighbors, exchanging bushmeat for produce or iron products. But as the forest frontier gave way to the banana, the Twa were pushed higher and higher into the mountains, where they were forced to adapt to a harsher climate and a less abundant larder.

Seven or eight hundred years after the Bantu expansion eastward across central Africa, a second great migratory wave spilled out of the highlands of northeastern Africa. This time the migrants were not farmers but pastoralists. As they spread to the west and south, they established the distinctive cattle-based cultures of the Fulani, Samburu, Maasai, Zulu, and Tutsi, among others. Some of these groups went no farther than the interlake highlands, where modern pollen studies show evidence of cattle grazing around mountain lakes and marshes by A.D. 1000. Outside of the highest mountains, however, these ancestral Tutsi found most of modern Rwanda settled by the Hutu, who were already organized into small kingdoms.

The Tutsi and their cattle avoided conflicts with the Hutu by moving into open ecological niches. Hutu farmers had developed sophisticated ways of using the lower, middle, and upper slopes of the hillsides that dominated the Rwandan landscape; Tutsi herders were drawn to hilltops and valleys. Some African pastoral groups had to move hundreds of miles in search of forage on a seasonal basis, but Rwanda offered much shorter altitudinal migrations. During the long rainy season, cattle could graze on grassy hilltops, then move a few hundred yards down into nearby valleys to browse in lush wetlands during the dry season. In neither instance did the Tutsi compete with the established Hutu for prime farmland—an ecological separation that facilitated political coexistence.

Over the next several centuries the Tutsi settled permanently in the region. Their tall, thin frames distinguished them from the stocky Hutu and much shorter Twa. Yet the Tutsi took the remarkable linguistic step of dropping their ancestral Nilotic tongue in favor of the Bantu language Kinyarwanda spoken by the Hutu. The Tutsi exchanged milk and other cattle products for Hutu farm produce. Cattle were allowed to browse on crop residues, leaving manure to enrich the soil in return. Ultimately, these systems of use and exchange evolved into more formal relationships between patrons and clients. Tutsi were apparently better able to convert their cattle wealth to arrange clientage relationships in their favor, although arrangements were flexible and many Hutu were patrons, too. By the fifteenth or sixteenth century, the minority Tutsi had either bargained, fought, or maneuvered their way into control over a significant part of Rwanda. Again, ecology played a role because the Tutsi did not have to follow long and variable migrations on a seasonal basis, like most pastoral societies, but could instead remain in one place and consolidate their power. Under a succession of Tutsi kings, or *mwamis,* a highly centralized and hierarchical kingdom evolved. Combining the human and agricultural resources of the Tutsi and Hutu, this kingdom aggressively expanded to conquer much of modern Rwanda. It also entered into a series of shifting alliances with neighboring states in what are now the nations of Uganda, Congo, Burundi, and Tanzania.

Geography helped certain groups resist Tutsi domination. In southwestern Rwanda, the rugged mountains of the Congo–Nile Divide and the dense Nyungwe forest limited contact with the central kingdom and allowed for a certain degree of local autonomy for those living on the western side of the divide. In the northwest, the same chain of mountains rose to meet the Virunga volcanoes, providing a double barrier to Tutsi penetration. There, thick forests afforded safety, shelter, and sustenance in times of trouble. A breakaway band of Tutsi, known as the Bagogwe, evolved a unique form of pastoralism within the Gishwati forest. But the vast majority of those who inhabited this mountain refuge were the ancestors of today's northern Hutu. And they fiercely defended their independence from Tutsi domination until the *abazungu* appeared.

THE GERMANS ARRIVED in Rwanda in 1894. They found a landscape that had already been radically transformed by its inhabitants. One of the first travelers through central Rwanda described "a vast area with neither tree nor bush, only banana trees . . . it is incredibly populated . . . and very cultivated." Over a period of less than two thousand years, Hutu farmers had cleared more than 75 percent of the country's montane rain forest. Traditional shifting cultivation had long since given way in most areas to more permanent and intensive forms of agriculture. Only the practice of naming places after local tree species provided a record of where the forest had once grown. Thus the settlement of Miyove stands to this day in mute memory of the *umuyove*—a giant hardwood of the *Entandophragma* genus—which once grew in a rain forest that extended almost forty miles southeast of the Virunga forest. So, too, the hill called Kiyumba, just west of Kigali, recalls the stand of *Prunus africana,* or *umwumba,* that once covered its crown, but from which endless farmlands now stretch in all directions. It is a long way from Rwanda to the American suburbs, but the cultural practice of naming places after that which we destroy is not so different in Cherry Hill, Pinesdale, Buffalo Springs, or Redwood City.

Hutu farmers were not the only agents of conversion. In Rwanda's drier regions, Tutsi cattle and fires had also converted much of the natural wooded savanna to low-diversity grasslands. But in the Tutsi kingdom, the Germans found a political structure through which they could rule their new colony. One early envoy described the Tutsi as "graceful" people with "noble traits" who compared favorably with European aristocracy. The existing Tutsi monarchy would provide the mechanism for indirect rule by a small, but powerful clique of resident Europeans, or *abazungu,* as the white foreigners were known in the local Kinyarwanda language.

One of the first uses of German power was to bring the mountain kingdoms of the northern Hutu under the authority of their new masters. Western weaponry accomplished in a few years what the Tutsi kingdom had failed to accomplish in centuries. Pockets of resistance were quickly eradicated. When Hutu rebels led by Ndungutse and Bassebya killed the first Catholic priests sent to establish a mission in the Virunga region, a combined force of German and Tutsi military was dispatched to the north. Ndungutse was killed and Bassebya was summarily hanged with several accomplices at the site of Ruhengeri's current outdoor market. More fundamental change soon followed, as ancient Hutu land tenure and land use practices were supplanted by the increasingly despised Tutsi system that favored the cow over the farmer. Traditional Hutu leaders were replaced by Tutsi chiefs approved by the Germans, but with no local standing.

The Germans took direct control of very little land in Rwanda. Unlike in Kenya, where most of the productive highlands were transferred to British ownership under the process known as alienation, the vast majority of land in Rwanda remained in native hands. The Germans did intervene, however, to declare almost all of the remaining montane forests as *kronland,* to be managed under control of the Crown. The first area acquired under this policy was the Virunga forest, soon after von Beringe's "discovery" of the mountain gorilla. By 1911 most of the remaining natural areas along the Congo–Nile Divide were declared forest reserves. These were not parks, but were instead managed to assure that the benefits of commercial forestry would accrue to colonial business interests. Maintaining tree cover in the mountainous reserves also protected downstream watersheds: a concern of German foresters, who were well ahead of their time in this regard.

The chief administrator of Rwanda was the remarkable Dr. Richard Kandt. A medical doctor by training, Kandt was also a poet and avid naturalist. He traveled widely across the country, making observations and collecting new species, several of which—like the golden monkey, or *Cercopithecus mitis kandti*—still bear his name. Kandt also recognized that the limited German presence could not stop Hutu farmers, who continued to clear the forest at a steady rate despite official proclamations. He approached the reigning *mwami,* Musinga, to present the problem as one worthy of royal intervention. Musinga listened to his arguments, then advised Kandt that "the forest was there already when we were born, and it will still be there when we both die." It was an accurate prediction, based on understandable perceptions. But Kandt was thinking beyond his lifetime and remained an active advocate for forest conservation throughout his tenure.

Richard Kandt was also quite farsighted in his analyses and prescriptions for Rwandan agriculture. He called for the development of export crops to defray colonial costs and correctly identified the highland farming zone as an ideal environment for growing arabica coffee. To meet the subsistence needs of an expanding population, he called for increased production of the recently introduced white potato. World War I would see the Germans retreat from Rwanda, surrendering their combined colony of Ruanda-Urundi to the Belgians. Many of Kandt's ideas and initiatives, however, would be implemented under Belgian rule.

The awarding of Ruanda-Urundi to the Belgians was a contentious decision by the young and ill-fated League of Nations. Belgium's King Leopold had ruled the neighboring Congo as a personal fiefdom in which the most basic human rights were violated and perverted in the cause of unchecked economic exploitation. Critical press accounts, and the publication of Conrad's thinly fictitious *Heart of Darkness,* stirred popular revulsion at Leopold's abuses. Under public pressure, the Belgian legislature voted to expropriate management of the Congo prior to World War I. Many in the League of Nations still distrusted Belgium's record of harsh exploitation. Yet mindful of the need to throw a bone to the Belgians for their role in the war—while England and France feasted on the richer parts of the German colonial carcass in Tanzania and Cameroon—the League awarded its mandate over Ruanda-Urundi to Belgium.

Conservation was a high priority under Belgian rule in Rwanda. Leopold's son, Prince Albert, had visited Yellowstone and became an advocate of national parks. At Carl Akeley's urging, and with his father's support,

Africa's first national park was created in the Virungas in 1925 for the specific purpose of protecting its mountain gorillas. Expanded into neighboring Congo in 1929, the Albert Park would eventually cover more than four thousand square miles. Later, in 1934, the Akagera Park was created to protect the diverse wildlife assemblages of a savanna-wetlands complex that covered one thousand square miles of eastern Rwanda. The major forest reserves along the Congo–Nile Divide also received renewed attention, although the Belgians aggressively exploited the most valuable hardwoods.

Across the settled landscape of Rwanda, the Belgians instituted other policies and practices that would ultimately intensify human pressures on protected areas, especially in the mountains. Cash crops were promoted to increase local revenues and foreign exports. From the first 1,200 acres under Kandt, coffee production expanded to cover more than 100,000 acres; tea was planted on another twenty thousand acres in the highlands; and pyrethrum was introduced in the 1940s to produce insecticides for the war effort. Rwandan farmers were forced to cultivate cash crops, under threat of penalties, for which they received little compensation. But they latched on to the white potato as a subsistence crop. Potato production exploded from three thousand acres in 1930 to 85,000 acres in 1942, almost all of it in the higher-elevation zones of northern Rwanda. The following year, however, a fungus infestation known as the "black blight" wiped out the entire crop. As in Ireland a century before, a large population dependent on the potato faced starvation. Yet with the colony cut off from Belgium during World War II, there was no system in place to help. Even if help were available, northern Hutu were not a favored constituency under Belgian–Tutsi rule. More than 100,000 northerners died during the potato famine of 1943–44.

After the war, the Belgian administration introduced new blight-resistant strains of potato, and highland farmers steadily, if cautiously, renewed their production. Throughout the postwar period, per capita food production in Rwanda remained stable, contrary to trends in most other African colonies. This was largely a function of clearing new lands, especially wetlands, rather than improved productivity. Forest lands, too, were cleared. In 1958, the Belgians took the extraordinary step of converting more than twenty thousand acres of Virunga parkland to farmland. This was one of the world's first instances of removing protected land from an established national park, setting a dangerous precedent within Rwanda and beyond.

The Virunga parkland conversion was the direct result of human population pressure. The average Rwandan's health improved enormously under the Belgians. Life expectancies rose and infant mortality declined in response to modern medicines dispensed through a modest network of hospitals and clinics. The combination of improved health programs with increased food production resulted in dramatic human population growth. From one million at the turn of the last century, the Rwandan population surged to more than three million by the end of the Belgian era in 1962. Hutu made up 85 percent of this population.

Throughout their time in Rwanda, the Belgians continued the practice of indirect rule through the Tutsi. With several thousand administrators, technicians, and private citizens in the country, however, the Belgians played a far more direct role in day-to-day management of the colony's affairs than the Germans before them. Still, they maintained the façade of the Tutsi monarchy. In return for this charade, they blatantly favored the Tutsi through preferences in commerce and education. Eventually, however, the Belgians also provided an outlet for Hutu grievances. Beginning in the 1950s, Hutu leaders from southern and central Rwanda began circulating tracts critical of the preferential treatment of Tutsi with regard to educational and work opportunities. Meanwhile, outside of Rwanda, anticolonial sentiment was building on the twin pillars of Gandhi's nonviolent victory over the British in India and the military victory of the Vietnamese over the French. As the decade unfurled, demands for better treatment of Hutu increasingly merged with manifestos calling for independence. The logic of the former and the inevitability of the latter resonated with portions of the Belgian public. Socialists and Catholics eventually formed a surprising alliance in favor of transferring power to the Hutu majority before independence—which they imagined to be a distant, and therefore manageable, reality.

JEAN-PAUL HARROY spoke with great animation. A short, round man in his eighties, he repeatedly poked his stubby fingers into the air as he recounted his singular role in Rwandan history. First warden of the Virunga sector of the Albert Park. Author of *Rwanda: terre qui pleurt (Rwanda: The Land That Weeps)*, a heartfelt if prematurely dramatic account of the country's losing battle with resource degradation. Last governor-general of Rwanda-Urundi. After independence he would be named secretary-general of the International Union for the Conservation of Nature. In 1980, however, he was professor emeritus at the Free University of Brussels, where he met with Bill in his apartment.

During his tenure as governor-general, Harroy was told to promote a revolution: to allow the Hutu to overthrow the Tutsi. To complicate matters, this was to happen only in Rwanda, despite a comparable Hutu majority in neighboring Burundi. In his own recounting, Harroy discreetly passed this

Meadows mark the saddle region between the Karisimbi and Visoke (pictured) volcanoes. The Karisoke Research Center was located near the base of Visoke, to the right of the meadows in this photo.

The border of the Parc des Volcans is unmistakable, with farm fields cultivated to its very edge.

Midday siesta for Group 5 (*left to right*): Tuck, Pablo, Marchessa, Muraha, Poppy, and Beethoven.

Play is a central element of gorilla life, especially for youngsters like Shinda and Muraha.

Beethoven carefully folds sharp thistle needles into the center of his leafy handful before eating.

Mountain gorillas live in a rich "salad bowl," consuming as much as fifty pounds of leaves, stems, and roots per day, as Puck does here.

Many gorilla foods require preparation. Here Beethoven peels wild celery, a food that provides him with a significant source of water.

Karisoke was a safe haven from hunters for black-fronted duiker.

We lived in this cabin of corrugated tin, which provided little protection from near-freezing nighttime temperatures—or falling trees.

Amy holding Mweza, who was terribly ill when we reclaimed her from Congolese authorities.

Effie was the grande dame of Group 5, dominant over all other females. She conferred much of her confidence and standing onto her seven offspring.

The pattern of ridges and lines above a gorilla's nostrils distinguishes her from others, much like a fingerprint. We came to recognize Quince by her radiant face.

Poppy at age three, displaying the curiosity typical of young gorillas.

Pablo's boldness and keen interest in humans, in this case Bill, surpassed that of all others in the family.

Pablo eating decomposing wood as a four-year-old, not long after he was abandoned by his mother, Liza.

Eight-year-old Ziz sits out a cold, heavy rain.

Icarus with Amy; despite his insecurity as a subordinate silverback, he accepted us into his life.

Icarus towered over us as he beat his chest to warn off competing silverbacks or other intruders.

Pablo at twenty-six years old, now the elder silverback in the largest known gorilla family, comprising forty-four individuals.

Nyungwe is exceptionally rich in wildlife, home to at least thirteen types of primates, including the mountain monkey.

Uncle Bert was killed and beheaded by poachers in August 1979. His murder triggered a series of deaths that decimated Group 4.

Nyungwe Forest is one of the largest and most diverse mountain forests remaining in Africa, yet it remains little known.

information on to key Hutu leaders, mostly from southern Rwanda, who then seemed paralyzed by the opportunity. They had become adept at writing political manifestos, but after centuries of Tutsi subjugation they were apparently unprepared for decisive action. The northern Hutu had no such problems. Hearing that the Belgians would not aid their Tutsi clients, they sharpened their machetes and spears, poured out of their mountain settlements in Ruhengeri and Gisenyi, and began killing. Tens of thousands of Tutsi died in the brief but bloody uprising of 1959. Hundreds of thousands fled into neighboring Burundi, Uganda, and Congo as the Hutu consolidated their control over Rwanda. A Tutsi counterattack a few years later was brutally crushed by Hutu soldiers, now armed by the Belgians and under the direction of a young major named Juvénal Habyarimana. Jean-Paul Harroy smiled as he remembered Habyarimana as an unassuming but bright young man who left seminary school for better opportunities in the military. He would take advantage of those opportunities to play a leading role in the fight for independence, and eventually become president of Rwanda.

DESPITE THE CRUCIAL ROLE of the northern Hutu in overthrowing the Tutsi, the Belgians handed the reigns of power to a southern coalition at independence in July 1962. Grégoire Kayibanda emerged from this group as the new country's first president. A moderate who preached Hutu unity, Kayibanda led a government that practiced systematic discrimination against the Tutsi. The government also directed a disproportionate share of the nation's limited foreign investment to the favored central and southern provinces. After ten years of perceived regional neglect and national decline, the northern Hutu again took matters into their own hands. On July 5, 1973, a bloodless coup installed Juvénal Habyarimana as president of the Second Republic—the event that caused Idi Amin to hold us captive as Peace Corps volunteers in the Entebbe airport. Kayibanda was placed under house arrest, where he died of chronic alcoholism several years later.

The politicians of the northern Ruhengeri–Gisenyi axis were a dynamic group. Once they felt comfortably in control of internal affairs, they opened their borders in the mid-1970s and rolled out the welcome mat for foreigners. Belgian, Swiss, French, and German donors were impressed with the seriousness of their hosts. The image of Rwanda as the "Switzerland of Africa"—stable, hardworking, beautiful—began to take hold. The strict quota system, under which 10 percent of all government jobs and school openings were allocated to Tutsi according to their ethnic representation in the population, should have been repulsive to most Westerners. Yet even this was held up as an enlightened policy toward minorities in comparison with practices in most African countries at that time. Habyarimana was seen by many as the kind of benign dictator the continent needed to move forward. The Rwanda story warmed the hearts, and loosened the purse strings, of foreign investors. By 1978, even the United States Agency for International Development was hiring staff to open a full office in Kigali. As development assistance flowed at unprecedented levels, the rising tide lifted all regions. But the biggest, wealthiest projects were generally steered toward Ruhengeri and Gisenyi.

The Habyarimana government was not just a passive recipient of foreign assistance; it undertook some much needed initiatives. Nearly half of its budget was dedicated to agriculture. Following a post-independence period of neglect under the First Republic, Rwandans were again attentive to their land resource base. Six thousand miles of terraces and hedgerows were built or restored in the government's first five years, providing increased erosion protection to almost one quarter of the existing arable land base. Starting in 1975, millions of trees were planted each year. Both activities were supported primarily by Rwandan financing and labor. When we arrived in early 1978, fewer than one hundred miles of paved road existed in the entire country. Within two years, a grid covering all major arteries was planned and financed by a curious consortium of donors that included the Italians, Libyans, and Chinese.

Second only to agriculture, the new government invested heavily in education, spending more than one quarter of its annual budget in support of schools and schoolchildren. Still, fewer than half of all eligible Rwandan children attended primary school by the late 1970s; only one child in fifty attended secondary school. Part of the problem was a lack of classrooms, part of the problem was a lack of teachers. The colonial powers had not invested in this sector, certainly not for disempowered Hutu students. A radical program of educational reform was launched in 1979 to increase classroom access and make the curriculum more relevant to Rwandan needs. Without any setbacks or interruptions, the new program's first graduates would emerge from high school and college in the late 1990s.

Public health spending was more modest. Only 6 percent of the annual budget was dedicated to this sector, virtually none of it for family planning. In 1978 nearly one fourth of all Rwandan children died before the age of five—barely better than the 29 percent mortality rate for young gorillas. Still, fertility rates for women of reproductive age approached the biological maximum and the overall human population was growing at 3.7 percent per annum. Rwanda rose toward the top of lists of the world's most densely populated countries, with nearly five hundred people per square mile. If considering only agricultural lands, the density rose to more than 1,300 per square mile: an extremely relevant figure in a nation where 95 percent of the people lived off what they could produce on shrinking rural land holdings. The average farm size was down to an acre, farmers were becoming intensive gardeners, and further subdivision of most family plots through inheritance was out of the question. In a small landlocked nation with few minerals and an overdependence on coffee, alternative development options were limited.

As one of its first acts, the Habyarimana government created the Office Rwandais du Tourisme et des Parcs Nationaux (ORTPN) in 1974. This removed conservation from the Ministry of Agriculture, which had approved the 1969 conversion of 25,000 acres of Virunga parkland for pyrethrum production. It also gave new emphasis to development of the tourism sector of the economy. Yet by 1979, the rising tide of human settlement again lapped at the Virunga shore. Land-hungry farmers cast covetous looks toward what was left of the Parc National des Volcans. Northern politicians were all too ready to oblige their constituents' interests, as well as their own. And Western donors had money to spend.

Chapter Twelve

Filters and Perspectives

YOU'D HAVE TO BE the son of a white woman to think I'd grow bamboo when I can get all I want free from the park."

It was an honest answer. Asked if he would be willing to grow his own supply of bamboo for personal construction needs, the grizzled old man stated the opinion of most farmers living near the boundary of the Parc des Volcans: Why waste good farmland when a traditional free source was not far away? The response wasn't encouraging for conservation. But Bill took it as an indication that his survey was capturing heartfelt local opinions about the park and gorillas.

Rwandan attitudes and perceptions represented some of the key pieces in the conservation puzzle we were trying to assemble. It is never easy to penetrate foreign cultures, especially through the filter of multiple languages. We spoke French and Swahili, but no more than a few hundred words of richly complicated Kinyarwanda. Our approach was to seek information from a variety of sources: discussions with Rwandan co-workers and colleagues, classroom exchanges with students and teachers, and a formal survey of local farmers.

TOWARD THE END of our first year, Bill began a series of visits to all but two of Rwanda's twenty-nine secondary schools. When Dian reneged on her promised use of a Karisoke vehicle for this purpose, we took $3,000 earmarked for airfare home and purchased a used Renault 16 to fill the gap. The tour was well worth the money we ultimately lost on the car. At each school, Bill showed a National Geographic film about Dian and her work with the gorillas. It was not the most appropriate source of information, because it had little background on the country and less on its people. It was also in English. But Bill turned down the soundtrack and provided his own more conservation-oriented commentary in French. The Rwandan students, who had never before seen anything about the world-famous species in their midst, were fascinated by the mountain gorilla.

Social and behavioral aspects of gorilla life generated the most student interest. Polygamous family units were part of traditional Rwandan culture, and analogous gorilla arrangements intrigued the students. How many females per male? The boys in the class nodded at the notion of a three-to-one ratio—at least those who saw themselves as successful silverbacks. How long between births? The girls nodded in appreciation of the gorillas' four-year birth interval, which was beyond anything their mothers had known as they gave birth to eight or more children in church-approved, birth-control-free, single-partner relationships. The humanlike qualities of gorillas appealed strongly to students: a nine-month gestation period, perfect fingernails and toenails—*their eyes.* All of these attributes reinforced the more abstract message that we share at least 99 percent of our genes with our gorilla cousins. This genetic link, in fact, was a barrier to be overcome: a legacy of insults— *Monkey! Gorilla!*—hurled at Rwandans and other Africans by white racists in the colonial past.

Our conservation message was harder to choreograph with the film footage, which focused on gorilla and human personalities. Still, students seemed genuinely concerned that only twenty-eight families of this special creature remained in the wild and that the fragile species might soon disappear. Some asked why we didn't save gorillas in America. They were surprised and proud to learn that Rwanda—their little country, so little known to the outside world—held the key to saving the mountain gorilla. This was a breakthrough observation for us. Rwandans were a proud people who sought greater respect. Congo and Angola grabbed more attention in the Cold War; Kenya and Nigeria wielded more economic clout. But Rwanda was home to the mountain gorilla. This unexpected angle—conservation nationalism— would open new opportunities for us in the months and years ahead.

Traveling around Rwanda to speak in schools, Bill was exposed to the elite of the next generation. Fewer than fourteen thousand students were enrolled in secondary schools across the entire country. They would have privileges and opportunities unknown to the vast majority of their peers; many would rise to positions of authority. Traveling also opened a window on the lives of those less fortunate. Midway between Kigali and Ruhengeri was a dusty commercial pitstop named Gakenke. It offered a convenient stopping point for truck and *taxi brousse* drivers at the base of the long and dangerous escarpment road to the north. Here, Bill always stopped at an open-air stand to buy a Coke and an omelet, then rolled the omelet in a burrito-like wrap called a *chapati*. Both the omelets and chapatis were stacked on plates by the cash register and usually sold cold. Bill called the town Chapati Village. And the frail little boy with polio who lived there he called the Chapati Kid. The boy first stared at Bill from the background, leaning on his knotted wooden staff as other boys approached more closely to watch the *umuzungu* eat. Personal space is an unknown concept in Rwanda, whether one is white or black. But foreigners are a never-ending source of unabashed, in-your-face fascination for young Rwandans. Over time, the Kid moved closer, his wide eyes softened by long eyelashes. Bill bought an extra omelet chapati for him one day, but the Kid turned it down under the withering looks and laughter of the other boys. On his next visit, Bill walked up and handed him the chapati just before reboarding the taxi. As the Toyota pickup pulled away, he could see the boy's defenseless body crumple under the mass assault of the other boys. Now, with his own car and no time constraints, Bill made sure that the Chapati Kid sat at his side to eat his meal. He never spoke a word and was possibly mute. But the other boys didn't disturb this arrangement, at least not while the big *umuzungu* was around.

T HE SCHOOL TOUR was a great success. Students learned about the mountain gorillas, the park, and the potential benefits of both to Rwanda. In a curriculum dominated by a strong Eurocentric focus, they finally heard about something of international interest and value in their own country. In return, we learned a great deal about the students' perspectives on the world around them, their lives, and their values. Their questions challenged us to answer why parkland was more important than farmland, whether the gorillas were more important than the local people, why all researchers were white foreigners. If we were going to propose solutions, we needed to know—and be able to answer—their questions. And change some of our own thinking.

The school tour required us to spend more and more time apart. In addition to the school visits, Bill took advantage of each passage though centrally located Kigali to meet with government and foreign officials. The cost of these layovers would have been prohibitive, especially since Rwandan officials repeatedly postponed meetings. Fortunately, European and American friends helped out with free room and board. They also provided another source of information and different views on life in Rwanda—especially the Europeans. Still, the seven- to ten-day separations were longer and far more frequent than any in our seven-year-old marriage.

At the end of one trip, Bill finished his work in Kigali by early afternoon and felt the urge to return to our Karisoke cabin before nightfall. He made a quick shopping run, then hurtled north in the Renault, skipping his customary stop in Chapati Village. He reached Ruhengeri in record time, then drove more carefully along the final lava track to a small parking area at the base of the mountain. By 5:30 he was ready to climb. He knew he needed to shorten the usual fifty-minute climb to our cabin if he was going to arrive before dark. It was certainly doable. In any event, he didn't have a sleeping bag or tent.

Ten minutes up the trail, all plans went to hell. An enormous black cloud surged out from the saddle between Karisimbi and Visoke, snuffing out what little daylight remained. Rain soon followed. Bill picked up his pace and began to run up Camp Trail. With the premature arrival of complete darkness, he turned to his memory of the trail he had climbed dozens of times before—even as he berated himself for not carrying a flashlight. An irregular smoker at that time, he carried about a dozen matches. These he rationed to shed light on critical junctions, sheltering the ridiculous—but effective—little torches within his cupped hands. By the trail's midpoint, though, he was out of matches, the rain was torrential, and the Cape buffalo were beginning their nocturnal promenades. Bill could no longer feel the trail beneath his feet, which were numbed by the cold water coursing down the pathway. He then dropped to his hands and knees. Mud oozing between his fingers confirmed that he was still on the much used trail; vegetation meant he was off course. Every stream, every log, every large rock was a marker of progress. Every stretch without familiar landmarks brought on disquieting thoughts. Far worse were the thoughts of buffalo that go bump in the night. Bill began to sing at the top of his lungs, alternating the Rolling Stones and Grateful Dead with oldies and show tunes. Anything to make noise, maintain sanity, and not think about the cold that had already numbed all feeling in his hands. Finally, the distinctive rock outcroppings around Bone Ravine reassured Bill that he was on course. From there, he knew that Camp Stream would soon appear on the left, channeling the trail as it passed within thirty feet of our cabin. He couldn't get lost now. In another five or ten minutes, Bill rose up again on two feet, climbed the steps to the cabin, and pushed open the door. There, caught in dinner delicto, Amy and David Watts could only look in disbelief at the shivering, mud-covered nightcrawler in the doorway. Shared laughter warmed the heart; the body followed more slowly. This was not the

last late-day adventure in the Virungas, but in the future Bill always carried a
flashlight.

WALKS ALONG THE PARK boundary presented a study in stark con-
trasts. Neat rows of white potatoes were planted right up to the sparse
line of exotic cypress trees that marked the boundary. Frequently, the cypress
roots were hacked away on the downward side by the hoes of aggressive
women farmers. Above that line, rain forest vegetation thrived in unrestrained exuberance. It was a binary world: all fields and people to one side,
all forest and wild animals on the other, with no transition zone to buffer in-
fluences in either direction. Our sympathies usually lay with the forest and
its denizens. It was from the forest that vast areas had been repeatedly carved
and cleared. It was the forest that suffered from a thousand smaller wounds
inflicted each day, from cut bamboo to wire-snared animals. People always
gained in this relationship, the end of which for the gorillas seemed tragically
clear. But there were also days when we reversed our perceptions and sought
to understand how local people looked at the park; days when the cold statis-
tical findings of Bill's surveys could be woven into a portrait of hard human
existence along the Virunga frontier.

The houses of the Virunga region looked like half-buried acorns from a
distance. Round, dark, windowless walls of stick and adobe construction
were topped with thick thatch shaped into a cone. At its tip, the acorn's stem
was replaced by an upside-down clay pot, which covered a central vent in the
roof. On most cool mornings, smoke seeped steadily from every roof. Inside,
the thatch was blackened, and eyes reddened, by the constant smoke of an
open fire. We could only imagine the color of the occupants' lungs. In towns
like Ruhengeri and surrounding areas, people were switching to rectangular
houses with windows and tin roofs said to be better ventilated and better
sheltered from rain. Near the park, the traditional house was still the norm
and its form blended better with the backdrop of volcanic cones. Yet people
knew that change was coming, in their houses and their lives.

Bill's surveys showed that most of the farmers who lived around the park
produced enough food to support their families' basic needs. Yet a majority
felt that they didn't have sufficient land to subdivide with their eldest sons, as
Hutu tradition required. Almost all were aware that Rwandans outside of the
Virungas faced even more severe land shortages. Asked their solution to the
lack of land, many noted the ancient release valve of emigration, though
without specific ideas of where to emigrate. In fact, residents of Ruhengeri
were already returning in the late 1970s from a failed resettlement program

in southern Rwanda. Others threw up their hands and called on God or government for help. Almost no one mentioned birth control.

Asked specifically about the park, more than half of all local farmers thought that they could cultivate its land. This was an interesting perception, since none of their crops were adapted to cultivation above the park boundary elevation of 8,800 feet. Forty percent further felt that local people needed to hunt or cut wood in the park, while recognizing that both activities were illegal. As for the natural forest itself, most local farmers couldn't cite a single value of its existence if it couldn't be cut. The same held for wildlife, if animals couldn't be hunted. A notable exception was tourism. Notable because more than one third of those surveyed thought it was possible; even more notable because there was no viable tourism program in the Parc des Volcans at that time.

The mountain gorilla fared well in Bill's surveys. Informed that only 260 were left in the entire world, more than four out of five local farmers thought the gorillas should be protected. The connection between protecting gorillas and protecting their habitat was not well understood, however. More than half of those same farmers thought that the entire park should be converted to agricultural purposes.

Other surveys of urban residents and university students in Rwanda showed greater support for both gorilla and park protection. Yet the farmers living closest to the park were the most critical population. Even if some of their beliefs were ill-founded or inaccurate, their perceptions represented their reality. They were also reinforced and enhanced by government policies and regional politics. Rwanda's tourism advisors at that time focused on the Akagera Park, which fit the East African model of vehicular tourism in savanna parks. Hiking through Green Hell did not fit their model and so the Parc des Volcans was ignored. Not only was there no tourism, there were only eight poorly paid, poorly trained, and poorly equipped guards for the entire park. This policy of neglect resulted in the loss of more than half of the original Virunga parkland between 1959 and 1973: losses funded by European donors in the name of development. Rumors persisted of plans to clear more land.

It was easy to walk through their postage stamp fields, past their smoky, cramped houses, and understand the views of local farmers. They saw little value in the forest or its wildlife, and they desperately needed more land to feed their growing families. They had nothing against the gorillas, which they didn't perceive as dangerous predators or even serious pests—they just thought that gorillas could be saved somehow outside of the forest, perhaps in a zoo. In the past, big development projects had paid to clear the forest and provide land for more people. So now that northerners dominated Rwandan politics, why not ask some foreign donors to pay for another round of park-land conversion? It made sense to local people. It made sense to northern politicians. It apparently made sense to the European Development Fund, which supported another such scheme in early 1979.

Chapter Thirteen

The Cattle Are Coming

CONSERVATION MEANS DIFFERENT things to different people. For some it is akin to a religion that attracts a passionate mix of true believers, missionaries, and zealots. For others, it is a more secular belief system that calls for human action to reinforce the intrinsic value of wildlife and wilderness. Many see in the term a mandate to use, but not waste, valuable natural resources. As they define conservation in ways that cover the entire spectrum from strict preservation to sustainable use, conservationists set forth an equally broad range of related rights and responsibilities. This can be confusing for those who prefer their definitions and labels clear; for others, the flexibility of its use reflects the power of the concept.

Conservation is also a science. It is an applied science that seeks to understand and resolve problems that diminish biological diversity and degrade natural ecosystems. As we enter the twenty-first century, the field is firmly grounded in the biological sciences, yet is increasingly interdisciplinary in its scope. Ultimately, the success of conservation science depends on the ability of its practitioners to move from the collection, integration, and analysis of information to the identification and pursuit of concrete action steps: to move from problem analysis to conflict resolution.

In the 1970s there was nothing like the current field of conservation biology, nothing approaching an interdisciplinary conservation science. Most field research focused on animal behavior rather than more applied issues in wildlife ecology. In Africa, most researchers concentrated on the savanna ecosystems of the east, while only a few individuals worked in the more demanding rain forest environments to the west. Almost no one studied human factors in conservation, wherever they worked. In our effort to take an interdisciplinary approach to science—one that looked at both people and wildlife—and then use our results to formulate a plan of action in the forests of Rwanda, we were on our own.

By the end of our first eighteen months, we had turned over many new pieces in the gorilla conservation puzzle and revealed a much clearer picture. Inside the park, conditions were mixed. The total mountain gorilla population of only 260 individuals was highly vulnerable to extinction, especially under the extreme poaching pressure that we had witnessed firsthand. No more than thirty silverbacks made up the current male breeding pool, and these alpha males were primary targets for trophy hunters. On the more positive side, the steep population decline of the 1960s had been halted and the percentage of young gorillas was the highest it had been since Schaller's initial study. The forest habitat retained a remarkable capacity to sustain the gorilla population despite the loss of nearly one third of all Virunga parkland since 1958. Food resources even seemed sufficient to support a much larger population, though the gorillas would have to live on less land and at higher elevations. Although this mixed image gave us some hope, the picture outside the park was far more troubling.

More than 100,000 farmers lived within five miles of the Parc des Volcans. Behind them, millions more scratched out a living on shrinking plots of tired land. The park and its gorillas held no value to local Rwandans, who saw only potential farmland under the green blanket of forest that towered above them. In endless battles with the powerful Ministry of Agriculture, senior ORTPN staff were hard-pressed to defend the value of a park that earned less than $7,000 in total entry fees in 1978. Left unchanged, these conditions would only encourage continued poaching and conversion. Our preoccupation was to change this equation, so heavily weighted toward extinction.

IT WAS NOT OUR EXPERIENCE that conservation proceeds in a tidy progression from research, through analysis, to action. From our earliest days in the park we were actively engaged in anti-poaching efforts around Karisoke. We had little understanding of who the poachers were, what their motivations might be, or why a market existed for their grisly goods. With time, we would learn the answers to some of these questions. But first we occupied the front line of defense in a deadly struggle to save the gorillas. So we burned poacher huts, confiscated weapons, and cut traps wherever we found them. We chased poachers as far as our lungs would last and with little thought to what we would do if we ever caught them. We herded gorillas away from known poacher zones. And we did our futile best with limited means to save those who could not escape the lethal grasp of traps.

On his frequent trips to Kigali, Bill was asked for advice on many issues concerning the park. Detlef Siebrecht was a German tourism advisor to ORTPN who felt that much more should be done with gorillas. Unfortunately, one of his first ideas was to capture a small group of gorillas and keep them in an enclosed area where tourists could see them. Detlef would then quickly add that the enclosure should be within the park boundaries, "of course, *mon ami.*" His smile was engaging and his logic straightforward: *we need to make enough money to protect your park and the other gorillas.* But his idea was frightening and Bill made it a priority to quash any such thinking in discussions with Detlef and other ORTPN personnel.

On one trip to Kigali, Bill learned that the secretary-general of education wished to see him. This followed soon after Bill had proposed a speaking tour of Rwandan secondary schools and he hoped that this meeting would result in a letter of authorization. He was asked instead if he would help to design an environmental studies program for grades one though twelve as part of a national curriculum revision. It was difficult to say no to the second most powerful man in the ministry, but Bill begged off claiming— correctly—that he didn't have the necessary training for that kind of work. Two months later the secretary-general caught him in a more optimistic moment, though, and Bill agreed to consult with Gilles Toussaint and François Minani, a team of professional educators within the ministry already assigned to work on the revised environmental curriculum.

Our early involvement with anti-poaching, tourism, and education was based more on inspiration and effort than on knowledge and training. Our understanding of these issues and their interrelationships grew steadily, however, until a meeting with Jean-Paul Sorg put everything into focus. Sorg was a forester assigned to the Ministry of Agriculture, or MINAGRI, as it was known. Like most of his Swiss colleagues and compatriots in the Direction des Eaux et Forêts, he took a very practical view of his work. Growing and cutting wood was his business. But Sorg also saw forestry as a way to save Rwanda's last natural forests, about which he cared deeply. When Bill learned of his interests, he scheduled a meeting at an old colonial research station outside of Butare to discuss ideas for protecting the Parc des Volcans. Bill was just beginning to appreciate Sorg's singsong French, when the bearded Swiss stopped and stared blankly at a question about the park. *Alors, vous ne savez pas?* No, Bill certainly didn't know. Sorg dropped the bombshell news that the government had agreed to clear another 12,500 acres from the park for a cattle-raising project. There had been many rumors, but nothing definite until Sorg pulled out a recent government publication outlining the proposed scheme in considerable detail. Five thousand cattle on five thousand hectares. One third of the park. European funding. Sorg apologized for bearing such bad news, but Bill thanked him for the warning and hurried north to Kigali.

Alain Monfort of ORTPN appeared to be caught off guard by the news. Bill wondered if Alain was unwilling to acknowledge the park service's impotence in the face of MINAGRI. Or perhaps, as a Belgian *fonctionnaire*, he simply couldn't go against powerful government and European interests. Monfort said he would learn what he could. Bill set up meetings with government officials for the following week and returned to Karisoke to spread the news.

Our personal relationship with Dian at that point ranged from negative to nonexistent. On work-related matters, though, we made a persistent effort to keep communication channels open. And even though she was essentially a recluse, Dian knew she needed some contact with the outside world, at least with regard to matters affecting her gorillas. If we brought news of poaching, Dian reacted viscerally. She was animated and in command. *Call up Vatiri. Organize patrols. Nemeye, take a gun.* At the end of the day she would want a full debriefing from everyone—usually separately—in the main room of her cabin. Her reaction to the news of another major parkland conversion, however, was strangely passive, almost indifferent. She certainly was unaware of the plan, and Dian didn't like being out of the loop. Perhaps she knew that events at this scale were beyond her control, that she, Vatiri, and Nemeye were powerless to halt the forces of destruction. After all, she had been at Karisoke when 40 percent of the park was cleared for pyrethrum in 1969, and she couldn't alter the outcome, except to recover a few acres around the Visoke parking area. Now, ten years later, Dian was also a different person from the dynamic woman who first staked her claim high in the Virungas. Perhaps the loss of another one third of the Parc des Volcans simply fit with her own increasingly dark view of the future. *They're all going to die* was all she could say.

~~~

WE WERE TOO ENERGIZED for passive acceptance and too young to believe that we couldn't change the course of events. Our experiences and research results also convinced us that we were closing in on an approach that might turn the Rwandan equation in favor of gorilla and park conservation. First we needed just a little more information about the opposition.

Votre démarche m'épate! The Belgian advisor to the minister of agriculture certainly looked astonished. *How can you even think about gorillas in a country where the people are so poor? Farmers need the land, and if gorillas are on that land, then that is too bad for the gorillas. People's needs come first.* Bill listened to the arguments and imagined that it was rare for this middle-aged technocrat to become so impassioned about a subject. Later in the discussion, the advisor was less self-assured when asked why a cattle project was needed if what the poor farmers required was land. Bill was escorted down the hall to the office of the director of livestock husbandry for further information on that apparently tricky subject.

The director was a very thin Rwandan, with thick glasses and a crisply pressed white shirt and tie. He was proud of the project that would be run through his office and eager to describe its anticipated benefits.

One thousand calves would be born each year, resulting in 500 liters of milk produced per lactating cow. Taking 150 liters per cow for sale at ten cents per liter, we have a milk production value of $15,000 per year. Selling 900 head of cattle, at $0.60 per kilogram for a one-hundred-kilo carcass, would generate another $54,000 per year.

The total of nearly $70,000 per year was not insubstantial by Rwandan standards. But the director's calculations included no costs to set against his suggested benefits. Some of these costs would go toward salaries for the hundreds of herders, milkers, veterinarians, and others who would manage the herd. Would they be local hires? He presumed so. How would the five thousand cattle be controlled so they didn't graze in what remained of the park? The herders would control them. Who would control the herders? He laughed.

Alain Monfort added further information over lunch at his house. Helicopters had flown over the park and prime areas for conversion were already mapped. He said he couldn't get the maps, but that they included the entire bamboo zone and most of the saddle areas. The saddle area around Karisoke was excluded. This meant the loss of the gorillas' most important food resources, the fragmentation of the park into an archipelago of natural islands separated by pasture lands in the saddles, and the further isolation and concentration of the remaining gorillas in high-elevation enclaves.

The enemy was now clear. Poaching was almost a minor threat compared to the devastating cost to the gorillas if we were to lose the fight against the cattle and further forest destruction. It was time to act.

IN ANY CONSERVATION CONFLICT, there are factors that favor conservation and factors that work against it. In 1979, the equation in Rwanda was heavily weighted against the mountain gorillas. At the global scale, world opinion was strongly pro-gorilla, thanks to the widespread publicity surrounding Dian's work in articles and films produced by her financial backers at the National Geographic Society. Yet this foreign constituency had no way to exert influence in any tangible manner inside Rwanda. The thousands of protest letters that were sent in the wake of Digit's death included sharp criticisms, but no checks to help the Rwandan park service. The real benefits of gorilla conservation accrued to organizations like National Geographic through the sale of its films and magazines and to the foreign audiences that enjoyed vicarious contact with the gorillas through these outlets. Western science, too, cared deeply about the fate of the gorillas. Yet, again, the rewards of research accumulated in Western academic institutions, publications, and career advancement.

Within Rwanda, Western interests were greatly devalued when weighed against Rwandan needs and perceptions. Gorillas and rain forests were simply not important. Land and employment held the highest value for local populations. Politicians are expected to be attuned to the interests of their constituents, which they ignore at their peril. Dissatisfied guerrillas might overthrow the government, mountain gorillas would not.

If gorillas were to have any chance, the scales had to be rebalanced through a program of active intervention. Factors favoring the gorillas and their forest habitat needed to be amplified, negative factors reduced. Education could help in the long run to increase Rwandan knowledge and understanding of the gorillas. The practical value of the natural forest for local water catchment protection could also be stressed. Anti-poaching could be intensified, although such an effort would likely strengthen negative attitudes toward the park, at least in the short term. All of these actions were good. But the overwhelming need was for a major source of revenue and employment to offset the powerful interests of MINAGRI and to win the hearts, minds, and money pouches of those who lived around the Parc des Volcans. Politicians could be trusted to follow the money. Only tourism offered the potential to meet this need and radically alter the equation.

There was no such thing as ecotourism in 1979. Nature tourism was well developed in East Africa, but it was strictly limited to savanna parks and a few high mountains of interest to alpine climbers. We had seen firsthand a poorly managed program of gorilla tourism in eastern Congo, where the Kahuzi-Biega gorillas were clearly unsettled by a lack of control over visitor numbers and behavior. Now we had to improve on that program to make it work for mountain gorillas and for the Rwandan people. We also needed to convince a daunting array of often opposing interests of the value—and absolute necessity—of our plan.

The Mountain Gorilla Project

Chapter Fourteen

Crazy White People

F RIDAY WAS ONE OF TWO DAYS when porters delivered mail and food
supplies to Karisoke. Dian was kneeling on the floor of her cabin, surrounded by piles of correspondence that she received from around the world.
Most were letters from fans and supporters of her work. The letter in Dian's
hand was also intended to show support for her efforts. It was from Sandy
Harcourt and it stated that he and John Burton of the British Fauna and Flora
Preservation Society had established the Mountain Gorilla Preservation
Fund to raise money for gorilla conservation in the Virungas. This was in
early 1978. The brutal killing of Digit a few months earlier had been widely
publicized in Britain, and the Fund intended to convert the public's sympathy
and anger into money for action. Dian saw matters differently.

It's blood money. It's Digit's money! Dian's eyes were red and her face contorted as she uttered Harcourt's name like a curse, then spat on the woven
raffia mat on the floor beside her. *This money should come to me, to Digit. Harcourt will just spend it on himself.* During her trip to the U.S. in the wake of
Digit's death, Dian and some of her backers had established the Digit Fund
to channel contributions from America. Within a few months she was receiving a modest flow of money, but the Digit Fund was not the immediate
success she had hoped it would be. In contrast, the Mountain Gorilla Preservation Fund was launched with a major media blitz and was backed by the
Fauna and Flora Preservation Society, the most respected British conservation organization at that time. Worse, Dian saw the Fund as a creation of
Sandy Harcourt—someone she despised as a former "student" who she believed had turned against her.

Our experience with Sandy to that point had been quite limited, but very positive. He had provided helpful advice and encouragement when Amy called in late 1977 to discuss her proposed research. He and Kelly Stewart kindly hosted us for an overnight visit at their Cambridge home when we passed through England on our way to Rwanda. And while Sandy was generally very serious, Kelly had a riotous sense of humor. She, too, had been a researcher at Karisoke and had once been very close to Dian. When Kelly and Sandy moved in together, however, she joined him on Dian's blacklist. To their credit, neither Sandy nor Kelly used our initial meeting to make disparaging remarks about Dian, or otherwise seek to influence our opinions of her personality or professional abilities.

The Mountain Gorilla Preservation Fund lingered in the remote background of our lives for nearly a year. Meanwhile, back in England, the Fund was gaining energy from the support of people like David Attenborough. A distinguished British naturalist, Attenborough had filmed Group 4 just before Digit's death. His footage of the gorillas and his personal stature combined to mobilize the British people to open their hearts and wallets to the mountain gorilla cause. More than $100,000 was collected in this manner, almost entirely from small private donations. It was the most successful campaign ever launched by the Fauna and Flora Preservation Society and a very large sum for conservation from any source in those days. Yet the Fund was an empty vessel, a campaign with the best of intentions but no real plan of action. It was also a vessel grounded on the shoals of continued resistance by Dian Fossey. Her opposition had paralyzed Rwandan officials, who were otherwise all too ready to accept money from any source, but fearful of her wrath and condemnation. Other international conservation groups, like the World Wildlife Fund, were waiting for signs of Dian's approbation before they joined the cause.

Faced with five thousand head of cattle on the near horizon and the future of the park in clear danger, we were ready with information and ideas to guide the Mountain Gorilla Preservation Fund and put its money to sound use. But first we needed to convince Dian to acknowledge the Fund's existence, enter into some level of discussion, and preferably endorse a plan of action. Or at least not oppose the use of this money in Rwanda.

Dian remained contemptuous of Sandy Harcourt and dismissive of "his" Fund. On rare days when she could move beyond this level of personal enmity, another point of fundamental resistance emerged. The Mountain Gorilla Preservation Fund had declared its goal of working with Rwandan institutions to improve their ability to manage the Parc des Volcans and protect its gorillas. ORTPN was a certain recipient of support. This was anath-

ema to Dian. She barely acknowledged the government's sovereignty over
the park and dismissed ORTPN as hopelessly corrupt and inept. She was
partially correct about ORTPN, though ineptitude was a more fundamental
problem than corruption. How the park service could improve its performance without some source of revenue to hire, train, and equip more
guards, however, was not clear to us. Nor was it a concern of Dian's. The
Digit Fund sent money directly to her, some of which she used to fund her
own private anti-poaching patrols around Karisoke. These patrols helped to
limit poaching within a fifteen-square-mile area surrounding the research
station, but left the great majority of gorillas that lived in the remaining 150
square miles of the Virunga forest without protection. Our recently completed census confirmed that the Karisoke gorillas were faring comparatively
well, but it also confirmed continued declines in the outlying population.

News of the pending cattle project and park conversion forced the need
for decisions and action. The Mountain Gorilla Preservation Fund was sitting on substantial sums of money and wanted to show contributors that action was under way to save gorillas. The Fund decided that Sandy Harcourt
would return to Rwanda with the Swedish conservationist and author Kai
Curry-Lindahl. Curry-Lindahl had spent some time in the Virungas during
the colonial era and was selected as a senior emissary. Their mission was to
determine a plan of action, identify willing partners, and sign an agreement
to move forward. It was understood that only the Rwandan government
could sign an agreement, and cooperation with ORTPN was a precondition.
It was implied that other partners—namely, Dian—were welcome but not
required.

We had worked extremely hard to form good working relations with
ORTPN staff, convinced that the future of the park and gorillas ultimately
depended on a vast improvement in their performance. We were also convinced that this improvement would be a long-term process, with many ups
and downs along the way. In 1979, the park service was incapable of formu-
lating or managing an effective gorilla conservation program on its own. Belgian advisors like Alain Monfort were helpful but compromised by their own
and their government's much greater interest in the savanna wildlife of the
Akagera park in eastern Rwanda. The gorillas needed strong, independent
advocates for their interests: people on the ground in Rwanda who could
help plan the activities of the Mountain Gorilla Preservation Fund, work
with ORTPN and other groups, and assure the effective use of funds. We
were willing to play this role, but we had only local standing. Sandy was living in England and planning an academic career. Dian was the only one with
a global reputation who could endorse any plan and help to solicit greater

support. Up to that point, however, she had used her power only to veto any action.

We were aware of the Harcourt and Curry-Lindahl mission, thanks to Sandy and Alain Monfort. Knowing that mail was never private at Karisoke, Sandy wrote to Alain, who passed correspondence on to Bill when he visited Kigali. When the dates and objectives were set, we again approached Dian to request her participation. Playing to her known antipathies, we argued that the project would go forward no matter what, but that without her participation all decisions would be made by ORTPN, Harcourt, and Monfort. This was the most unholy trinity that Dian could imagine. She began to budge. What would her role be? We honestly believed that Dian was needed as a fund-raising icon for her global constituency, and that her views and ideas should be taken into consideration in developing a conservation plan. We also thought that she would be an impossible partner—especially for the Rwandans—in any day-to-day working relationship. With somewhat less candor, we stressed the need for her input and global leadership. Who would lead anti-poaching efforts? This would require an improved park guard force, we answered, adding that they would of course benefit from special training from the Karisoke staff. Dian countered that she was leaning toward hiring American and European guards through her Digit Fund. What about tourism? This was unacceptable to Dian. We were not unqualified enthusiasts, either, after our negative experiences with uncontrolled tourists in Kahuzi-Biega and a few Kenyan parks. Yet tourism offered the only source of revenue and employment that might offset the claims of the cattle project, as well as provide potentially sustainable funding for improved park security. We felt that tourism must be considered as part of any search for a solution. Our discussion ended on this subject. Dian reiterated that she would not meet with Harcourt or Curry-Lindahl. She ordered us not to talk with them, either, stating bluntly that we would be banned from Karisoke if we did.

The issues were too important to ignore, however, and we had too little faith in others' ability to formulate an effective plan. We decided that Bill should meet with Harcourt and Curry-Lindahl and bring them up to Karisoke, if possible, to see if Dian might change her mind. By then we had sold our used Renault, so Bill left camp early one morning and walked the four hours to Ruhengeri. From there he boarded a Toyota pickup and rode through an intense afternoon storm to Kigali.

MONFORT, HARCOURT, and Curry-Lindahl were already seated at La Taverne when Bill arrived. The ruddy-faced Swede introduced himself pointedly as the *chef de mission,* underscoring his position with an insistence that he pay for the meal. Bill readily accepted, though he could already imagine Dian accusing him of feeding on "blood money." The next several hours were spent discussing ideas for a mountain gorilla conservation plan. Bill reviewed our research and conclusions, which in turn informed our recommendation to attack the problem on three fronts: anti-poaching to halt the killing, education to change people's perceptions and values, and tightly controlled tourism to generate political support for the park and gorillas through foreign revenue and local employment. Each of the participants had some experience with the park and gorillas and each had his opinions. Alain lobbied for ORTPN support, but with some outside control over finances. Curry-Lindahl questioned the value of long-term education in the face of such immediate problems. Everyone debated the double-edged potential of tourism—and the lack of any model for foot-based nature tourism in a rugged jungle. As the evening progressed, however, agreement coalesced around our original three-part program. Curry-Lindahl concluded that he could make the case to Dian, whose support would then create a unified front for a final proposal to ORTPN.

The next morning, Bill and the others joined Alain Monfort as he careened his Peugeot 504 north to the Virungas. Bill had never experienced the rutted mountain track in such comfort. Arriving at the base of Visoke, Bill climbed up alone while the others returned to wait in nearby Ruhengeri. We reviewed the Kigali meeting, then went to discuss the matter once more with Dian, hopeful that she would relent and allow the others to come up to Karisoke. But Dian again dismissed our arguments. She instead wrote a rude note to be carried down to the base of the mountain telling the team to stay away when they arrived the next morning. We went to bed dejected, yet determined to try one more time. The next morning, we returned to Dian's cabin to restate the case for cooperation with the Fund. We offered no radically new arguments, but Dian seemed more attentive and clearheaded. After no more than twenty minutes, she changed her mind. Amy and Rwelekana then raced down the mountain in a record twenty-two minutes, arriving just in time to intercept the porter bearing Dian's first note. In its place they delivered a more conciliatory, if limited, invitation: Curry-Lindahl could come up to Dian's cabin, but neither Monfort nor Harcourt was welcome anywhere near Karisoke. We had done our part.

Kai Curry-Lindahl spent that evening with Dian. We were not invited to join their deliberations. The next morning Curry-Lindahl stopped briefly at our cabin to report that he had had "a most successful visit." He then said that Dian was "a remarkable woman," adding that she had moved close to him after dinner and ran her fingers through his hair. He went on to tell us that we didn't understand her, that "she just needs love." With no further comments on the conservation mission that brought him to Karisoke, Curry-Lindahl departed to walk down the mountain. We were left with his apparent admonition to remind us that no good deed goes unpunished. Certain that romance was not an option we wished to pursue in our relationship with Dian, we were nevertheless encouraged. The project might now move forward.

AFTER THEIR DEPARTURE, Harcourt and Curry-Lindahl were quick to publish an article titled "Conservation of the Mountain Gorilla and Its Habitat in Rwanda." The paper contained a recapitulation of our ideas with no reference at all to our work, not even an acknowledgment of our discussions in Rwanda. It was discouraging to see the results of three combined years of hard work, analysis, and recommendations appear under the names of people who had asked for our suggestions and whose visit we had helped to facilitate. We published our basic recommendations for gorilla conservation in an East African journal, but their article was printed in the much more widely distributed *Environmental Conservation*. At that point in our lives, this slight hurt—especially from Sandy Harcourt, whom we considered a friend and supporter. Looking back, the issue of attribution seems exceptionally unimportant in comparison with the higher goal of seeing our ideas applied to saving the gorillas. In that respect, the Curry-Lindahl mission proved to be a useful vehicle for conservation.

In the weeks after his visit, we would learn that Dian's deal with Curry-Lindahl was sealed with more than a kiss. On the mission's return to England it was announced that the Fauna and Flora Preservation Society would be joined by a new partner, the African Wildlife Leadership Foundation. Dian had insisted on this addition. AWLF was a relatively small American organization with a regional focus on East Africa and a concentration on training and education. The major attraction of AWLF, however, was Dian's friendship with its president, Robinson McIlvaine, a former U.S. ambassador to Kenya. Through the power of her position at Karisoke, and supported by Curry-Lindahl, Dian convinced the others that AWLF should take the lead in what was now being called the Mountain Gorilla Project. Jean-Pierre von der Becke, a former colonial park warden in Congo, was named project director.

We were satisfied with the new arrangement, although it did seem to marginalize the role of the Fauna and Flora Preservation Society in the project that they had initiated through their fund-raising campaign. We knew and respected the work of the African Wildlife Leadership Foundation★ and had benefited from excellent advice from its Nairobi-based director, Bob Poole, before we began work in Rwanda. Poole had specifically encouraged us to make the effort to work with African institutions and individuals, as he was convinced that Africans were ultimately responsible for their own parks and wildlife. Within months of our meeting he was killed in a tragic car accident, and East Africa lost an important voice for conservation. We had also met Jean-Pierre von der Becke during an earlier visit to Karisoke, and he seemed heartfelt and earnest in his commitment to conservation. With his silver hair, attractive features, and multiple language skills, he would be an impressive team leader. Now we needed to know where we fit on the team.

Our interest lay in the education and tourism components of the Mountain Gorilla Project. We had the necessary experience with both gorillas and schools, excellent contacts, language skills, and familiarity with how things worked—and didn't work—in Rwanda. We had also laid the conceptual groundwork for the whole project, invested considerable time and energy in its promotion, and were ready for action. So, when Robinson McIlvaine visited Rwanda to formalize arrangements for the MGP, we looked forward to our first meeting with him. After a brief presentation, though, McIlvaine cut us off. *AWLF will only support anti-poaching work. The other activities are unnecessary.* This was Dian's view, but it was surprising to hear it from the head of an organization that had built its reputation on local education programs and had seen firsthand the power of tourism to help save East African parks. But then came the real shocker. *It is my understanding that you two are very selfish people and that you just want to stay on to become famous.* We knew that the ambassador had spent the prior evening talking with Dian, but we still weren't prepared for this unfounded assault on our character and motives. Putting his rudeness aside, we returned to our ideas and justifications for the education and tourism aspects of a three-pronged conservation program. We noted that we could work on these issues independently with continued support from the Wildlife Conservation Society, but we thought that a cooperative approach would be more productive for all concerned. McIlvaine dropped his personal attacks and became somewhat more engaged in the discussion.

★ The African Wildlife Leadership Foundation is now known as the African Wildlife Foundation. The Fauna and Flora Preservation Society has changed its name to Fauna and Flora International.

Ultimately, he came to support the idea of a three-pronged Mountain Go-rilla Project in which we would manage the tourism and education compo-nents, while acknowledging Jean-Pierre von der Becke as project director.

One final hurdle remained before the MGP could move forward. We needed ORTPN to approve the details of the tourism program. Dismas Nsa-bimana had surprised everyone in the preceding months with his opposition to any tourism program centered on the gorillas. It was widely suspected that his changed position resulted from a bribe by Dian. But he defended his stance on the laudable grounds that the gorillas should not be disturbed. With half of the gorilla population dead and the announcement of the cattle proj-ect, however, such arguments were meaningless. Nsabimana's views were rendered moot when he was abruptly replaced by Benda Lema as director of ORTPN.

Benda Lema was a short, balding man with piercing eyes and a predilec-tion for fine suits. He had no background related to conservation or tourism, unless one counted his degree in economics from Patrice Lumumba Univer-sity in the Soviet Union. It was rumored that his surname, Lema, was short for Lenin Marx. If Benda Lema had any Marxist tendencies, however, they did not emerge in our discussions of gorilla-based tourism. His only ques-tions concerned how much money could be made from the gorillas. The questions were certainly relevant to ORTPN, which needed to generate rev-enue to cover its costs. They were also central to Benda Lema's ability to make the case against the Virunga cattle project, which was still pending within the Ministry of Agriculture.

We offered Benda Lema the best possible answers to help his cause, with-out promising results that we could not deliver. Here we occupied very shaky ground, though, as we had no experience at all with the market for this new kind of tourism. As tourists ourselves in East African parks, we felt that park entry fees as low as $2 grossly undervalued wildlife and wilderness. In Rwanda, we had seen fourteen hundred tourists—most of whom were resi-dent expatriates—visit the Parc des Volcans in 1978, each paying only $5 for a seven-day pass. We knew the market could bear a much higher price—especially if we could guarantee a view of gorillas. But how large was the market and what price would it pay? Needing an answer, we put our "oppo-sition research" to good effect: if five thousand cattle were projected to bring in $70,000, we would claim that three thousand tourists would pay $25 each for a total of $75,000. To up the ante, we stated our conviction that more peo-ple were sure to come and pay even higher prices if the program were well run. Benda Lema did not seem moved by the need for quality control, but he was dazzled by the numbers. *There are that many people interested in the gorillas?*

Yes, we assured him, there were thousands of crazy white people out there who would pay a lot of money to hike through the cold rain and steep terrain to sit with wild gorillas. The director laughed at the notion. *Beaucoup d'abazungu foux?* Yes, that much we could vouch for: the world was full of crazy white people.

Chapter Fifteen

Lee

As the Mountain Gorilla Project began to take shape in the spring of 1979, we were pressing hard to finish our work at Karisoke. Dian never followed through on her threat to expel us, but we had much to accomplish as we approached the end of our planned seventeen-month stay at the research station. We were also keen to begin work on the Mountain Gorilla Project as soon as possible.

Amy spent her final months, then days, with Group 5. The weeks of vegetation sampling provided an unintended form of emotional severance, the first time since our experience with Mweza that Amy passed more than three field days in a row away from her adopted family. She also introduced Beethoven and his family to Peter Veit, a new researcher who would begin his own studies of sexual cycles among Group 5 females as soon as Amy finished her work.

Amy knew that we would be living and working nearby on the Mountain Gorilla Project in coming years, but it was sadly uncertain whether we would be allowed to visit Karisoke. So each hour with the gorillas took on added meaning. How much longer would Beethoven reign? Icarus seemed an increasingly improbable successor, more likely to leave the group and go off on his own. Ziz was next in line: an impressive blackback with a seemingly strong network of relationships within the group. Social and political skills were surely part of gorilla leadership, too. Then there was Pablo. It was hard to imagine the Clown Prince—with his crossed eyes, crooked grin, and penchant for trouble—as an adult, let alone with the mantle and authority of a silverback. Yet it was sad to think of him outside of the family setting, banished to the social life of a solitary male. Amy hoped he would find a role that would let him remain with the family. For poor timid Shinda, Amy could only hope that he might someday find courage, like the lion in *The Wizard of Oz*. And what of the dominant matriline? As the alpha female, Effie held major leadership functions within the group. Would her oldest daughter, Puck, carry on those roles? Would she compete or cooperate with her sister, Tuck? Or would Tuck transfer to another family before she bred?

There were so many questions with so few answers. There was also too much to do if we were going to move on from research to the more important conservation work of the MGP. If we could help save the mountain gorillas, there would be time to answer some of our questions. If not, those questions didn't really matter.

B ILL WAS SPENDING almost half of his time in Kigali during this period, laying the groundwork for the tourism and education components of the MGP. In camp, he monitored outlying groups when he wasn't helping with the vegetation sampling. By that time we had also convinced Dian to hire our old Peace Corps friend and housemate from Bukavu, Craig Sholley, to help with the non–study groups and other essential work that full-time researchers could not manage. During our final months, Craig was trying to habituate Nunkie, or at least the rest of his family, to the daily presence of human observers. David Watts also went to observe Nunkie's Group for research purposes. Nunkie was still uneasy in the presence of humans and gave every indication that he would stay that way as long as he lived.

Others in Nunkie's Group were more approachable. The most relaxed were two adult female transfers from Group 4, Papoose and Petula, and their youngsters, N'Gee and Lee. N'Gee was a four-year-old male, one year older than his half-sister. Lee was still occasionally nursing and rarely strayed out of her mother's sight. In the presence of human observers, Lee's soft eyes offered an invitation to approach more closely, Petula's calm demeanor a signal that it was okay. Papoose and N'Gee were similarly approachable. The two mother-child pairs were welcome sources of calm in an often volatile family kept on edge by Nunkie's nervous nature.

One day in May, David returned to camp early and angry. As he had approached Nunkie's Group on fresh trail from that morning, he found Lee sitting beside the trail. Petula was nearby, but something didn't look right. As David moved closer, Lee stood up and the problem was clear. She was caught in a trap, with the wire noose cinched tightly around her ankle. David looked to see if he could remove the wire, but was stopped by a scream and bluff charge from Nunkie. Back at camp, David discussed the problem with Dian. She told one of the trackers to approach the group, then shoot a gun so that the noise would scare away the other gorillas. In theory, this would give David a few moments to remove the trap. When they actually fired the gun, however, the noise so frightened Lee that she ripped the snare from its spring mechanism and fled with the rest of her family.

For the next several days, Craig watched helplessly as Lee limped along, dragging the wire behind her. He could see where the force of ripping free had tightened the metal noose so that it cut deeply into her skin. Craig was driven by a profound empathy for animals and a deep revulsion toward suffering. Six years earlier, during the massive flooding from Hurricane Andrew in his native Susquehanna Valley, he had spent several days rescuing wildlife and domestic animals trapped by the rising waters. He had originally hoped to be a veterinarian. Yet Karisoke had no plan or method to deal with the recurrent problem of injured gorillas, despite the increasing availability of veterinary darts and drugs to immobilize animals at a growing number of field stations. Without the means or authority to act, all we could do was wait. And plan a rescue if Lee were ever left alone.

Bill joined Craig on what seemed like a slow-motion death watch, as Lee steadily declined. During the day she would sit still for hours, her head tipped forward on her chest. When she stood, she would wobble unsteadily for a few steps, then sit again. Petula would carry her at times, but mostly she just stayed by her side or close by. The rest of the group would move away to feed, but then return to rest and nest near their weakest member. Occasionally the young gorilla would make a muffled moan, a plaintive cry of pain. Anger mixed with frustration at her plight. Finally, Petula began to spend more time away from Lee. Bill and Craig concocted a plan for which Peter Veit was also recruited. At the first opportunity, Craig would grab Lee and run at full speed back to camp. Bill and Peter would remain as a rear guard in case Petula—or Nunkie—came after him. Peter asked how a silverback would respond to a cough grunt; we said we'd never tried it before.

Arriving at the group that morning, it appeared we were too late. Lee lay still in the night nest she had shared with her mother, yet Petula was nowhere in sight. We moved closer and stared for signs of life. Craig said he saw Lee's chest heave and we sprang into action. Heart pounding, Bill ran past Lee to take up his post, with Peter a few steps away. Craig yelled "Got her!" and fled—just as Nunkie broke through the surrounding vegetation. In a brief terrifying moment Nunkie lunged forward, stopped just short of the arm-waving, shouting white apes, then whirled 180 degrees in retreat. Bill and Peter knew this was no time to linger. Stepping backward, staring intently at the spot where Nunkie disappeared, we quickly joined Craig in high-speed flight.

Back at camp, nervous laughter took some of the edge off our adrenaline-fueled state. Yet Lee was in terrible shape and we had to act fast. Dian was not at camp. In any event, we had learned with Mweza that Karisoke stocked no drugs of the kind needed. Earlier in the week we had alerted the doctors at the French hospital in Ruhengeri that we might appear at any time with an emergency patient. So Craig and Bill took turns carrying Lee down the mountain, then used a Karisoke van to drive Lee the final fifteen miles to Ruhengeri.

Pierre Vimont was *chef de mission* of the French hospital in Ruhengeri. Though it served the general population, this was a military hospital and Vimont had a military commission as well as a medical degree. He was a short, solid man with a quick sense of humor and a passion for wildlife. We had come to know Pierre through common friends and had earlier discussed the need for emergency medical care for injured mountain gorillas. He told us to bring him any gorilla anytime, at work or at his home, and he would do all he could to help. Arriving with Lee during Vimont's post-lunch siesta at home, we put his offer to the test. Pierre responded quickly, assembling a team of assistants and operating first to remove the wire and then clean the massive infection that had spread around it. A full course of topical and internal antibiotics followed. Then we waited. Vimont opened a guest room at his home where Craig and Bill took turns staying and sleeping with Lee. But to no avail. On the second day, Lee died of her systemic infections.

We remained at Karisoke for a few more weeks. We said our teary good-byes to the living gorillas, but Lee's death marked the true emotional end of our stay. Another gorilla was dead. An African poacher had set the trap. The world's preeminent gorilla research station had absolutely no means or backup system to help an injured gorilla, leaving us to jury-rig a dangerous rescue of Lee, who could have been easily saved by willing French doctors even a few days earlier. It was time to move on and take more control over circumstances affecting our lives and those of the gorillas.

Chapter Sixteen

Moving the Mountain

JEAN-PIERRE VON DER BECKE cut a dashing figure in his military uniform. He was not a spit-and-polish fanatic—a few years of study at Berkeley in the late 1960s had taken the edge off most of his colonial tendencies. Jean-Pierre was most comfortable in a loose sweater and blue jeans. Yet there he was in a neatly pressed uniform of uncertain origin, white hair and beard flowing as he paced back and forth, exhorting his new charges.

The score of Rwandan park guards facing their new boss were a decidedly motley crew. No two uniforms were the same and few guards had even a matched set of shirt and pants. Barely half wore boots. None owned waterproof rain gear to work in a park where it rained ten months of the year. The weapons on their shoulders looked like relics from World War I: single-shot, heavy rifles that were scarcely better than flintlock muskets. Still, as Jean-Pierre barked commands, the guards struggled nobly against all odds to look like a reasonable facsimile of a professional corps.

In late August of 1979, Bill watched this theater on many mornings before going out to follow the gorillas he had selected for the MGP tourism program. If he returned by mid-afternoon, the guards and Jean-Pierre were gone. Lyre-horned Ankole cattle had taken their place, lazily grazing the lush grass in front of the dilapidated building that passed for the headquarters of the Parc National des Volcans. A solitary *Markhamia* tree, its lower limbs amputated for firewood, added to the desolate atmosphere. Bill could imagine Jean-Pierre back at his house at Gasiza, pouring a healthy scotch and wondering what he was doing here. The question would have been entirely reasonable.

If Jean-Pierre had any doubts, his Rwandan counterpart appeared to have none. Camille the park *conservateur,* showed no interest in drills and training, preferring to remain in his warden's office conducting whatever official business he could create in a park that averaged four visitors per day. After the drills, Camille would emerge in his crisply pressed uniform to strut before his troops. If visitors were present, he would announce for all to hear: *The gorillas have nothing to fear. If I find a poacher in the forest I will personally execute him!* It was a bold pledge, repeated frequently with full confidence. Fortunately for local poachers, Camille almost never entered the forest. By midday, however, he could usually be found killing beers at one of Ruhengeri's many watering holes—until he rolled the park's only Land Rover returning from one late-night binge.

～⌒～⌒～⌒⌒～⌒～

OUR CIRCUMSTANCES CHANGED dramatically with the debut of the MGP in the summer of 1979. It was one thing to think of ideas to help the mountain gorillas, another to sell those ideas to reluctant listeners. Now we needed to deliver on our proposals in a climate of high expectations, with equally high potential for failure.

Our work began earlier in the summer with a whirlwind trip through Europe and the U.S. We had lost money on our education vehicle purchase and sale. With almost no money left, we bought tickets from Aeroflot at considerably less than the legal market rate. Of course, the flight from Nairobi to Paris took almost twenty hours with stops in Cairo and Moscow along the way. Alain and Nicole Monfort, on vacation in Europe, then kindly drove us to London, where we met with Sandy Harcourt, Kelly Stewart, and John Burton of the Fauna and Flora Preservation Society. We were asked to describe conditions in Rwanda and outline plans for the Mountain Gorilla Project at a reception for invited supporters and journalists at the London Zoo. Afterward, we continued our private planning discussions in more detail. We even had time for a pleasant day in the countryside with Sandy and Kelly.

Unfortunately, our London stay was cut short by the announcement that Sir Freddie Laker's air service across the Atlantic was suspended indefinitely. In the 1970s, Laker had built a highly successful cut-rate service across the Atlantic on the strength of his own personality and the wings of a second-hand fleet of DC-10s. But several recent accidents on other airlines had precipitated a global ban on flying any of these sturdy workhorses. Too embarrassed to borrow money or ask for a loan, we were counting on marshaling our last few hundred dollars to fly home with Freddie. Fortunately, British Airways announced that it would make available a limited number of half-price fares to help stranded Laker passengers on a first-come, first-serve basis. We hastily bid farewell, grabbed our bags—including two large trunks full of research data and plant specimens—and made a mad dash at the height of the hectic evening rush hour through the Underground. Arriving at Victoria Station, we found ourselves in a rapidly swelling crowd in front of the British Airways office, where officials announced they would not sell any tickets until the next morning. So we claimed our place in line and took turns sleeping on the sidewalk next to our belongings. One day earlier, *The Times* of London had solicited our views and opinions on the fate of the mountain gorilla. We were wined and dined by our distinguished hosts. But as midnight struck, our ballroom gowns turned into rags on the cold pavement of Victoria Station. There, all that mattered was our proximity to the front of the queue. The next morning the cutoff for half-fare tickets fell seven places behind us and we smiled as we boarded our flight home.

TWO LARGE TRUNKS distinguished us from other travelers in the U.S. Customs line at JFK. The customs agent looked over the battered trunks while also studying our appearance: young, longish hair, beard, jeans, well-worn field boots. He asked to open the first trunk, then stood in admiration at the revealed stash: hundreds of small plastic bags filled with dried plant matter, mostly leaves and stems. *And what have we here?*

"Gorilla foods," Amy answered. "Dried gorilla foods. I study what gorillas eat."

Another agent who had joined us chimed in. *We have gorillas at the Bronx Zoo. You could study what they eat there and save the trip to Africa.*

"Actually, the Bronx Zoo is part of the group that paid for my research."

Amy produced Rwandan export permits stating that none of her more than one hundred plant species was endangered. This would have been impossible for anyone to confirm, given their dried, crumbled state, and many of the plants did share a marked resemblance to marijuana. But after a few good laughs over the bags of dried dung that accompanied the plant samples, the agent decided that we were unlikely drug runners and allowed us to leave.

A LETTER FROM THE Wildlife Conservation Society awaited us on our return to Madison, Wisconsin. We had informed WCS of our intention to return to Rwanda and submitted a very modest request for renewed support. Their response was a shocker. We were congratulated on our successes, thanked for our efforts, and told that WCS would not fund our proposed work with the MGP. A phone call to New York changed nothing, though an official did explain that the Society saw its role as supporting research, not applied management. We had hoped to convince WCS to take its rightful place in the Mountain Gorilla Project consortium; now they wouldn't even fund our own work. Our families were frankly relieved at this news. Three and a half years in Africa was enough. They hoped that we would finish our degrees and get "real" jobs, preferably in the U.S. We were crushed. We called Sandy Harcourt immediately and explained our predicament. He was very concerned that we would not be able to return to Rwanda, or at least not in a timely manner. After forty-eight nervous hours, Sandy called back with confirmation that the Fauna and Flora Preservation Society would fund our activities. They would even pay us the princely stipend of $500 each per month, up from $200 during the preceding eighteen months. We were relieved and elated.

Actually, Bill had more reason to be excited. Amy was unable to do any of her analyses under field conditions in Rwanda and we still needed her final results for management purposes and to meet our obligations to ORTPN. We agreed that she would register for the fall semester and finish her work in Madison, while Bill returned to Rwanda. We were twenty-eight years old and had been together for nine years, our entire adult lives. Our earlier work in Rwanda had required many separations of a week or two. Now we were looking at four months. After one last emotional night, we set a date for Christmas and headed our separate ways.

IN 1979, THE RHINO RONDAVEL was a common sight in East African parks. Manufactured in Kenya, it offered low-maintenance housing under rugged conditions. It was made entirely of relatively cheap sheet metal that was easily transported and quickly assembled. The Belgian aid program had purchased several of these huts for ORTPN, one of which was given to Bill upon his return. The round structure was thirteen feet in diameter, with a dirt floor, one small window, and a conical metal roof. This was much smaller than the cabins at Karisoke, but still contained enough room for a bed, small desk, and two or three trunks, which served as combination chairs, closets, dressers, bookshelves, and food boxes. The real problem with the hut was the low roof that made rainfall oppressively loud and hail deafening. When the sun appeared, the entire structure became a high-efficiency solar oven. These deficiencies encouraged long hours working outdoors.

There was much work to do. Based on his census findings, Bill had made a preliminary selection of gorilla families that might serve the purposes of tourism. The Karisoke research and peripheral groups were off-limits. This allowed research and monitoring to continue without the repeated disturbances that we had experienced from illegal tourists. Group 6 was also ruled out, despite its proximity to a road, because of Brutus's well-known temper. Other families were too far away or too small. Three groups in the central part of the Virunga range offered the greatest potential for habituation based on their accessibility, their calm reaction to observers during the census, and the presence of young gorillas. Our experience with the Karisoke groups had underscored the endless curiosity of these youngsters, and we believed that their presence might be a key to habituation.

From his hut near the park office at Kinigi, just south of Mt. Sabyinyo, Bill visited one of the selected groups each day. Though all of the gorilla work was on foot, he now had a motorcycle to travel around the edge of the park. Some days this increased mobility even allowed two site visits: following Group 13 or 15 in the morning on Mt. Sabyinyo, then shuttling over to see Group 11 on the eastern flank of Visoke in the afternoon. The Yamaha Enduro dirt bike also allowed Bill to take advantage of the many foot trails that crossed the region, though he learned the hard way that two-plank foot-bridges were not made for men on machines. Fortunately, only a broken mirror and bruised pride resulted from his plunge into a shallow ravine in front of a small group of wide-eyed boys.

Before we left Karisoke, Bill had told Craig Sholley about Group 11. This was one of the most tranquil groups observed during the census, plus it had four infants and two other subadults. During the summer of 1979, while we were in the U.S., Craig had taken the initiative to visit Group 11 several times. He, too, was impressed with their calm demeanor and even named the silverback Stilgar, after a powerful yet serene ruler from the novel *Dune*. When Bill returned, he began to visit Group 11 regularly and confirmed the family as a prime candidate for the tourism program. To the east on Mt. Sabyinyo, Group 15 appeared to be another good choice. These gorillas were more skittish than those in Group 11, but the family included three infants. The two broad canyons that seemed to constitute their core range also fell within easy walking distance of the ORTPN office in Kinigi where tourists would arrive. Unfortunately, Group 15 did not respond well to the presence of observers. Within two weeks of being followed, the gorillas fled their core area, scaling cliffs that would leave all but serious mountain climbers behind. Tourists could not be guaranteed a view, or their safety, under these conditions. And nothing was worth terrifying a family of gorillas.

Group 13 on Sabyinyo's western flank emerged as the next best contender. This family offered an attractive composition of two silverbacks, six other adults, and four young gorillas. With Group 11 and Group 13, we now had two potential tourist groups, which we thought was the minimum needed to meet demand and to allow for a great number of uncertainties. But the home ranges of both Groups 11 and 13 were near the border with Congo, and it was not known how the Congolese would respond to tourists crossing into their sector of the Virungas. Bill decided to proceed with habituation of these two families on an accelerated schedule, often visiting both groups in a single day.

In the agreement to create the Mountain Gorilla Project, the Rwandan Park Service had committed to the cooperation of guides already assigned to the Parc des Volcans. This was important for several reasons. Most significant was the project goal of creating a Rwandan capacity—national staff paid by national institutions—to manage MGP activities whenever possible. A more immediate goal was to eliminate the illegal tourism program in which guides accepted bribes to take visitors to see research Group 5 near Karisoke. In recent years, more than one group of tourists had video cameras ready to capture the gorillas emerging from the brush, only to film an irate Amy as she gave a tongue-lashing to their guides. After more than twenty such encounters, Gilbert Irandemba and Léonidas Zimulinda were inured to the lecture. Both were legal park guides, knowingly making illegal visits to Karisoke research groups. In their minds, Amy's lectures were a lower risk punishment than being charged by unhabituated gorillas. Now we could provide a legitimate alternative and leave the research gorillas—and researchers—in peace.

Gilbert and Zimulinda didn't see it the same way. They had a moderately lucrative business going and a sure thing in Group 5. They, like many others in the region, further believed that these gorillas accepted a human presence only because Dian Fossey had used her magic powers to control them. They found it highly unlikely that Bill had any such powers. Besides, it was much harder work to follow Groups 11, 13, and especially 15. When asked to work with Bill on habituation of these new groups, they refused. Bill met with the park warden and with ORTPN officials in Kigali to resolve the problem, but no one wanted to force the old guides to take on new responsibilities. Bill had the strong impression that all were just waiting for him to fail.

The short-term solution to the guide strike was to hire Big Nemeye, with whom Bill had worked for much of the census and who was currently unemployed. He spoke neither French nor English, but Nemeye knew the forest and the gorillas. Another local resident with forest skills, Jonas Gwitonda, was also hired as a tracker and apprentice guide.

The strategy for habituation of both gorilla families was the same. The hybrid system drew on George Schaller's practice of approaching the gorillas in an upright manner to allow them to see us clearly, then kneeling once the group was settled. From our Karisoke experience, we added belch vocalizations and mock feeding behavior that helped to calm the gorillas as close contact was being established. To first find the group, Bill and Nemeye would return to where they had left them the day before, then follow their trail until a visual contact was made. This contact was maintained as long as possible, or until the gorillas showed displeasure. In the first few days, even Group 11 had little tolerance for prolonged contact. Any attempt to follow the gorillas provoked silent flight. Then Stilgar began to assert himself, screaming and charging in full strut. Bill disliked the idea of disturbing the gorillas to this point, but knew that this stage should pass quickly. Schaller and Fossey had both written that the best response to a charge was to hold one's ground, a lesson each had learned from experience. Now Bill could learn the same way, since he always stood in front with Big Nemeye right behind him. On at least one occasion, he felt Nemeye's hand on his back, providing physical as well as moral support. After several days of aggressive charges, Stilgar suddenly stopped. This was a relief; the program could go on. Yet Bill was surprised to find that the abrupt ending to such high-voltage contacts also caused an emotional letdown.

Stilgar's next tactic was to lead his band to a ravine or other favorable site where he could position himself to stare at the intruders. There, the young gorillas would come out to peer at us, too, though they quickly retreated at the silverback's first cough grunt. Within weeks, all four youngsters were allowed to stay in full view. The process of habituation was working.

The habituation of Group 11 moved more quickly than we ever imagined, but Group 13 proved more resistant. In part, this seemed to be a function of its habitat. The gorillas' core range between the western flank of Sabyinyo and the low-lying crater of Muside was covered with dense bamboo. This required that Bill and Nemeye approach more closely to make visual contact. After an initial period of flight, the silverback began to stand his ground and scream, though he rarely charged. After about a month, the gorillas of Group 13 would tolerate a human presence only ten yards away. Good contacts, though, lasted only four or five minutes well into the second month. Still, Bill was ecstatic. He knew that habituation was not magic, but neither was it guaranteed.

The early success with Group 11 expanded in October to include "practice tourists." Friends from Kigali and Ruhengeri were eager to see the gorillas, and we began taking additional people into the forest whenever they visited Bill in the volcanoes. At first Bill took only one or two others, then built up to a family of four. Stilgar was not entirely happy with this change, but the young gorillas appeared delighted. Now there were more white apes, wearing more colors of rain gear, and carrying new objects. Each of the juveniles approached the visitors differently: some sneaked up behind us, others ambled slowly past. One preferred aerial reconnaissance from overhanging branches and bamboo. We were the morning entertainment for the young gorillas and with their attention, the habituation hook was firmly set. Now we began to see a new challenge: how to keep the gorillas and tourists apart. At Karisoke, close contact had never been discouraged and some researchers actively encouraged contact. But we couldn't trust the behavior of tourists, or their health status; nor could we guarantee that Stilgar would tolerate the same proximity as Beethoven. Bill decided that a no-contact policy of physical separation was essential. Cough grunts and positioning of the guides proved effective in maintaining an optimal viewing distance, though the young gorillas constantly tested this boundary.

All the "practice tourists" paid full park entry fees. Then one morning, Camille, the park warden, was waiting at the Visoke parking area when Bill returned from viewing Group 11 with friends. Camille had brought Gilbert and Zimulinda with him. The warden puffed himself up and demanded to know what these "illegal tourists" were doing in his park. Camille knew very well that the visitors had paid and that Bill always dropped off their money at the Kinigi office when he returned home at the end of the day. Still, he accused Bill of running a "private tourism program" from which he kept the profits. Zimulinda and Gilbert smiled at this reversal of accusations made so often—and accurately—at their expense, even as Bill produced the official receipts and money from an envelope in his knapsack. The theater continued after the embarrassed visitors drove away, but along more productive lines. Camille proposed that Bill work with the official park guides from now on and not with his private tracker, Nemeye. Bill immediately accepted this absurd restatement of his original request for cooperation, congratulating Camille on his excellent idea. But, of course, since Nemeye and Jonas were now familiar to the gorillas, he insisted that they also be hired by ORTPN as official guides. Camille didn't have the authority to agree to this, but he had gained what he wanted and the bars were waiting, so the deal was sealed with a handshake.

The reason for the end to the guides' strike was simple. Nemeye and Jonas were earning good tips from the visitors to Group 11, and old-fashioned jealousy was setting in. Gilbert and Zimulinda were jealous of others making money at the same time that their illicit income from Group 5 visits was diminished. They wanted in on the new program. The morning after the meeting with Camille, there were four gorilla guides waiting when Bill arrived at park headquarters.

Soon after resolving the guide issue in mid-October, Bill returned from a visit to Group 11. Walking through the fields along the park's edge, he encountered a solitary *umuzungu* with a shock of long red hair and a matching thatch that burst from his open shirt. This had to be an American. Asked what he was doing there, Mark Condiotti informed Bill that he was waiting to go out to the gorillas. *Which group? Group 5. No you're not,* Bill quickly replied. The short discussion that followed established that Zimulinda was still trying to make a few more illegal francs on the side, thereby earning the first of several warnings. Yet the event provided another breakthrough in the gorilla tourism program. Bill explained why Mark couldn't go to Group 5, then discussed his own work with habituation. Mark was passionate about gorillas and immediately offered to work as a volunteer for the MGP, even paying his own expenses. Over the coming weeks, he proved his mettle, and Bill convinced Jean-Pierre and the Rwandans to allow him to work as an unpaid project assistant. His official authorization was delayed for months, however, by Rwandan concerns that he was a CIA agent. Bill tried to assure the Rwandans that a real CIA agent would have all the necessary documents that Mark didn't have, and would definitely have better boots and rain gear. But Rwandan suspicions of foreigners still ran deep, especially of those who arrived suddenly from nowhere and then offered to work for free.

Mark took over the Group 13 habituation process, allowing Bill to concentrate on Group 11. Equally important, Bill moved his base of operations—and his Rhino Rondavel—to the parking area at the foot of Visoke. From there he could not only reach the gorillas more easily, but also control the flow of tourists up the east side of the mountain and block illegal visitors to the south side, where Karisoke was located. This move, however, led to a bizarre interaction with the American embassy. Bill received word that an embassy official wanted to meet with him on his next trip to Kigali. He honored the unusual request and joined the young American attaché, whom he had never met before, in his office. Later, they would become friends and the officer would go on to serve as a distinguished ambassador in several African countries. On that day, though, the attaché had been told to play the designated hit man. He accused Bill of "threatening Dr. Fossey by moving to the base of Mt. Visoke." He had a hard time saying what the threat might be, but he had been told that it had to do with a "takeover" of Karisoke. The performance would have been comical if it weren't so outrageous that U.S. diplomats were intervening in a matter of no geopolitical importance and about which they knew absolutely nothing. Bill informed the earnest young embassy officer that his strategic location at the base of Visoke actually helped keep unwanted visitors away from Karisoke, its gorillas, and its overly territorial director. More pointedly, he noted that the MGP was not a Dian Fossey project or a U.S. government project, but a Rwandan project funded entirely by private conservation groups. The project would make money and create jobs for Rwandans. He added that if the U.S. government really wanted to help, it could find more constructive ways to support conservation than by blindly subscribing to Fossey's paranoid view of the world. Riding away from the embassy on his motorcycle, Bill thought the lunatic factor was getting a bit out of hand, with the Rwandans imagining CIA agents and Dian calling on the U.S. embassy to block an imagined invasion of her camp.

T HE FIRST TIME THAT Bill rode into Chapati Village on his motorcycle, a crowd of young boys gathered around before he could even dismount. Their close inspection of the bike ended when he removed his helmet and visor. Several of the boys pointed in recognition of the earlier visitor to their town. One boy stepped forward and stared at Bill with obvious satisfaction. The Chapati Kid now had a benefactor who wasn't just big, but he had wheels! The boy still didn't talk and the rest of his body was almost as thin as his one shriveled leg. But he kept smiling as he finished his rolled-up omelet chapati. As Bill rode out of town, the Kid seemed to have new standing among the other boys.

The greater mobility afforded by the motorcycle also allowed a greater range of friends and contacts. On his passages through Ruhengeri, Bill often stopped to share lunch with Pierre Vimont and his wife, Claude. The food was always exceptional *chez* Vimont, but the doctor was also a truly good person. He had a quick mind and keen sense of humor, and he had shown his concern for the gorillas and the cause of conservation during his treatment of Lee. In his capacity as top-ranking medical officer at the French military hospital, Pierre renewed his offer of medical treatment for any sick or injured gorillas. Bill discussed the matter with Jean-Pierre von der Becke and all agreed that the MGP would break with Karisoke tradition and take advantage of this offer for any injured gorillas.

Thankfully, we never needed Pierre's services to help save any of the tourism gorillas. He did intervene, however, in a case he couldn't help. Late on a cold, blustery November day, Mark Condiotti sent word that the silverback of Group 13 had been shot and killed. Jean-Pierre arranged to have the body transported to Vimont's for an autopsy. As he watched Pierre cut into the massive body, Bill's thoughts returned to Uncle Bert. He wondered why the killers didn't take the silverback's head or the all-too-human hands. In fact, we never saw the hands cut from any Virunga gorillas, despite persistent rumors of the practice. As we milled around talking and consoling Mark, Pierre searched in vain for the bullet. Once again the projectile had entered from one side, shattering the shoulder before passing on through soft tissue, and exited the other side. Whoever was killing gorillas had a very powerful rifle. Alain Monfort believed it might be a FAL: a military weapon produced in his native Belgian town of Liège. Whatever its make or source, what was most exceptional was that someone possessed a gun at all. Most Rwandan poachers set traps, some used poison. But almost no one in the country out-side of the military had either pistols or rifles in those days. Doctors we knew remarked that they had treated innumerable machete wounds from bar fights, but never gunshot wounds. It wasn't that Rwanda was a more civilized place than Europe or America: there simply weren't any modern weapons in circulation at that time. With one deadly exception.

There was no motive for the Group 13 killing. The silverback's head wasn't taken, no infants were missing. Maybe Mark scared off the poacher, though he never heard a shot. All we could say was that one more gorilla was dead. The thought that someone might now be targeting the tourist groups was unlikely, yet disturbing. The only encouraging observation was that a second silverback was already in place to maintain control of the family and perhaps avoid the infanticide seen in the aftermath of the Group 4 killings. Mark returned to the group and we all placed our hopes in Mrithi, "the suc-cessor."

L ATER THAT SAME NOVEMBER, on another cold afternoon, there was a knock on the metal door of Bill's hut. Robinson McIlvaine stood outside in a light rain and asked if he could come in. The president of the African Wildlife Leadership Foundation looked worn. Bill offered him his chair and some tea, then listened as the elderly gentleman described a disastrous visit up the mountain to Karisoke. He had survived the climb and the elements remarkably well, but Dian had angrily rejected every idea for collaboration with the Rwandans. This was not unexpected. But then she turned on Jean-Pierre von der Becke and said that he was "too soft on the Africans, too weak to run the project." Finally, she accused McIlvaine himself, her old friend, of misusing her money. Not only was AWLF mismanaging the Digit Fund, she said in her dark state of mind, but McIlvaine was allowing his organization to expropriate funds for its own use. The charges were typically unfounded, but this was not the time for "I told you so." Bill just let him talk. After a short while, McIlvaine thanked him for the tea and rose to join his chauffeur in the waiting car. "I'm sorry for what I said about you before. We do appreciate what you are doing. Dian needs help, but not the kind we can give her." The apology was unnecessary, but appreciated just the same. If we were going to move the great mountain of inertia and resistance before us, it was essential that we all push in the same direction.

Chapter Seventeen

White Apes and Ecotourists

THE SUCCESS OF THE Mountain Gorilla Project ultimately depended on our ability to generate tourism-based jobs and revenue, which would then turn local attitudes and national politics in favor of the gorillas. The shooting of Group 13's silverback underscored the absolute need for improved park security. Jean-Pierre von der Becke was making progress in this critical area. Our education program was beginning to show promise in changing attitudes and creating a future constituency for conservation. But only tourism—especially tourism that brought in much needed foreign revenue to the Rwandan economy—could generate the political clout needed to stave off the cattle project and other looming threats to the integrity of the Parc des Volcans. We had promised that we could habituate gorillas and we were succeeding. We had also promised that we could deliver three thousand paying tourists, not just a few friends and resident expatriates.

The first nonresident tourists to visit the gorillas were the TransAfrica overlanders. We had first met these truckloads of young Europeans years earlier while in the Peace Corps in Bukavu. Twenty or more to a truck, they left from southern Europe, crossed the Sahara, and then traversed the Congo Basin. After months of grueling travel and cramped conditions, they arrived in the promised land of East Africa, where they would visit parks, lie on the beach, and enjoy fresh milk and honey before returning home. Rwanda was a pit stop on their way east—a brief but tranquil passage between its war-torn neighbors Uganda and Burundi—until gorilla tourism provided a reason to stay longer in the country. In fact, the TransAfrica groups were perfect for the early days of the tourism program. They were completely self-contained with food, water, and tents all neatly stored aboard their trucks. They were young and generally healthy. And, after the often hellish crossing of the lowland Congolese rain forest, they greatly appreciated the opportunity to leave their trucks to see and do something on foot.

Our biggest problem with the TransAfrica corps was arranging for so many visitors to see the gorillas. Group 13 was off-limits to large groups in late 1979, and we were trying to limit the number of people visiting Group 11 to six tourists in one group per day. This required that we set up a three- or four-day rotation for each truckload: most of the people climbed Visoke one day, some climbed all the way to Mt. Karisimbi's summit on the second day, some simply wandered the forest, and one group of six visited the gorillas each day. A typical gorilla trek involved a twenty-minute walk through the fields, another hour hiking in the park, perhaps a half-hour with the gorillas, and a leisurely return in less than an hour. One TransAfrica group leader had been crippled by polio. He desperately wished to see the gorillas, but didn't want to hold up the others. After hearing their stories of great contacts and relatively short hikes, he decided to join us on the third and last day of their visit. Of course, that day Group 11 bolted for some reason up and over the Ngezi crater, down the other side into Congo, only to circle back onto Visoke's eastern slope. We walked farther than ever before over very rough terrain, finally arriving at the gorillas after six hours of hard hiking. It took another two to return, with the TransAfrica leader painfully hobbled by the ordeal, even though Bill and Nemeye carried him over some of the more difficult passages. That night, however, there was no need for a fire. His smile lit up the entire Visoke campsite and might have been visible from high on the mountain. His physical ordeal had ended in an emotional high. The power of the gorillas to move and even transform those fortunate enough to spend time in their presence was clear.

Although the TransAfricans were great visitors because they didn't ask for much, they also didn't contribute significantly to the local or national economy. They paid their entry fees—now up to $20 for one gorilla visit and three more days hiking in the park—but spent little else. ORTPN and its tourism advisors wanted bigger spenders who stayed in hotels, rented cars, and feasted in restaurants. In their dreams, they saw organized tours of wealthy visitors streaming in from around the world, dependent on an array of Rwandan services, even though the latter were not yet ready for most tourists' demands.

Tourism advisor Detlef Siebrecht was bursting with pride at the news he conveyed to Bill. He had arranged for eight German tour agents to come see Rwanda's new gorilla attraction. Never mind that it was late October and the habituation process was only ten weeks along. Or that eight visitors was more than we wanted to take out at one time. These were influential people. Important people, as the porters could tell when they finally showed up with several loads to carry into the forest, including one cooler filled with beer. Matters quickly took a turn for the worse when a light rain began to fall, revealing that several of the Germans had failed to bring rain gear to the rain forest. Then, when the gorillas' trail became muddled in a thick stand of bamboo, Bill asked the tour agents to stay together while he and the guide sorted out the situation. The first beers were opened while they waited and soon the forest rang with the loud call of one especially obnoxious agent: *Gorilla! Gorilla! Komm hir, gorilla!*

Bill's spirits crashed. He had promised tourists, but were these agents a preview of the nightmare to come? The group eventually did see the gorillas, but Bill was left with new doubts about tourism. And Amy was ten thousand miles away, unable to share his misgivings or restore his hopes.

THE GORILLAS OF GROUP 11 were a calming influence. They revealed more of themselves and their characters as time passed. In deference to Craig Sholley, we continued to call the silverback Stilgar. The other gorillas were all given African names in either Kinyarwanda or Swahili which, in turn, reflected each individual's personality, behavior, or personal attributes. Some were quite pedestrian, like *Ndume*—meaning "male"—for the emerging second silverback. One adult female was called *Mkono,* or "the hand," in reference to the mangled fingers that rendered her left hand nearly useless. She was almost certainly a trap victim, as was a younger female named *Kosa* for her completely "missing" hand. *Kalele* earned her name with constant "noisy" vocalizations. And when it became clear what agitated this young mother, her baby was named *Sababu,* or "the reason." The oldest youngster, who loved to play, was called *Furaha,* or "joy," while his best "friend" and playmate was named *Nshuti.* The last of the young gorillas to be identified was named with the number *Tano* to mark her place as "fifth" in line. The guides named most of the gorillas and took pride in their responsibility to confer appropriate African names.

Interactions with the members of Group 11 were becoming more interesting and complex. On one occasion, two young French women accompanied Bill and Nemeye. The gorillas on that day were secluded at the bottom of a shallow crater, so Bill decided to wait above until they emerged in better view. A fallen *Hagenia* trunk provided an excellent vantage point from which to observe individuals as they appeared in the crater's grassy clearing. Stilgar was nowhere to be seen, but the other gorillas appeared calm. Then, as Bill and the two women were leaning on the log watching the scene below, two immense black hands rose up and slapped the trunk right in front of them. Stilgar pulled himself up to peer over the log, his huge head a few feet from their faces. Bill gave out an involuntary cough grunt—a somewhat aggressive vocalization—which Stilgar fortunately ignored. Instead, he calmly surveyed the human apes before him for a very long minute. Apparently satisfied with his reconnaissance mission, the silverback lowered his powerful frame and retreated as silently as he had appeared. Bill looked around to find both women on the ground. One was in tears. The other was grinning in full appreciation of the moment.

As more and more tourists came to see the gorillas, their dress—or lack thereof—was a steady source of entertainment. One visitor in five seemed to arrive without any rain gear at all, though some apparently thought that they could walk through the dense vegetation under an umbrella. Few tourists brought sweaters or vests to wear under their rain gear to fight off the cold. There were no guidebooks to Rwanda's high mountain regions in those days and, despite our warnings to tour agents, visitors failed to appreciate how cold and wet the Equator could be at ten thousand feet during the rainy season. Dozens of women appeared in short dresses; one complemented her skirt with matching shoes with two-inch heels. A hardy minority of both sexes hiked through the mud and nettles in sandals. Among those who did wear rain gear, there was a marked preference for bright yellow and orange—colors that seemed to attract the younger gorillas. So, too, did the diversity of visitors and the variety of their appearance. Bill was convinced that the arrival of the white apes provided a daily dose of gorilla entertainment.

Siafu! The word was always shouted and visitors quickly understood its meaning. We had learned the hard way that driver ants were much more common in the bamboo zone than at the higher elevation of Karisoke. Several times a week we would stumble on their bivouacs, disturbing their military columns and triggering an instant attack. As smaller worker ants remained in formation, half-inch-long soldiers with oversized heads would swarm over any foreign body, their sharp mandibles primed and ready to pierce our skin. More disturbing, they would often wait to bite until they had climbed up our legs or dropped into our shirts after we brushed an infested branch. The result was the *siafu* strip: a rapid shedding of shirts and pants to brush off the ants, if possible, or unhook them, if they were biting, before they could inflict any more damage on the body's more sensitive parts. There were times when it was an advantage to wear less clothing.

Tourists did not have to remove their clothing in order to bond with each other. The shared experience of spending time with a free-ranging family of mountain gorillas was sufficient. Sitting around the fire, drinking hot tea at the end of the day, visitors recounted stories of the gorillas. Many said they felt like a gorilla by the time they arrived at the group, after crawling over logs and through the mud. Then other recollections of their trek through the Virunga forest would come to mind. The smell of the vegetation and rich, rain-soaked earth. The silverback's musty scent. The taste of wild celery, the sounds of unseen creatures. The skill of tracking and the thrill of discovering night nests. Finding warm dung and knowing it meant the gorillas were nearby. The childlike joy of getting wet and dirty. The physical exertion of climbing and slipping. Even the sting of nettles and ants. Visitors to the Virungas came alive in ways that Bill had never seen in East Africa, where tourists were carried though the wilderness and past the wildlife in zebra-striped minibuses and protective Land Rovers. Gorilla tourism was completely different: an intense physical, sensory, and emotional experience. It would be a few more years before the term *ecotourism* was coined, but in late 1979 we were already seeing the attraction of entering the forest on foot and meeting animals on their own terms in the Virungas.

VISITORS ALSO PROVIDED a distraction for Bill, though he was generally happier when they didn't spend the night at the base of Visoke. He liked being alone with the mountains, their sounds and silences. The clean air and cold temperatures at 8,800 feet combined to heighten all senses and clear the mind. Sometimes the senses were preoccupied with nagging knee pains, or the mind with thoughts of never-ending challenges. But time to think without external distractions was a luxury that Bill greatly appreciated.

For the purposes of the MGP, the steady flow of visitors was encouraging. Local employment was on the rise, both around the park and in nearby Ruhengeri. The old Hotel Muhavura was renovated, as were several local restaurants, to serve visitors on their way in and out of town. The growing number of foreign tourists also increased revenues and political support for the park service in its battle against the MINAGRI cattle project, which was now on hold, but not dead. Gorillas became a hot topic of conservation among resident expatriate visitors, many of whom returned from a day with Group 11 to become effective conservation advocates in their various roles as government advisors and technical assistants. They were an important group of insiders.

Yet increased visitation had its downside, too. Bill had lobbied for limits on the number of daily tourists, suggesting that six was the optimal group size for observing gorillas at a single time. The number wasn't chiseled in stone, but it did reflect Bill's practical experience with the adult gorillas of Groups 11 and 13, who were much calmer if all the visitors remained within sight. Six also represented the maximum number of visitors that he felt the guides could control and keep separated from the gorillas. This group size further assured a quality experience for each visitor. People who had traveled thousands of miles and spent thousands of dollars to see gorillas didn't want to view their subjects from behind the heads of fellow tourists. ORTPN, however, was extremely reluctant to establish any limits. In part, this reflected a very real need for increased political clout based on revenue, in part it reflected ignorance and greed. *You said that you could make money from the gorillas, Monsieur Weber. So let's make money.*

One December morning in 1979, Bill heard the distant drone of two Volkswagen vans grinding their way up the final stretch to the base of Visoke. This auditory early warning system was the only way of knowing who might show up each morning. With three overnight campers already scheduled to go out, the two vehicles signaled a probable overload, but Bill was not prepared for the thirteen tourists who emerged from the vans. They were all members of an Air France crew that had rented the vehicles, purchased their tickets, and insisted on seeing gorillas. Bill tried to persuade half the group to return the next day, but they were adamant about staying together. With a recent order from Benda Lema not to turn anyone away, Bill had no alternative. He assembled the sixteen visitors and told them to stay together, keep quiet, and follow instructions.

The group set off with Big Nemeye in front, Bill in the middle of the caravan, and two porters in the rear. Within a half-hour Nemeye found fresh trail from that morning, not far from where they had left Group 11 the day before. Bill noted that it was not unusual for the gorillas to follow the trail of departing tourists, sometimes shortening the path to recontacting the group the next morning. He hoped it was a good omen. That day's shortcut, though, led straight up the mountain. Over the next hour and a half, the band of twenty humans became increasingly dispersed as altitude and exertion forced stragglers to the rear. Bill stopped the group several times, reminding them to stay together and remain quiet. Still, the gorillas moved steadily higher up Visoke's eastern flank. As the gorillas neared the treeline, a female and two youngsters finally appeared in the notch of a *Hagenia* tree about twenty yards away. Bill was in front and gathered five or six of the tourists around him as he sought to identify the gorillas.

*M**ERDE! PUTAIN* . . . mais merde!

The cries came from a Frenchman who had fallen in a small ravine and was stung by nettles. He was not pleased. Neither was the silverback who screamed a single *wraaagh* in response to the Frenchman's curses. The other gorillas fled from sight, leaving Bill to contemplate an unappealing set of options. Had he been alone or with Nemeye, he would have simply left the gorillas in peace and returned home. But tourists complicated the equation. They had rented their vehicles, paid their fees, hiked a long way, and all wanted to go on. Most had behaved well under difficult circumstances. Bill gathered the group together and told them to sit quietly and wait with Nemeye and the porters while he went ahead. If the gorillas crossed a ravine where they could be seen without further disturbance, he would call the others forward. If the gorillas showed signs of fear and flight, they would call it quits. Everyone nodded as Bill set off alone.

Bill climbed only a short distance before the gorillas' trail led into one of Visoke's many ridge-top tunnels of vegetation: three- or four-foot-high openings maintained by the steady passage of animals. He crawled in on all fours, then came to a quick halt. Inside the tunnel, only four or five yards away, a silverback stared down at him with pursed lips, swaying back and forth. Just as Bill was thinking that this wasn't Stilgar—and that he didn't look happy—the unknown silverback screamed and charged. Bill knew this was not a bluff and braced for a hit. He never felt the blow, just a warm burning sensation as the gorilla's giant canines sank into his neck. There must have been more to the tussle, since he was dragged or rolled about thirty feet down into a small ravine. But Bill was knocked unconscious after the bite. When he came to, Nemeye was standing over him. *Bwana, uko sawa?* No, bwana wasn't okay and he wasn't happy. All he knew was that he was lying on his back, his shirt was ripped, and his eyeglasses were missing. And where was the big black truck that ran over him?

Nemeye quickly found the glasses and Bill stomped back to the tourists, who were huddled only thirty yards away. They had heard the gorilla's scream and commotion. Now they stared in stunned silence as Bill curtly told them the visit was over and that they were all going home. He quickly led the way down the mountain, silently berating himself. *You're here as the great gorilla expert and you end up trashed by a gorilla.* He felt terrible that he had forced a gorilla to react so violently. After a while, one of the Air France pilots spoke up. *Tu saigne, mon vieux.* Ah, yes. Blood. The neck bite. Adrenaline and shock had erased the memory of the attack, but it all came back as Bill inserted his little finger into a surprisingly deep gash—more than an inch wide and nearly as deep. A shallower puncture wound lay just to the other side of the spinal column. The location of the bites made Bill shudder. One of the stewardesses offered him an Air France lemon-scented towelette that stung as he cleaned the wound. The experience began to take on elements of the surreal. As he continued walking, the pain in Bill's ribs also became much sharper. He began to realize that he was badly hurt.

The return to the Visoke parking area was a blur. A bench seat in one of the tourist vans was cleared to take Bill to the Ruhengeri hospital. The French are never far from their beloved red wine and a bottle was quickly offered as a palliative. Neither alcohol nor the intense nicotine of a Gauloise cigarette, however, could dull the pain of the jarring descent down the fifteen-mile rocky track to Ruhengeri. At the hospital, a doctor first examined the two neck wounds. He then depressed two of Bill's ribs like piano keys. An X-ray confirmed that the ribs were broken. Bill was given a bed, where he tried to collect his thoughts until an angry nurse yelled at him for lying in bed with his dirty clothes and boots on. The offending items were stripped off and he was ordered to take a shower—down the hall and outside. When the first cold water hit his neck, Bill passed out. He was found unconscious in the shower, then brought back to his bed where he apparently passed the cleanliness test.

When Bill awoke, Pierre Vimont was standing next to him in full evening wear with Claude in a shimmering gown at his side. Pierre had heard of the attack while at a party in Kigali and set a personal record for the sixty-mile drive up the escarpment road to Ruhengeri in the dark. Claude's pale appearance was probably due more to Pierre's driving than the sight of Bill and his bandages, but it was reassuring to see the Vimonts' faces and concern. Jean-Pierre von der Becke, too, appeared at the hospital and was pleased to learn that the injuries were not life-threatening. He was also relieved to hear that Stilgar was not the assailant. Bill was increasingly convinced that it must have been Brutus—his old nemesis from the census—who had bushwhacked him in the tunnel. The ambush style of attack fit with Brutus, as did the bright silver hair on his saddle.

Tourism was suspended for the next several days as Jean-Pierre and the guides went back over the trail and surrounding area to determine what happened. They found that Brutus's Group 6 had indeed moved north to interact with Group 11 late on the day before the attack. The extent of flattened vegetation indicated an unusually intense amount of displaying by the silverbacks. The interaction may have continued in vocal form throughout the night, as the two groups had made their nests within one hundred yards of each other. The next morning, Group 11 moved off to the north, while Group 6 briefly followed our tourist trail before turning west to climb to the safety of higher elevation. This was the trail that Nemeye and Bill followed, unknowingly subjecting an agitated group of unhabituated gorillas to even more harassment. When Brutus saw Bill crawling after them in the tunnel, he did what any self-respecting silverback would do. He attacked, drew first blood, and moved on. Brutus 1, Bill 0. End of match.

<center>~~~~~</center>

O N THE MORNING OF DECEMBER 23, Amy was waiting for her connecting flight in the transit lounge at Belgium's Zaventum Airport. Many of the people around her were going to Rwanda for the Christmas holidays. She could not help overhearing one group discussing a planned visit to the gorillas. *But we're not sure that we'll be able to see them. Why not? One of the guides was attacked by the gorillas.*

Amy had heard enough to jump into the discussion. *Was it a Rwandan guide?*

No, an American. A big man with a beard. I've seen him, but I don't know his name.

How badly was he hurt?

They say he was bitten in the neck, but I don't know how badly.

Amy knew that Mark Condiotti was the only American besides Bill working with the gorillas, but didn't know if he was big or bearded. Nor did the Belgians know the extent of the injuries. They said the attack happened at least two weeks earlier, but Amy had heard nothing from Bill or anyone else. She flew on to Kigali, uncertain of what lay ahead.

Meanwhile, Bill was released from the hospital after a few days. His broken ribs kept him off his motorcycle and out of the forest, but Pierre Vimont arranged for a French pilot friend to fly him to Kigali. There he worked on Mountain Gorilla Project business and waited for Amy. On the day of her arrival, her plane landed earlier than expected and Bill was not at the airport. She suspected the worst as she searched the waiting crowd. Maybe he would arrive in a wheelchair? Worried, Amy also felt somewhat foolish since she had worn a big red bow as Bill's Christmas present. Then Bill appeared with a large bandage on his neck and his arm in a sling to protect his broken ribs. It was a huge release for both of us, after months of separation, an attack that could have been much worse than it was, and an anxious twelve hours for Amy. When Sabena Airlines announced that they had again lost Amy's bags, it was the least of our worries.

Back at a friend's house, Amy discovered that the deeper of Bill's two bite wounds was badly infected. She proceeded to rip open the lightly healed gash and gouge out the infected tissue. Rarely has so much pain been administered so lovingly. In the process, she shuddered to think how close the bites were to either side of Bill's spinal cord. Later that evening after a shared shower, we realized that the misplaced luggage contained Amy's birth control products. We joked that if a baby girl resulted from our reunion we would name her Sabena.

By the time the lost bag was delivered a few days later we had already decided to conceive our first child. Bill's brush with mortality was part of the reason for our decision, but more compelling was the realization by each of us during our time apart that we simply cared for each other more deeply than we had ever imagined. Now we wished to share our lives in a new, more meaningful way. This was the right time and place. We spent the spring of 1980 working on the MGP, following gorillas, and enjoying the first signs of life of our son, Noah, who would be born that fall.

Chapter Eighteen

Limits and Reservations

IF BILL'S BANDAGES and sling evoked any sympathy from Benda Lema, he gave no outward sign. The ORTPN director rose from his desk, shook our hands, and gestured where we should sit. He welcomed Amy and patronizingly added that he was glad she was there to help her husband. He also stated his satisfaction that the recent attack was not committed by *un gorille de tourisme*. In Benda Lema's worldview, this meant that tourism could go on and that the gorillas would continue to generate revenue. The cash cows would produce a steady flow of milk.

We were encouraged, too, that it was the maverick Brutus and not the familiar Stilgar who had attacked Bill. It would have been much more difficult to bring tourists to view a gorilla that had shown such a violent side. Still, there were many other issues to be addressed before the tourism program could really move ahead. The most basic need was for an absolute limit on the number of tourists who could visit the gorillas at one time. Following the mysterious death of Adrien deSchryver—perhaps by poisoning— Congo's Kahuzi-Biega Park had abandoned all limits on gorilla visitation. Dozens of tourists in a single unruly group now went to see the Kahuzi-Biega lowland gorillas on many days. This resulted in constantly agitated gorillas, dissatisfied tourists, and an increasingly bad reputation for the park and its program. We had made arguments against overcrowding before, but to no avail. This time Benda Lema was more sympathetic. Maybe the bandages were working. Maybe he actually understood the idea that fewer people would pay more money for an exclusive, high-quality experience. He may have been trained as a Marxist economist, but the profit motive was foremost in his mind. Benda Lema agreed to a maximum limit of six visitors at one time.

Limits on daily visitation would only lead to a growing number of frustrated tourists if there wasn't a system of advance reservations. Benda Lema thought this was a fine idea, but his European advisors disagreed. They felt that the available technology was insufficient for our needs. International telexes to Kigali were fairly reliable, but the daily radio contacts between the ORTPN office in Kigali and the park headquarters in Kinigi were too uncertain and there were no telephone connections. In private, some of the Europeans stated that the Rwandans were incapable of the organization and attention to detail required for advance reservations. We disregarded this dismal belief in African inferiority and moved ahead with Benda Lema's blessing.

We sat down with a group of ORTPN receptionists to make the first six-month reservation calendar by hand. Using legal paper, we drew boxes large enough for ORTPN staff to write in individual names, agents, and contact numbers for each day's scheduled visitors. We left the master set in Kigali, dropped one copy at the park headquarters in Kinigi, and took another to our hut. Reservations were to be made through ORTPN-Kigali, then communicated by radio each morning to Kinigi and Visoke. Eventually, a fourth link was established at Mark Condiotti's base camp near Mt. Sabyinyo. Names were often butchered by mispronunciations and the Donald Duck quality of our radio communications, but the numbers and dates usually came through. MGP and ORTPN staff traveling between Kigali and the park learned to bring copies of the most recent version of the central calendar. Word of the system quickly spread among residents and major international travel agents, who welcomed the greater certainty of reservations, especially for top-dollar package tours. Africa's legions of backpackers and independent travelers were more likely to be caught off guard, but most of them had the flexibility to wait several days, if need be, for an opening. Sadly, those most affected by the new system were our first supporters. The TransAfrica and other overland trucks never knew when they would arrive in Rwanda after several months in transit, and their passengers were generally too numerous to wait for standby opportunities. Most shifted their focus to the "no limits, no reservations" program in Kahuzi-Biega.

Our final request to Benda Lema was to hire more guards to patrol the areas around Groups 11 and 13. Greater security was always a good idea, although we had already noted a significant decline in the number of traps found in areas regularly used by tourists. Equally important in the wake of Bill's encounter with Brutus, we needed a better idea of which other gorilla groups might be near our habituated groups. Group 6 was almost the same size as Group 11 and no tracker could be expected to distinguish one large group's trail from another. By monitoring the surrounding areas, though, we would have advance warning of the potential for interactions, nervous gorillas, and mistaken contacts. The benefits of an expanded monitoring program were clear: increased park security, employment and training for new trackers and guides, and an insurance policy for the tourists, guides, and ourselves. Benda Lema again gave his approval.

IT TOOK A FEW months to work out the glitches in the reservation system but the result was a much better organized program that benefited visitors, authorities, and advisors alike. During this transition period, one short-lived compromise to accommodate extra visitors without changing the "Rule of 6" was to take a second tourist group to Group 11 on overflow days. The gorillas seemed to tolerate this imposition, but none of us was happy with the situation.

One day in late January, we had seven people waiting to go out, including an elderly German couple in their seventies, the man walking with a cane. Bill had recently decided his broken ribs were field-worthy, so he offered to go out with the first five visitors. Amy, who had been doing all of the gorilla work for the past month, took a more leisurely route with the older couple. The first contact went well, although the terrain and vegetation conspired to limit viewing opportunities and the gorillas never completely stopped moving. When a tracker arrived after about forty-five minutes to tell him that Amy was nearby, Bill moved off with his group. Amy waited a few more minutes, then approached the gorillas on the main access trail used by the others. As she reached the base of a small rise she saw Stilgar in full strut above her. He was used to Amy by this time, but something about a second group of visitors set him off. Amy dropped to her knees as he charged downhill and she signaled the other two to do the same. She had a terrible sense of déjà vu and prepared for a Brutus-like attack. Stilgar roared, ripping a sapling from the ground as he charged, then delivered a glancing cuff to Amy's head as he shot past. He turned to glare at her, then strutted back to his family. Amy caught her breath, afraid to look around. The old man was lying on his side. He was a lighter shade of pale than when he started, but otherwise okay. His wife was still standing, beaming broadly, blue eyes glistening as bright as a little girl's. *I have never seen anything like this. It is wonderful!* Amy smiled back, wondering about the woman's sanity and thinking that the charge could have easily been the last thing the woman ever saw. She then explained that this was definitely not a good time to follow the gorillas. The couple agreed, apparently satisfied with the intensity, if not the duration, of their experience. It was the last time we would ever take two tourist groups to the gorillas.

———✵———

OUR LIFE AT THE BASE of Visoke was simple and satisfying. With Amy's arrival, we placed a second foam mattress on the matted floor of our metal hut. A borrowed folding chair doubled our stock of furniture. We worked outside as much as the rain permitted, sitting on a wooden bench that faced the eastern mountains of Sabyinyo, Gahinga, and Muhavura. Ephemeral patterns of cloud and light played across the mountains, constantly recasting their wild character. To the south and east, the Virunga piedmont spread for more than twenty miles. This broad apron of rich lava soils—domesticated rock—supported one of the world's highest agrarian population densities. A thin tongue of rain forest vegetation curled around our hut as it followed a ravine, buffering us from the endless fields below.

Above our heads the Virunga forest rioted on the sheer flanks of Mt. Visoke. At night, tree hyraxes from within this sanctuary conspired to multiply their individual calls into unearthly crescendos of sleep-defying screams. With daylight came a more subtle, yet steady chorus of bird songs and calls, perhaps punctuated by the *piao!* of a resident band of golden monkeys. Higher up the slopes, Brutus patrolled a territory that no others seemed to desire, his presence announced by the occasional *pocketa-pocketa* of his resonant chestbeat. From time to time his trail passed close behind our hut, leaving Bill to wonder how his old wrestling partner was doing.

Our home was on the edge of a small parking area at the base of Mt. Visoke. The relatively level expanse of volcanic cinders had been created by an earlier quarry operation. Solitary robin chats and small flocks of black and white fire chats thrived along its edges. A mongoose once darted out of the forest to cross the open space in a daring dash to the other side. But the Visoke parking was really human habitat. By 6:30 every morning, gorilla guides and porters would walk up to gather around the night watchman's stillglowing fire. Those with tobacco shared with those without; most rolled their own in a motley assortment of used paper. By seven, the sharp smack of playing cards slapped onto a rough-hewn log announced the beginning of a raucous game of cards. The rules were never clear to us, but the flair and sound with which one slapped down his cards appeared to be central to the game's essence. So, too, were the accompanying curses, charges, denials, and laughter. Big Nemeye loved the game.

Tourists usually arrived between 7:30 and 8:00, though their vehicles could be heard ten minutes earlier struggling up the rugged path until they reached the road's terminus at *le Parking*. The arrival of visitors brought an end to the card game as the players took on the more serious challenge of securing employment. The primary guide for that day was assured of work, as was at least one tracker or guide-in-training. Another pair of guides was on call for Group 13 at a meeting point several miles to the east. Any other guides were left hoping that someone wanted to climb the mountains rather than see gorillas that day. Porters had their own dominance hierarchy for the right to carry tourists' packs or camera equipment into the forest. All we asked was that they work this out ahead of time and not argue over clients. By 8:30, we were on our way to see the gorillas. Semitoba, a former Karisoke woodcutter whom we hired as our housekeeper, stayed behind to wash the morning dishes, sweep out the hut, and do a little laundry if it wasn't raining too hard.

A day with Group 11 was usually finished by early afternoon. Visitors were welcome to stay and share a pot of hot tea while they replayed the day's events. Some stayed longer and invited us to savor a meal along with their memories. A few even prepared elaborate celebrations, complete with freshly baked bread, imported cheese, and good French wine. This was a welcome relief from our standard fare of beans, rice, cabbage, and potatoes. The personal pilot of President Habyarimana came to see the gorillas on a day when no other tourists appeared, so he and his wife ventured out alone with the two of us. Back at our hut, muscles fatigued from a steep climb and pants caked in mud, the pilot produced a bottle of Dom Pérignon and a jar of beluga caviar as we sat outside on our rough *Eucalyptus* bench. It was the first time we had ever tasted caviar—and the first long discussion we had ever had with a French military officer. He said that he liked his work for President Habyarimana, but he grew more passionate about his love of hunting on a reserve near the Akagera National Park. We saw the world differently but enjoyed our time together. This experience compared favorably with the day Amy told a Belgian military officer that he couldn't go out to see Group 11, even though he had a reservation, because he was two hours late and the other visitors were already out with the gorillas. His threats to return with a gun made for a few nights of uneasy sleep.

But on most days tourists exited quickly, eager to change out of their cold wet clothes. As they packed their vehicles, the guides and porters lingered uneasily, waiting to be paid. In the first few months of the tourism program this was a time for loud exclamations by the guides, safe in the obscurity of their native tongue. One frequent comment was that the *abazungu* were in terrible shape and couldn't keep up with the Rwandan guides. A second common assertion was that the same foreigners were tight with their *amafaranga,* paying precious little in tips. Bill listened to their proclamations for several weeks, then suggested a connection between the two correctly observed phenomena. Yes, the foreigners were often exhausted by their efforts, even the young and apparently strong. But this was at least partly a function of their transition from lowland America, Europe, Australia, or Japan to the high elevation of the Virungas. Grasping the concept of elevation-induced debility required a great leap of faith by most of the guides, however, who had spent their entire lives above eight thousand feet and had lungs like soccer balls. They preferred the equally accurate explanation that Westerners had easier lives than Rwandans, who were used to working harder. Bill's second point was that exhausted tourists who felt mistreated by their guides were less likely to pay big tips. This remark was greeted with initial skepticism, but some of the Rwandan guides began to change their behavior. They tried slowing their pace to that of the group's weakest or oldest member. Some even deigned to take the hand of struggling tourists to help them over ravines. Best of all, many of the guides started taking brief rest stops along the way, during which they would talk about the forest and the gorillas. As a result, tips rose sharply from appreciative visitors. Soon guides were making $5 to $10 per day, far beyond the dollar-a-day earnings of the average Rwandan worker. A few tourist groups gave $50 tips—a small fortune in the local economy. Several of the men bought more farmland with their earnings. Some built new houses. Big Nemeye paid the dowry for a second wife. All saw their status rise, and several became important advocates for the park in their communities.

The dramatic rise in guide earnings provoked some strong negative reactions, however. Some of the park guards fell back on old habits of questioning why guides should be making more money than other park workers. Benda Lema then entered the fray, supporting the guards. It was unclear whether he was motivated by concerns for forced equality or by a desire to capture greater profits for ORTPN. Whatever the reason, he issued an order banning all individual tips, requiring instead that tourists purchase a "guide tip" ticket at park headquarters. Profits from this $2 fee were to be split between the individual guide and the park service. This angered visitors, who were forced to pay for a service before it was rendered, regardless of its quality. It greatly angered the guides, who lost a significant source of income and much of the motivation to perform their jobs well. It angered us because it added an unnecessary element of contentiousness to our program. Bill went to Kigali to meet with Benda Lema and Alain Monfort. He explained the need for performance-based pay for work that was very demanding, at times dangerous, and otherwise poorly remunerated. He made the further point that while most tourists would never meet someone as educated as the ORTPN director or other *important* Rwandans, they would spend several hours with our guides. It was essential that these "ambassadors" be proud—and happy. Alain supported Bill, and Benda Lema rescinded his order.

Our ambassador-guides needed a great many skills to do their jobs. Afternoons, when the tourists were gone, we turned to hands-on training. Nemeye and Jonas had never attended school and Swahili was the only language they needed to work with us in the forest. Learning French was essential, however, for them to move beyond tracking to lead groups of tourists by themselves. We took the basic vocabulary we used for "gorilla Swahili" and created an effective French primer for use by the guides, all of whom could read. Zimulinda and Gilbert were both proficient in French, so they moved on to "gorilla English." *Here is a nest. Follow me. Don't move. He is the silverback. The mother has a baby.* Equally important for all the guides were lessons in dealing with foreigners. Rwandans were not obsequious, but they were certainly deferential to pale-skinned foreigners. This was an unfortunate legacy of the colonial era. It was also a dangerous syndrome for a program where the misbehavior of tourists could not only undermine our objectives but even threaten the lives of the gorillas we were trying to protect. Guides needed the confidence to tell an *umuzungu* that he was too close to an infant, or that her cough was a danger to the gorillas. Although private contempt for foreigners came naturally, the persona of authority figure came more slowly to the guides. With time, however, they proudly assumed this role.

By late afternoon, the parking area was ours to enjoy in peace—unless some visitors were staying to camp. Most overnighters were polite and respected our privacy. But there were exceptions. A few lost souls arrived without rain gear *or* any camping equipment and had to borrow one of our tents. Some thought there would be food at the end of the trail and ended up joining us for dinner. A disgusting few who remembered to bring food nonetheless thought nothing of leaving their garbage strewn around the campsite and our hut. And then there was the *choo,* or latrine. Bill's Christmas present to Amy was a custom-made seat for our outhouse, which had previously consisted of just an open hole in the ground. The new throne, made from a sturdy wooden box, offered a comfortable respite with a discreet view of the valley below. We didn't advertise our fine facility, but neither did we turn down any requests for its use. One day, after a group of young Europeans had ended their visit and driven away, Bill visited the *choo* only to discover muddy Vibram boot prints on either side of the seat's opening. Behind the hole was a pile of vile excrement that the misguided bombardier had failed to drop with any precision from his elevated perch. Our guest's apparent concern for his own hygiene certainly compromised ours, and Bill's next addition to the *choo* was a sign that said "Please Sit Down" in four languages.

At one point, we tried to reduce public use of the parking and stimulate local business at the same time. Our closest neighbor, Bidele, lived only eighty yards away and had protected a nice patch of bamboo on his property just outside the park boundary. We were already renting about sixty square feet of garden space from Bidele for fresh vegetables. Now we suggested that he turn his bamboo stand into a rental campsite. He liked the idea and immediately set both of his wives to work preparing enough room for two or three tents and a small fire pit. He also arranged for wood and water delivery by his children for a modest fee. Bidele was soon making several dollars per night as an eco-entrepreneur. Then Camille, the park warden, appeared with a cease-and-desist order all the way from Kigali. Learning of Bidele's venture, Benda Lema realized there was more money to be made, even from those who didn't stay in hotels, and he again wanted ORTPN to capture that revenue. Not only was Bidele's operation shut down, but all camping was banned at the Visoke parking and redirected to the Parc des Volcans headquarters at Kinigi. Never mind that Kinigi was eight miles away and that most overnight campers were on foot. Never mind that one of the Mountain Gorilla Project's goals was to generate local employment and revenue. Big Brother had a better idea and this time it stuck.

We felt badly that Bidele had been cut off from his extra income. He had a large family with four or five children always dressed in scanty rags and with chronically runny noses and rheumy eyes. If we said anything about the need for warmer clothing, Bidele would laugh and mutter something about crazy *abazungu*. One day, Amy tried a more direct approach when his second wife stopped to talk on her way back from gathering water. Amy gave her a 1,000 franc note and suggested she buy the children some clothes when she went to market. The woman bent deeply at the knee, despite the heavy water jug balanced perfectly on her head, and thanked Amy profusely as she accepted the gift. When Amy next met her, the woman had a swollen face where Bidele had hit her before he took the money. We were rudely reminded of the limits to our well-intentioned interference in another culture. Not that wife-beating wasn't an ugly fact of life in the Western world, too, but Amy hadn't considered this result in a land where all money was controlled by men. In the future, she purchased the clothes herself and gave them directly to Bidele's children. As far as we could tell, he never resold them.

We had no nearby expatriate neighbors. Mark Condiotti lived about six miles to the east, Jean-Pierre von der Becke almost twenty-five. A single American woman in her sixties, Rosamond Carr, lived twenty miles to the west on the slopes of Mt. Karisimbi, where she had managed a pyrethrum and commercial flower plantation since before independence. A small community of French doctors and teachers lived nearly fifteen miles to the southeast in Ruhengeri. Everyone around us was Rwandan. The administrative zone around the foot of Visoke was called Bisate and was home to several thousand farmers and a handful of merchants, almost all of whom were Hutu with ancient roots in the region. We never locked our door, unless we left for several days. Yet only one object—a Swiss Army knife—was ever stolen in the time we lived there. As a member of the community as well as our houseworker, Semitoba took the loss personally and arranged for the knife's anonymous return the next day. The families who lived along our access path to the forest came to know us by name—*Ami* and *Biru,* in their pronunciations; their children routinely greeted and stared at each group of tourists as they walked past their *rugos.*

We witnessed the daily cycle of rural African life. About half of the local children walked off to school each morning in color-coded clusters of girls in blue dresses and boys in khaki shirts and shorts. Their peers who couldn't find or afford a place in school set about their equally gender-specific chores: boys tending goats, girls gathering firewood and water. Babies were strapped to the backs of their mothers, who set out to hoe, weed, or harvest their fields. The men owned the land, but the women worked the land. The most common Rwandan greeting from a man to a woman in the field acknowledged this fact: *urakoze,* or "thanks for working." Come late afternoon, the women returned to their homes, where a grandmother, aunt, or oldest daughter would already have the pot of beans boiling for the evening meal. Potatoes and onions might be added, some goat meat on a special occasion. The men of the region sought salaried jobs: generally as laborers, sometimes far away. Often they would gather to drink and talk on their way home. By six o'clock every night it was dark. If it wasn't raining, the sounds of hundreds of conversations—and a few arguments—would rise up with the breeze and drift over our little outpost, competing and mixing with the sounds from the forest behind and above us.

Everyone was friendly to us, but few Rwandans sought out contact. Nemeye, Semitoba, and other former workers from Karisoke such as Rwelekana were our closest friends and our windows on the world around us. They could be warm and humorous. But as Peace Corps volunteers we grew accustomed to the overly social Congolese, who seemed to want nothing more than constant contact with foreigners and each other. Rwandans, or at least the northern Hutu, were much more insular, almost antisocial, even among themselves. We appreciated the solitude.

O UR RELATIONSHIP with the gorillas of Group 11 was different from that which we had with Group 5. This was a family that we had habitu-ated from scratch, which meant that we had witnessed the entire behavioral evolution from flight and charge to acceptance and mutual curiosity. We had seen the rich potential of the acceptance phase at Karisoke, and Amy had stretched the tolerance of Group 5 to include close observers as they moved. But we had not participated in the earlier stages of Group 5's habituation process. Our role in selecting Group 11 and changing the lives of its gorillas through habituation also meant that we felt even more responsible for their welfare. Of course we cared deeply about Group 5 and would have been dev-astated by any harm to its members, even though we did not make the origi-nal decision to expose that family to a human presence. But we felt even more directly accountable for any negative impact of tourism on Stilgar and his family.

Limiting the number of visitors was an important safeguard for the goril-las. If the "Rule of 6" was not chiseled in stone, it was accepted gospel as long as we were involved. Limiting the time that visitors could spend with the go-rillas was also very important. Our visits lasted no more than one hour and were designed to approximate the gorillas' typical mid-morning rest period. In fact, Group 11 seemed to choreograph *its* schedule to fit our visits. Soon after we arrived each morning, Stilgar and the adult females would find an appropriate area with good visibility. Then they would settle down to rest while the younger generation played around them. Forty-five minutes to an hour later, they would begin to move off, signaling the end of our visit. This pattern held with remarkable consistency, whether we arrived at 9:00 A.M. or noon.

Our final effort to assure the gorillas' well-being was to maintain a space of at least five yards between the gorillas and tourists. We intended to limit both excessive familiarity with visitors and possible exposure to human-borne diseases. Most visitors arrived in peak health. Tourists who were sick were not supposed to go out to the gorillas, but illness could be difficult to detect and so this rule was sometimes hard to enforce. It should have been easier to maintain a safe distance between people and gorillas. Yet this was a constant battle, largely because of the young gorillas' insatiable curiosity and their own desire for close inspection, if not direct contact. And while most visitors actively tried to avoid close encounters of the wrong kind, a few proved unable to resist the attraction of direct contact. We worked constantly with the guides on this issue and they took their responsibilities seriously.

Our understanding of the gorillas of Group 11 was limited by the amount of time we actually spent in their company. One hour per day didn't expose us to the full range of the gorillas' lives. This was especially true when the gorillas spent most of that hour at rest, even if this was prime time for social behavior such as play and grooming. Another limitation was that we almost never spent time alone with this family of gorillas. Visitors depended on us or the guides as their source of expert information about the gorillas, their lives, and the scenes unfolding before them. If we weren't whispering to someone, we were looking to see who we could help move to a better vantage point. Many times, we found as much interest in watching people's reactions to the gorillas—the wonder, surprise, and joy in their faces and in their body language—as in watching the gorillas themselves.

Despite the limits and distractions, we were fortunate to have the opportunity to know another set of individual gorillas and to observe their group dynamics. In many ways Group 11 reminded us of Group 5. Stilgar was an older, darkly silvered patriarch who looked much like Beethoven. Ndume was a young silverback, perhaps a few years younger than Icarus, though he occupied a more central role in family life. Five females completed the adult contingent in Group 11. They were initially shy and we didn't get to know them as well as Effie, Marchessa, Pantsy, Puck, and Tuck. And there were young gorillas everywhere. More numerous and outgoing, the seven subadults were at least as playful as the youth contingent from Group 5. They lacked the wild card personality of Pablo, however. Perhaps with greater time and more familiarity they would show an equally mischievous side. Perhaps Pablo was one of a kind.

One striking difference in Group 11 was not behavioral but physical. Three of the gorillas had maimed or missing hands: the silverback Ndume, the adult female Mkono, and the juvenile Kosa. This high incidence within a family of fourteen was unlike anything in the Karisoke study groups; it was powerful evidence of the high intensity of wire snare poaching within Group 11's home range on the northeastern slopes of Visoke. When we began the habituation process in that area, we discovered and cut trap lines every third or fourth day. Within a few months of daily tourist visits in the same vicinity, however, we almost never saw another trap. Poachers apparently decided that the rate of discovery was too expensive and too risky to continue their illegal trade. Mark Condiotti noticed the same effect around

Group 13. This decline in trapping was one of the most immediate and rewarding impacts of the MGP gorilla tourism program. It also allowed Jean-Pierre von der Becke to concentrate his guards' anti-poaching efforts in other sectors of the park.

We never stopped thinking about what the unintended effects of exposing more gorillas to human visitation might be. Behaviorally, socially, and ecologically, the tourism groups showed no obvious differences from habituated groups at Karisoke. In fact, they had twenty-three hours per day away from people: far more time without human exposure than Groups 4 and 5. Yet we had undeniably intruded on the natural lives of free-ranging, independent gorillas. Was the benefit to the gorillas worth the price of our intrusion? Yes, if a few individuals were disturbed so that we could save the entire population. But only time would tell if that ambitious goal could be realized. Meanwhile, the stumps at the end of Ndume's and Kosa's arms and Mkono's mangled fingers stood as stark reminders of the failed alternative of years past.

Chapter Nineteen

Across the Virungas

IN MAY OF 1980, Amy set off with three Frenchmen to hike the full length of the Rwandan sector of the Virunga chain of volcanoes. Up to that point, she had spent almost every day in the park with gorillas, either at Karisoke or with Group 11. These experiences gave her intimate knowledge of about 10 percent of the forest, but now she wanted to see the rest of the park—including the three easternmost mountains that she had watched peering in and out of the clouds since her return from the U.S. The trek had a work-related purpose, too, which was to explore and map a network of trails that would cross the Parc des Volcans from one end to another. Access trails already existed for each mountain, but no trail had ever linked the different summits. Such a link was important for visitors who wished to experience the mountains as well as the gorillas. With gorilla tourism growing rapidly, we wanted to disperse this pressure and demonstrate that the entire park was important for conservation, not just the few areas where tourists or scientists followed gorillas.

Didier Blaizeau taught secondary school in Kigali. He had visited the gorillas several times with us, but he spent much more time exploring the park whenever he could free himself from teaching responsibilities. He was a native of the French Alps and passionate about mountain climbing. The idea for a Trans-Virunga Trail grew from a dinner discussion at Didier's home in Kigali, where Bill lamented the lack of connecting trails during his census work. Didier had recognized the same problem, though the lack of trails rarely stopped him. We all agreed that a network of linked trails could make the Virungas an important international backpacking destination—and help to justify its full protection. Over the next two months, Didier and Amy planned their reconnaissance mission, which two other teachers decided to join.

The expedition set off from the eastern flank of Mt. Muhavura, near the Uganda border. In this section of the park, Amy entered a world unlike any she had seen around Visoke. Bare lava flows, some as recent as the late 1950s, covered most of the ground. Lichens adapted to holding their own moisture in an environment without soil colonized the lava surfaces, while mosses grew in moist nooks and crannies. Sparse pockets of grasses and shrubs were widely scattered. Brightly colored petals of red-hot pokers stood in sharp contrast to the black rock background. Amy was struck by the realization that this was probably the same landscape the first gorillas crossed to arrive in the Virungas.

Current thinking holds that gorillas originated in the lowland rain forests of west-central Africa. As the population grew and some migrated east, they followed the curve of the Congo, remaining on the north side of that vast river. At one time, thousands of years ago, a continuous population of gorillas stretched from Africa's Atlantic coast to eastern Congo. Climate change then created colder and drier conditions that split the rain forest in two separate blocks, isolating the eastern gorillas from the main gorilla population to the west. At some point, the eastern gorillas reached the highlands around Lakes Kivu and Albert. The Virunga volcanoes would have been in sight to the east. Yet to reach the rich forests of the interior Virungas, the gorillas would first have to cross the extensive lava fields of the still active Nyiragongo and Nyamuragira volcanoes blocking passage from the west. Standing at the opposite end of the chain, Amy imagined the combination of courage and curiosity—perhaps necessity—that drove the first gorillas to cross such a bleak landscape on their way to the unknown beyond. At least one band survived the crossing and later reproduced. They were the founders of the mountain gorilla population we know today. Those who remained to the west of the Great Lakes became the subspecies we know today as Grauer's, or eastern lowland gorillas.

At the top of Muhavura, a cloud cap lifted to reveal stunning views in all directions from the 13,540-foot summit. Beyond endless fields and countless lakes, Amy and her companions could see Uganda's Bwindi Impenetrable Forest thirty miles to the north. There lived the only other population of mountain gorillas, one that had once been in direct contact with the Virungas. By the late 1800s, however, African farmers and herders had cleared all but a few relict tracts of forest between the two sites, leaving two island populations to struggle for survival in a sea of humanity.

Leaving the summit, the four hikers clambered through Muhavura's dense cap of *Senecio,* then descended a series of heath-lined lava fins—narrow ridges above steep ravines where a misstep could be costly. That night they camped in the saddle between Muhavura and Gahinga. The next morning, the group took less than an hour to reach Gahinga's summit. There, at an elevation almost two thousand feet lower than that of Muhavura, a carpet of everlasting flowers and orchids greeted the climbers. Amy had seen the tight silver balls and feathery leaves of the everlasting flowers at much lower densities on the summit of Visoke. But nowhere in the Virungas would she ever see clusters of the same bright pink orchids on display. Resting on Gahinga's rim, Amy stared down into the two-tiered crater. She wondered what animals made and maintained the well-worn trails that crossed the larger of the two marshes. It was easy to imagine elephants browsing leisurely below. Leisure was a foreign concept to Didier, however, and the group quickly mustered to move on. With no established trail to follow, they bushwhacked across Gahinga's western flank. Recent elephant trails released the musty odor of freshly disturbed soil as they wound through uniform stands of enormous bamboo. The group arrived in a high meadow to the east of Sabyinyo. Bill and Nemeye encountered hyenas at this spot during the census, but no such surprises disturbed Amy's deep sleep. She would need her rest.

The expedition had always planned to ascend the second highest of Sabyinyo's jagged teeth. The notion sounded exciting in Kigali; it grew disquieting as Amy emerged from Sabyinyo's thick bamboo to stare at the summit in question. Their position on a narrow ridge, overlooking a bottomless gash that cleaved the former crater in half, didn't help. On the other side of that canyon, a solid wall rose straight up until the forces of erosion carved out the distinctive teeth that gave Sabyinyo its name. The last thousand feet were more directly vertical than anything Amy had ever climbed—and Didier had said that ropes weren't needed! She decided to follow directly behind Didier and copy his movements. Toward the top, she kept one eye on Didier's hand-and footholds, the other on her own tenuous grip on the mountain. The modest bulge of her four-month pregnancy felt like a much larger mass, pushing her away as she tried to hug the rocky fin they were climbing. As she reached the top with the wind in her hair, the adrenaline rush and clear views made it all seem worthwhile—at least for a few minutes. Then darker clouds closed in and Didier ordered a retreat. Amy again followed Didier for the descent, but she couldn't see what he was doing below her. She couldn't even see her own feet, as loose rock crumbled beneath her toes. Almost immediately, she felt that she was recklessly risking her own life—and that of the child she was carrying inside her—to go any farther. She remembered that

Didier's own pregnant wife had stayed at home for this trek. Swallowing her pride as the only woman on the expedition, Amy told Didier that she needed help. The sturdy Frenchman climbed back up to her and carefully placed each of her feet as she descended the next several hundred feet of the most difficult terrain. He then returned and did the same for each of the others. At the bottom, the other two Frenchmen thanked Amy for asking for help and Didier apologized for the needless risk. Sabyinyo's highest peaks would not be part of the proposed Trans-Virunga tourist trail. But in subsequent years at least two climbers died attempting the same route scaled by Amy and the others.

A visit to Group 13 had been planned to provide a break at the midpoint of the trek. After the experience on Sabyinyo, the break was even more appreciated than anyone had imagined at the outset. Mark Condiotti had made considerable progress with Group 13, especially in light of the killing of the dominant male only six months before. A younger silverback named Mrithi had held the group together and was apparently breeding with at least one female. Now Mark was waiting to show his family to Amy and her friends. Unfortunately, the gorillas were in very thick bamboo on that day and they never settled down to rest. Mark was disappointed, but Amy did her best to cheer him up, and the others seemed happy with their fleeting glimpses. There are no guarantees with wild animals.

The cabin at the eastern end of Ngezi was even more dilapidated than on Amy's last visit. Built by Dian several years before, it was intended to provide shelter and gorilla viewing opportunities for tourists on the other side of Visoke. There was no follow-up, however, and the structure fell into disuse except by giant rats and poachers. Still, the site offered a beautiful view of Visoke's summit, which reflected on the surface of the narrow crater lake that gave Ngezi its name. The next morning's climb up Visoke passed a fifty-foot waterfall just below the treeline. Near the top, Amy and the others explored two large side craters. Giant *Lobelia* and *Senecio* dotted the crater bottoms and Amy stopped to admire the specialized adaptations that allowed these plants to survive the extreme cold of the alpine zone. *Senecio* does not drop its dead leaves, but rather folds them around its ten-foot trunk, where they join with resident mosses to form a thick layer of insulation. The *Lobelia,* too, droops its dead bayonet-shaped leaves from its crown, but otherwise leaves its tall hollow stem unprotected. Amy appreciated both adaptations even more after a cold, driving rain swept over the mountain.

Descending just north of Karisoke, so as not to upset Dian with unwelcome visitors—she had shot a pistol over the heads of some Dutch tourists a few months earlier—the group crossed the saddle in a hailstorm. The water level in the marsh was knee high and freezing. Amy noted the absurdity of not stopping briefly to warm up at Karisoke's fire pit, but they pushed on. Slogging up the flank of Karisimbi, the group climbed into the subalpine zone, where porters had dropped off tents and food supplies. As darkness set in and the rain continued to pour down, Amy had no interest in food and instead went straight to bed. Curled up and still shaking in her sleeping bag, she wondered how the equally soaked but less-well-equipped porters would survive the night. At some point, the subfreezing temperatures numbed her mind as well as her body and allowed a restless sleep.

At 14,797 feet, Karisimbi is Africa's fourth highest mountain. For traditional Rwandans its summit is home to their most powerful god, Ryangombe. The mountain's name, however, derives from *isimbi,* the cowrie shell. The cowrie came to the African interior from the Indian Ocean coast as a valued form of currency for trade—and Karisimbi's often snow-covered crown recalled the shell. Even though precipitation declines above 11,000 feet in the Virungas, the previous night's steady downpour dropped almost four inches of fresh snow on Karisimbi's summit. Amy and Didier arrived first at the top and couldn't resist a snowball fight. The Rwandans were unfamiliar with this custom, but everyone quickly joined in the fun. It was a good way to shake off the chills, discomfort, and fatigue of the preceding twenty-four hours.

Karisimbi marked the western terminus of the expedition, a cone-shaped bookend to match that of Muhavura to the east. Partway down Karisimbi's southern face, however, is a side crater larger than any other crater in the entire chain, including Visoke. Almost two miles across and nearly a thousand feet deep, it is a spectacular formation. Amy wished they had time to explore, but they were running late and a vehicle waited for them at the base of the mountain. This was obviously a prime tourism spot. Yet equally obvious was the potential for damage from uncontrolled use. Dozens of giant *Senecio*—a slow-growing woody shrub that lives for hundreds of years—had already been hacked down for firewood, and blackened fire pits dotted the rim. Attracting more visitors to this site without strict regulations and enforcement would only further degrade a highly sensitive alpine ecosystem.

On the final descent of Karisimbi's southwestern slope, Amy's thoughts again turned to gorillas. There weren't any. Yet unlike Muhavura's lava landscape, here there was ample food of high quality: celery, thistle, *Galium,* and more. As Bill had discovered on the census, there were large areas of the park with good gorilla habitat, but no gorillas. Poaching had to be the cause. Could increased tourism help reduce poaching pressure and allow gorillas to return? If so, how many *Senecio* is one gorilla worth?

T HE BLAIZEAU-VEDDER expedition demonstrated that the Virungas could be crossed from end to end and that all of the summits, save Sabyinyo's, could be reached without special climbing skills or gear. Backpackers, with or without porters, could now join day hikers on all of the mountain summits, yet still spend their nights in the saddles. Amy and Didier mapped new trail possibilities and we argued the case for a Trans-Virunga Trail to ORTPN. The response was not positive. The Rwandans seemed little interested in backpackers, claiming that they spent less than other tourists. This was true while they were in the park, but they would still require hotels and restaurants in Ruhengeri and Kigali. More surprising, Mark Condiotti opposed the idea. He argued that increased hiking in the park's interior would create opportunities for more illegal viewing of the tourism gorillas. He was especially concerned that hikers might encounter Group 13 and follow them for an extended time or otherwise disturb them. Mark's concerns were legitimate, but we felt that since backpackers also required guides, better guide training could solve the problem.

We believed that the park was still at risk of conversion to farmland or pasture. This was especially true of those areas away from the center of the park that were not used for gorilla tourism. Increased hiking and backpacking would add value to those areas and decrease the incidence of poaching. Ultimately, these activities might help prepare the way for a growing gorilla population to return to viable parts of its former range. For these reasons, we supported the idea of an expanded trail network. But in the end, ORTPN simply let the plan die. There were only so many battles we could fight and there was higher priority work for the Mountain Gorilla Project to complete.

Chapter Twenty

Why God Created Gorillas

THE FIRST CATHOLIC missionaries to arrive in Africa had sharp eyes for real estate. Backed by the expropriation powers of the colonial government, they picked prime waterfront and hilltop sites for settlement whenever possible. Rwanda offered many lakes, endless hills, and magnificent views with a backdrop of volcanoes for the discerning cleric. The mission at Rwaza combined all three.

Rwaza was a collection of nondescript, old brick buildings with spectacular views of Lakes Bulera and Ruhondo to the west, and Mt. Muhavura towering to the north. Inside the compound stood the church itself and working, living, and eating quarters for its priests, nuns, and friars; a school with dormitories and a cafeteria; and a cigar factory that used locally grown tobacco and provided work to more people than any other nongovernmental employer in the Virunga region.

Bill's morning presentation at the school was well received. The students were an impressive group: the elite of northern Rwanda—the future of the country under its current political alignment. They were bright, attentive, and inquiring. Bill extended the question-and-answer session as long as possible, then joined several of the Belgian priests and brothers for lunch. Later that afternoon, preparations began for an evening show for the local population. Bill brought a projector and a generator, but wondered where they would project the film for the expected large outdoor audience. The answer was a sheet strung between two tall poles: an ingenious system that created a nearly 360-degree amphitheater for viewing from both sides of the sheet. By sundown, it was clear that they would need all the space possible. The priest estimated the crowd at over two thousand and joked that he should switch to nighttime masses with a movie to attract parishioners.

By this time Bill had replaced the National Geographic films with recent films more appropriate for education purposes. Their soundtracks were also in French, which was the language he used for secondary school presentations. For our work in the forest we spoke almost entirely in Swahili. But most rural Rwandans spoke only Kinyarwanda—the complex local language of which Bill had only limited knowledge. He was therefore grateful when Camille, the warden of the Parc des Volcans, offered to translate the film for the general public at Rwaza. Following a prepared slide show with recorded commentary in Kinyarwanda, Camille took over. The sparse text of the French filmmakers was overridden by a torrent of commentary in Kinyarwanda that included admonitions, exhortations, and an unprecedented degree of animation. Camille worked the crowd like a master politician, his oratory commanding silent attention at times and provoking roars of laughter at others. The three-ring circus drew exuberant applause at its end. As people filed out, many stopped to give their personal thanks. One woman held her young baby up to Bill and said *just like a gorilla,* then walked away laughing. Camille stayed to talk with anyone who would listen, still high from his performance.

When Bill arrived home after an hour-and-a-half drive from the other end of the Virungas, we talked about the experience. He was excited by the turnout and the opportunity to reach so many people, but wondered what message they took home. He certainly couldn't follow all that Camille had to say. Did the crowd learn something about gorillas? About their threatened status? Were there more people like the woman who saw in gorillas something close to our own species? Or had we turned these most admirable creatures, the focus of our own deepest respect and commitment, into clowns or targets for laughter? Was entertainment an acceptable first step in the learning process? In such a large crowd, there were certainly many different responses, some more positive than others. At a minimum, two thousand more people had seen gorillas in action and been prompted to think about their lives and their right to exist. We talked long into the night and came to no solid conclusions except that education was too important not to succeed.

THE MGP EDUCATION PROGRAM aimed at multiple targets. Local adult populations were very important because of their proximity to the park, their history of forest use and clearing, and their active interest in access to new agricultural lands. Bill's surveys also showed that although local residents voiced few negative attitudes toward the gorillas, they knew little about the animals and nothing about their conservation status or needs. More surprising, only a minority of neighboring farmers saw any connection between the health of the mountain forest and their own well-being. Yet the natural vegetation cover of the Virungas' steep slopes inhibited otherwise disastrous erosion and flooding of the surrounding farmlands. The same vegetation and soil cover also acted like a giant water tower, holding, filtering, and releasing a steady flow of clear, clean water to streams and wells throughout even the driest times of the year—a luxury to most Rwandans living in areas already cleared of forest cover. One of Bill's more revealing findings was that those local farmers who had traveled more widely recognized the severity of water problems in the rest of Rwanda. The MGP education program needed to bring that same understanding to those who never left the shadow of the Virungas.

Bill targeted secondary school students because they rode the fast track to future leadership positions. Less than 2 percent of all young Rwandans attended secondary school in 1978; less than half of them would graduate. These fortunate few—two thirds of them males—were destined to be the government officials, teachers, businessmen, and military leaders of the future. For this reason, we believed that it was worth a nationwide effort to make presentations about gorilla conservation at each of the country's twenty-nine secondary schools. With only 47 percent of all children enrolled in primary school, even sixth-grade graduates were likely to earn positions of some privilege. However, since there were too many primary schools to visit—and because their graduates were more likely to remain in their native areas—we decided to focus on primary schools within the Virunga region.

Beyond direct presentations, we provided information for use within the formal Rwandan school curriculum. In a review of all written science materials used in grades one through twelve, Bill found two references to gorillas. The first described their dental structure. The second inaccurately stated that they were crop raiders. There was no reference to the gorilla's kinship with the human species, no mention that mountain gorillas were unique to the Virungas. Not even a statement that they were highly endangered. Future generations would not be any better able to address the critical problems facing gorillas if they weren't exposed to more comprehensive and accurate information about gorilla behavior, ecology, and conservation needs. This realization led Bill to accept the secretary-general of education's earlier invitation to work with the Bureau Pédagogique to reform its biology and environmental studies curricula.

MONSIEUR TROG was a short, balding, humorless man. As director of the Bureau Pédagogique, he was responsible for implementing a complete revision of the core teaching materials and content for the entire Rwandan school system. The revision was long overdue, as the Belgian books and courses taught in Rwanda were already out-of-date in Europe and had little relevance to African students. Trog faced a difficult task. He was also obliged to humor the secretary-general when that official pulled tricks like suggesting that a young, untrained American could help advise on developing the country's new environmental curriculum. Trog politely met with Bill, then introduced Bill to his science staff. Gilles Toussaint was a former marine biologist by training, who settled into the Belgian teaching establishment in Rwanda when he found no gainful employment in his chosen field. He led a pleasant life with his Tutsi wife and children, but his passion for biology could still be rekindled by a challenge such as the national curriculum revision. His colleague François Minani was bright and ambitious, in a generally constructive way. He wanted to learn more to do a good job and advance as far as possible in his field, or in any other field that might offer greater opportunities.

Nicole Monfort also was asked to help with the environmental curriculum revision project. Nicole was a savanna ecologist with a lifetime of African experience, a strong commitment to conservation, and a great sense of humor. Like Bill, she was an advisor to the formal team of Toussaint and Minani, and the four worked well together. From the beginning, Bill knew that he didn't have the training or the time to give the curriculum initiative the necessary attention. So he approached the newly established Peace Corps program in Rwanda and asked whether a volunteer could be assigned to the Bureau Pédagogique. The position was approved on condition that the government officially ask for it and provide an office and support for the position. Trog could live with independent advisors, but he didn't like a foreign presence within his bureaucratic kingdom. Caught between his Rwandan superiors and a hard place, he turned to his Belgian superior for help.

Belgium was Rwanda's primary foreign donor and paid for much of the curriculum revision. Trog therefore invited Bill to explain the value of a Peace Corps volunteer to a Belgian Ministry of Education official who was visiting Rwanda. As soon as the meeting began, however, it was clear that Trog's goal was to scuttle the planned collaboration. Turning to the Belgian official, he stated that *Monsieur Weber has a mission to save gorillas and he would like us to help with that mission. What he might not understand is that Rwandans have*

many more important needs and we need to direct our science teaching to those needs.
Trog could not hide a slight smile as he concluded his subversive introduction. Bill responded that he did indeed care about the gorillas, but that curriculum revision discussions had focused almost entirely on the need to
teach basic ecological principles for people more than wildlife. This was in
line with our belief that Rwandans would need to address their already
strained relationship with their land, forest, and water resource base—of
which the gorillas' endangered status was but one symptom—long before
they would have the luxury of putting their scientific training to work on
other subjects. The Peace Corps position was approved.

THE MOUNTAIN GORILLA PROJECT also provided equipment for the
education program. John Burton of the Fauna and Flora Preservation
Society had arranged with a contractor from World Wildlife Fund–UK to design and equip a vehicle for school and public presentations. It was an expensive item, but promised to be helpful and much more professional-looking
than our earlier efforts. It did look better, especially the dynamic silverback
logo of the MGP that was painted on both sides. And the audiovisual material inside was a great improvement for mobile presentations. Yet the vehicle
arrived six months late and was poorly adapted to Rwanda. A basic Renault 4,
its motor had nowhere near the power needed to cope with Rwandan topography. Every time Bill brought the vehicle back to our camp at the base of Visoke, local men and boys would hear the engine's whine and run out to help
push the small car up the final few hundred yards. It became a challenge to
local standards of manhood to see who could make it to the top without tripping or slipping on the rough track. The low suspension of the R4 also
turned this and other rocky roads into dangerous, muffler-eating shoals. Inside, the vehicle had two front seats and a windowless rear section that contained projectors, films, a screen, a generator, display panels, and about a
dozen metal tubes used to support the panels and screen. After less than a
week of back country travel, though, the tubes and display panels came loose
from their attachments and made a deafening racket over the rough roads
that covered most of Rwanda at that time. But rattletrap or not, the Goril-
lamobile looked cool and it gave the MGP education program an identity.

For school presentations, Bill would usually arrive in mid-afternoon and
park in a conspicuous spot. After introducing himself to school authorities
and assuring that all was in order for the evening talk, he would return to set
up the display panels on the side of the vehicle. These had pictures of gorillas, the forest, and farmlands, with accompanying written texts in French for

secondary schools, Kinyarwanda for primary schools. The panels were mag-
nets for students, teachers, and curious passersby, providing an opportunity
for open discussion. Each of these discussions gave Bill insights into how
Rwandans looked at the conservation issues we were trying to address. Most
students lived at the schools they attended, so the main presentation usually
occurred after dinner, avoiding any conflicts with classes or sports. As in his
earlier school visits while he was still at Karisoke, Bill was impressed with the
students' general level of interest and thoughtful questions. Conservation issues provoked a certain amount of discussion, but gorilla behavior and social
organization dominated the question-and-answer sessions. Often the latter
would go on for an hour after the talk concluded, or until the school superin-
tendent called an end to the evening.

Primary schools required a different approach. François Minani worked
with Bill to translate a fifteen-minute talk about the gorillas and key conservation issues into Kinyarwanda. This text was then recorded by a commenta-
tor at Radio Rwanda. Combined with about forty slides, the result was an
automated audiovisual presentation that covered the essential information.
After this introduction, a French language film was shown to give students an
idea of what gorillas were really like. Then elementary school teachers could
help translate questions from the students and Bill's answers. After a while,
Minani began to accompany Bill on his tours and soon took over the primary
school presentations. In addition to his native language skills, he had an excellent manner with the students and the value of the interaction improved
dramatically.

Local primary school presentations usually took place during school
hours, since most students returned home after classes. Often, the presentations were followed by an evening program for the local population. None
ever attracted the thousands who had attended the Rwaza show, nor did Minani ever try to match the high-octane theatrics of Camille's presentation.
But these evening events were always well attended and extremely popular.
Over time, Minani learned to combine content with showmanship in his
presentations. When Bill repeated his survey of local attitudes several years
later he would find that attendance at one of these presentations was a very
important factor in changing local people's attitudes toward more positive
views of gorilla and park conservation.

M OST OF RWANDA'S secondary schools were run by the central government; several operated under the auspices of the Catholic church.
By 1980, however, American missionaries were beginning to establish a

competing network of Protestant schools. These served the country's insatiable need for more schools and attracted many sudden converts to whatever denomination controlled admissions.

In northwestern Rwanda, the Seventh-Day Adventists built an attractive new school complex high on a grassy ridge to the south of the Virungas. Bill arrived at the school in the late afternoon and was met by the American director. The man was very polite and invited Bill to join him and his family for dinner. The food—all vegetarian—was exceptionally good, and the children were exceptionally silent. After supper they were herded off by their mother, who then tended to the dishes, leaving the men to talk. The director wasted little time in asking if the planned talk would say anything about evolution. Bill answered that the subject wasn't part of his formal presentation, except to note that one value of understanding gorillas was to help us better understand our own species. Students often asked questions about the genetic relationship between gorillas and humans, however, and Bill said that he would base his answers on current evolutionary concepts.

The director replied that he did not believe in evolution. He went on to describe the strict creationist curriculum taught at his school. Bill wondered what Trog would think about that, since creationism was not part of the approved national curriculum in biology, but he kept his thoughts to himself. He tried instead to deflect the subject, saying that his central message was the need to save gorillas from extinction, regardless of their origin. The director was not deflected. He asked if Bill knew why God had created gorillas. Not waiting for an answer he said, "I'll tell you why. God made gorillas and chimpanzees similar to people so that we could have something to make us laugh." Bill stifled an impulse to respond in kind about why God created some missionaries, but instead thanked the director for dinner and went off to prepare his presentation.

In his talk, Bill made a point of emphasizing the fact that gorillas and humans can trace at least 99 percent of their genes to a common heritage and that both species share an African origin. Otherwise, he stayed on message and emphasized gorilla behavior, ecology, and conservation. The director stood silently in the rear of the large meeting room, then cut off that evening's question-and-answer session after only about ten minutes. It was the last time that the MGP was invited to his school.

Chapter Twenty-one

Food, Cameras, Action

T HE JAPANESE FILM DIRECTOR listened intently, nodding frequently as we explained the Mountain Gorilla Project. At times he interrupted to translate particular points to his two colleagues. We then asked if he could capture not just the gorillas, but the main conservation issues in his film. More discussion, but no direct answer. The director asked how much time we could spend with his team. We replied that we had many other responsibilities and that it depended on how much our interests overlapped, again stressing the key point of his film's conservation content. More discussion. The director then asked the big one: Could Bill reenact being charged by Brutus? We laughed, and the Japanese politely joined in. This relationship would take some work.

Mountain gorillas attract film crews. The gorillas are charismatic, appealing, and extremely humanlike in their behavior and sociability. Most of all, they are exceptionally approachable. This means film crews that secure permission to film in Rwanda are guaranteed high-quality footage. National Geographic led the way with two pioneering films on the gorillas and Dian Fossey in the early 1970s. For almost ten years they maintained a monopoly on the subject, until the door opened a crack for others. The MGP expanded the number of habituated gorilla families through our tourism project, which pushed the door wide open for a greater number of film crews.

O UR FIRST EXPERIENCE with filmmakers came in 1978 while we were still at Karisoke. One day, without advance notice, a stream of more than fifty porters began to file past our cabin. Dozens of duffels, crates, trunks, and large silver boxes bobbed and balanced on the porter's heads as they continued toward Dian's cabin. In the rear was a giant basket carried awkwardly by four men, all of whom could have easily stood inside their load. Soon after the basket passed our window, the half-dozen white people who had led the column returned and knocked on our door.

Filmmaker Gérard Vienne was not happy. He explained that he was a French *cinéaste* and that he had permission from ORTPN to film the gorillas. Dian looked at his official permit, handed it back, and said that she would not work with him. Instead, she sent him to our cabin, knowing that our ORTPN authorization required that we help. Vienne was exasperated by the reception and reluctant to work with the "B" team of "Fedder and Veber," whose last names he never could get right. But he liked the fact that we spoke French, thought Amy would look *chouette,* or "cool," on film, and was frankly happy that his long hike did not end in complete disaster.

The result for us was a complete loss of three weeks of research for Amy and considerable logistical backup by Bill. But it became an enjoyable experience in most ways. Gérard Vienne was an award-winning filmmaker with a global reputation whose crew consisted of his son, two daughters, a son-in-law, a Salvadoran cameraman, and a French balloonist. They were an amiable group and we spent many pleasant evenings crowded into our cabin, happy to share their endless stock of good red wine while eating separate meals. The French found our one-pot mixtures of beans and rice with various sauces *trop dégoûtantes;* we thought their cans of *pâté de foie gras* equally disgusting. We found common ground in their Camemberts and Bries. Leaving our cabin each night, the Vienne crew returned in good spirits to their campsite in the wet meadow where Dian had required that they pitch their tents.

In the field the French crew demonstrated a high degree of professionalism, and an equally high tolerance for the disorder that seemed constantly to surround them. They were especially respectful of the gorillas, about whom they were passionate. Everyone likes gorillas, but most filmmakers are primarily interested in getting their shots and moving on. *L'équipe Vienne* adapted to the gorillas' pace and took a genuine interest in their personalities and histories. They also respected our request that they not touch or otherwise seek contact with any of the gorillas in Group 5, though Pablo and Company soon made a shambles of this agreement. Every camera lens was a mirror in which each young gorilla wished to see his or her reflection. The microphone on Amy's chest was a new plaything to be carefully inspected, licked, and ultimately detached—gently, of course—by Tuck. While Amy was filmed watching a grooming session in front of her, Ziz seemed to decide that the scene would be better if he was included. So he strutted up next to Amy, displaced his gaze for a moment, then sat down with his arm draped around her shoulder. As Amy turned to look at him, Ziz stared awkwardly away, like any young guy unsure of his next move. This great theater was duly captured on nature films shown widely on both European and American TV.

Vienne's hot air balloon also made it on film, though it never left its tether. He intended to launch the balloon from Visoke's summit, then film the other volcanoes and the Akagera Park as it flew east, before landing somewhere in Akagera. The prevailing winds did not blow in that direction, however, and any northerly error in the flight path would result in a potentially dangerous landing in unstable Uganda. Instead, the balloon was raised high enough to obtain some interesting footage of the Karisoke meadow complex without releasing it from its cable anchor. It also provided a once-in-a-lifetime experience for the Karisoke staff workers who braved this limited ascent inside the big basket—and much amusement for those who watched from below. Ultimately, Vienne used his artist's license to splice footage of the Karisoke "launch" and an equally artificial Akagera landing for use in one of his films.

The Vienne crew completed a series of thirteen half-hour films on Rwanda: three on the gorillas, including one with no humans, just straight gorilla behavior; eight on the savanna wildlife of the Akagera Park; and two others on the country and its people. These films were shown throughout Europe and some were eventually adapted and translated for American and Canadian TV. More important for our interests, the films were made available at no cost to ORTPN and the MGP for unlimited conservation education uses in Rwanda. National Geographic had brought gorillas to the Western world with its films, but these films were in English and only available on loan through the American embassy. The Vienne films were in French, and in the case of the three gorilla films, we wrote the accompanying French language scripts that Amy recorded. They became staples in our education work around the park and across Rwanda. And the Amy–Ziz encounter was a huge hit among Rwandan schoolchildren.

In early 1980, the Vienne films were given their world premiere at the French Cultural Institute in Kigali. It was a major social event for the elites of the capital, rendered more so by the presence and patronage of His Excellency, Juvénal Habyarimana, president of Rwanda, and his first lady, Agathe. Three of the films were shown, with *Un gorille nommé Beethoven* saved for last. As edited by Vienne, the film ended with Amy in Ziz's embrace. After the lights came on amid thunderous applause, President Habyarimana and his wife walked forward, first to congratulate Gérard Vienne, then to take their places at the head of an official receiving line. Bill took advantage of the only opportunity he would ever have to shake the president's hand, and even exchanged a few forgettable words. Amy never made it to the receiving line, as she was quickly engulfed by a host of newfound admirers. Her crowd was larger and stayed longer than that of the president, who seemed to take the competing attraction in stride, departing with his entourage while others stayed behind. Conservation is not a field that pays very well, but there are moments when one is rewarded with bountiful praise that can be heartwarming and even embarrassing.

BETWEEN OCTOBER 1979 and June 1980, three different film crews came to take advantage of new opportunities to film gorillas created by the Mountain Gorilla Project. The Belgian Radio and Television network arrived first. Their team was much smaller than the Vienne family affair, with only a producer, a cameraman, and a sound specialist. This, we would learn, was the more standard formula for nature films. The Belgians were also more businesslike in their approach than the French. They filmed for less than a week, and since they were filming the MGP as much as the gorillas, Bill was able to continue working while they filmed.

In the end, the Belgian Radio and Television film proved one of the best at capturing the essence of the MGP. Perhaps because the Belgian viewing audience was more familiar with conditions in Rwanda, this film included an in-depth description of the socioeconomic background to the project. Surprisingly, since the tourism habituation process was in its early stages, the Belgian team also captured enough gorilla footage to satisfy popular audience demand. The only drawback to the film was that a powerful communications union in Belgium would not approve its release for use in Rwanda without payment of unacceptably high commercial royalty fees.

In March 1980, the Japanese appeared with their ORTPN permit in hand. Once Bill resolved the fact that he would not reenact being charged by Brutus—or provoke a charge by Stilgar as a suggested alternative—we agreed to work together for an indefinite period. With their camp at the edge of the Visoke parking area, the Japanese did not have to transport their goods as high or far as the Viennes had. But if they had, they would have employed almost the same small army of porters. The protective aluminum boxes that are the stock-in-trade of film crews littered their campsite. An even greater number of trunks, however, contained their impressive stocks of imported

Japanese food and drink. It was becoming apparent to us that film crews live on their stomachs as much as on their technical abilities.

The Japanese distinguished themselves by wearing white gloves whenever they entered the forest. They also liked to shoot an annoying number of takes that no other crews had required and which we soon eliminated. Among the gorillas, they were mostly respectful, but their aggressive pursuit of certain shots required regular control. They also seemed more removed from the gorillas themselves than had other crews. Since the Japanese did not want to film tourists, they purchased all six tickets for Group 11 over a period of several days. Yet they were upset when told that they, like the tourists they replaced, had to leave after roughly one hour, or whenever the gorillas ended their mid-morning siesta.

One evening, a group of elder Rwandan men appeared at our hut to ask if Bill would discuss a *maneno,* or issue, with them. Bill agreed, expecting a variation on the usual theme of money. This time, however, the men had a question that had them clearly baffled: Were the Japanese *abazungu?* This was an opportunity to turn the question around and ask what Rwandans meant by *abazungu.* Was it just a catch-all term for white people? Or did it cover all foreigners, including other black Africans? The men quickly agreed that the term did not apply to other Africans, then proceeded to debate the remaining distinctions. In the end, they decided that the Japanese, like the Chinese, were different from other *abazungu.* Interestingly, the distinguishing attributes were cultural, not racial. From a Rwandan perspective, the Chinese road construction crews in their country worked like dogs, all wore the same outfits, smoked constantly, never mixed with others, and slept packed like sardines in flimsy tin huts. This earned them the label of *abashinwa*—the Kinyarwandan form of the French *chinois,* or Chinese. Similarly, it was decided by our ad hoc gathering that the Japanese preference for strange foods, eating with "little trees" (chopsticks), drinking clear alcohol (sake), wearing white gloves in the forest, bowing frequently, and speaking a mysterious language all distinguished our new visitors as *abajapone.* Bill was left to ponder what it was about the cultural similarities among the French, Belgians, Dutch, Italians, Swedes, Russians, and Americans that made us all *abazungu* in the Rwandan scheme of things.

One morning after nearly a week of working together, the Japanese director said that his team had been invited by Dian Fossey to film the release of a baby gorilla into one of the Karisoke study groups. We knew of the planned release of the young gorilla, named Bonne Année (Happy New Year) for the day on which Jean-Pierre von der Becke had recaptured her from poachers.

We also knew that a reintroduction was an extremely risky proposition that had only been tried once before, without success. The delicate dynamics of the operation would be unnecessarily complicated by the presence of a film crew, especially one that had already proven itself to be overly aggressive. We explained the situation and repercussions fully, then asked the Japanese director not to proceed since it could threaten the survival of the young gorilla. He replied that "Dr. Fossey approved the filming" and she wanted it done before she left for a trip to the States. With that lack of respect for the gorillas, we terminated our working agreement with the Japanese. Within days, word came of a complete fiasco at Karisoke. Dian had decided to try the reintroduction in Group 5, despite the fact that it was a large, stable family that had not accepted any transfers in years. It was also a family in which every adult female already had an infant, leaving no one to "adopt" the unrelated orphan. Group 4, having lost its females in the wake of the 1978 killings, was much more likely to accept the young female. But Group 5 was closer to camp, Dian was in poor physical shape, and she was due to leave the next day. As a result, the Japanese filmed a poorly planned release in which several Group 5 members attacked the terrified young gorilla dropped in their midst, before Bonne Année was picked up by Dian and carried back to camp. A few weeks later, Jean-Pierre von der Becke oversaw a successful reintroduction to Group 4, without Dian and without the Japanese film crew.*

IN EARLY 1980 we were contacted by Barbara Jampel, who was producing a film on mountain gorillas for the National Geographic Society. We explained that there might be problems since they sponsored Dian's work, but Barbara told us that National Geographic had recently ceased funding for Karisoke. When it was clear that she was serious about filming the MGP conservation story, we agreed to collaborate.

The National Geographic team arrived in early June, hoping to catch the very beginning of the dry season: after the rains, but before the dust and haze would dull the beauty of the landscape. They were a little early, but proved to be good sports about the rain. They also adapted quickly to the Visoke parking campground, where their multicolored tents added a certain flair to the black lava surroundings. The Geographic team of two Dutchmen and Barbara was less preoccupied with eating than the French or Japanese, and managed well on mostly local fare. After dinner, however, fine single malt scotch

---

* Bonne Année died of malnutrition later that year. Her six-month tenure in the wild is the longest to date of any reintroduced mountain gorilla.

and Cuban cigars appeared for all who cared to share. Life with film crews could prove addictive.

Moving around the forest with the Geographic team posed a problem. The Geographic camera was bigger than any we had seen before and the cameraman was significantly older. And heavier. Sweating profusely and taking strategic breaks, however, he proved his mettle as Group 11 led him over some of the more rugged parts of its range in the course of their two-week stay. He was also extremely skilled at his craft. So, too, was Barbara, whose interviews not only allowed us to tell the full conservation story, but also brought out deeper feelings about the gorillas and our motivations that we rarely shared with others. The result made up the final segment of *Gorilla,* an hour-long special that received the highest Nielsen rating of any nature film ever shown on American television when it appeared in 1981. It reran throughout the 1980s and remains on American video rental shelves to this day—where our sons can see their once skinny father and pregnant mother. Unfortunately, although National Geographic made a great profit on the film, it was never translated into French or otherwise made readily available for use in Rwanda.

WORKING WITH FILM crews was a distraction, but only in the case of the Japanese was it a waste of time. Though we were never paid, the resulting films were well worth our cooperation. The French, Belgian, and American film crews were highly professional and personally engaging. They brought variety to our drab diet and lubricated our evening discussions with high-quality alcohol. The films with which we cooperated were well made and helped project a fascinating and positive image of gorillas and Rwanda. Each served to create a larger pool of people around the world who cared about gorillas and no doubt contributed to the steady rise in tourism that started in 1980. All the films carried a strong conservation message, and two highlighted the work of the Mountain Gorilla Project. In this, their message went beyond a lament for the loss of nature to show a strong example of conservation in action. We especially appreciated the extra steps taken by Gérard Vienne to make his films freely and widely available to the people of Rwanda.

Chapter Twenty-two

Moving On

B Y THE SUMMER OF 1980, the Mountain Gorilla Project was more successful than we had dared imagine. In less than a year, there was visible progress in all areas: tourism, anti-poaching, education, and political support. Saving the mountain gorilla began to seem truly possible.

As the MGP picked up speed and power, its engine was tourism. More than one thousand visitors came to see Groups 11 and 13 during the first year, doubling the total number of park visitors for the preceding year. Almost everyone who set out to see the gorillas succeeded, except for an unfortunate few who were unable to meet the physical demands of the Virunga environment. Park revenues more than quadrupled as visitors paid significantly higher fees for a guaranteed session with the gorillas. As the year progressed, the percentage of visiting foreigners rose rapidly and the first formal tour groups arrived at Kigali's sorry excuse for an international airport. Hotels, restaurants, and auto rental agencies all felt the surge in business. Nearer the park, more guides and guards were hired, and dozens more local residents found full- and part-time employment in the growing service sector.

Park security was the most immediate challenge faced by the young MGP and here, too, there was remarkable progress. The silverback of Group 13 had been shot and killed in November, but even this one mysterious death stood in stark contrast to the bloodbath of the preceding year. The Parc des Volcans guard force had tripled from fourteen to more than forty. These guards—whose salaries were supported by tourism revenues—received greatly improved training and tactical oversight from Jean-Pierre von der Becke. Their deployment in the park was further aided by the provision of appropriate field equipment—such as boots, rain gear, and tents—by the MGP and the Belgian government. This equipment helped both their performance and their morale. Tourism was even helping by driving local trappers away from those areas visited by tourists on a regular basis.

The MGP education program started less dramatically but was making steady progress by the end of its first full year. It had an institutional home and a competent government counterpart, or designated colleague, in François Minani. A local schoolteacher from Bisate named Elie Sebigoli had taken the lead in establishing a student nature club that organized trips to see the gorillas and explore the park. The belated arrival of the mobile education unit allowed an expanded program of visits so that more than twenty thousand students and villagers had heard talks by Bill and Minani before the end of the school year. The Vienne films enhanced these presentations greatly. Late in the year, Minani, Nicole Monfort, and Gilles Toussaint formed Nature et Environnement au Rwanda, a nonprofit organization dedicated to the conservation of Rwandan wildlife. The organization adapted the pied wagtail—a Rwandan symbol of good fortune—as its logo and mailed its newsletter to every secondary school teacher in the country.

The MGP's success greatly improved the political calculus for the gorillas. At the local level, discussions of the park became less negative and voices of support began to be heard. Some of the strongest voices came from families of those who had found employment as guides or guards, or in the tourism-related private sector. The relative wealth of the gorilla guides, in particular, increased their community standing and influence. The bourgmestre of Kinigi, where gorilla tourism was concentrated, was increasingly supportive of MGP activities—though his colleagues from neighboring communities that received almost no financial benefits from tourism were more restrained. At the national level, gorillas and the MGP were rising stars. ORTPN saw its own stature rise in tandem with the project's success, and the park service's emphasis began to shift from the Akagera Park to the Parc des Volcans. Higher level government support was less certain, but the sharp rise in nonresident foreign tourists—long an unattained goal of the Rwandan government—certainly brought favorable attention to the gorillas, the park, and their institutional overseers within ORTPN. Most notably, the success of the MGP put an end to the proposed cattle project in the Virungas. Internationally, the global support network grew with exposure through films and through the powerful experiences of those who made the pilgrimage to sit among mountain gorillas in the wild. Among foreign governments, only Belgium provided direct assistance to wildlife conservation in Rwanda. But others were watching. Some Swiss foresters began to wonder if they could do more for conservation in the forest reserves of Nyungwe, Gishwati, and Mukura, for which they were responsible. The director of the U.S. Agency for International Development (USAID) invited us to his house to discuss Bill's idea for a parallel project to help the people who lived around the park.

Overall, the gorilla conservation outlook was much more positive than when we began our work. Poaching was down, economic benefits were up, local attitudes were improving, and national support was strong. Gorilla numbers were uncertain, though we knew of more births than deaths. A later census would confirm that the period after the MGP's inception was the turning point in the population's recovery.

Still, there were persistent problems. Unabated population growth intensified land pressure outside the Parc des Volcans. At the 3.7 percent annual rate that this population was swelling in 1980, even the unthinkable step of clearing the entire Parc des Volcans would meet the land hunger of only one third of one year's population increase. No amount of tourism development could ever provide alternative means to meet all these people's needs. Population and land use were inextricably linked, and both issues had to be tackled for the gorillas—and the Rwandan people—to have a secure future. The economic success of the MGP might forestall clearing of the Virungas, but other natural areas were increasingly vulnerable. The World Bank had paid to clear the Bugesera woodlands in southern Rwanda for settlement earlier in the 1970s. In 1980, the Bank was preparing to support another project to convert the Mutara Hunting Reserve, adjacent to the Akagera Park, for cattle grazing. Rwanda's other montane forests, which had no tourism and no international lobby for their protection, were next. Increasingly, we saw these forests as dominoes, with the Parc des Volcans as just the last in line.

THE SUMMER OF 1980 was a time of great hope for the MGP, but it was also a time to move on to address other problems. We had lived almost four years in Africa and had spent the last seven focused entirely on gorillas. Our goal in Rwanda was to help understand the gorillas' needs and the threats against them, then take action to reduce those threats. We had accomplished our initial goal, yet found much more to be done. Gorillas, in some curious ways, were easy. The challenges of human population growth, land use, and saving forests without a charismatic superstar were much more daunting. To take on those challenges, we needed far more knowledge and experience than we had attained by the ripe age of thirty. So we decided to finish graduate school and broaden our real-world education through work in other parts of Africa, with the goal of returning as soon as possible to expand our mission in Rwanda.

We believed that personnel changes could benefit the MGP by providing new ideas and energy. The challenge was to find good people. Our temporary departure from the gorillas, the Virungas, and Rwanda was eased by the discovery of excellent replacements. Conrad and Rosalind Aveling visited us on their way home to England after working on an orangutan conservation project in Indonesia. Rosalind spoke excellent French and was interested in education; Conrad had a Ph.D. in biology, spoke Swahili, and was more inclined toward park and project management. Both were captivated by the concept of the MGP; both were attractive and articulate. After several days together—including a late night songfest that Rwandan Friends told us carried far out over the surrounding farmlands—we knew we could leave the tourism and education programs in their capable hands.

M ADISON, WISCONSIN, was the perfect base for our interlude. The University of Wisconsin offered a variety of graduate programs that met our diverse academic needs, as well as a strong conservation tradition. George Schaller had walked the corridors of Birge Hall, where Amy's office was located; Schaller's advisor, John Emlen, was still active as an emeritus professor in the Zoology Department. Bill's program in Land Resources had recently been established as one of the country's first interdisciplinary programs in environmental studies. And although it led to a doctoral degree, Land Resources was intended primarily for conservation practitioners. Bill took full advantage of the freedom to take graduate courses in biology, anthropology, resource economics, tropical geography, and African history.

When we started graduate school in the mid-1970s, we had returned from the Peace Corps with the general idea that we wanted to prepare ourselves for conservation work in Africa. When we returned to grad school to finish our Ph.D.'s in the early 1980s, we had several more years of tough practical experience behind us and dozens of more detailed questions that we needed answered. We turned every course into a personal seminar and every paper into a topic relevant to our return to Africa. We joined a group that met weekly to discuss key issues in conservation from across the tropics. Seeing a void in the curriculum, we formed a group of graduate students who designed our own interdisciplinary course combining issues in both conservation and development. We then worked with interested faculty to obtain university approval for the course and served as its first teaching assistants. It was a heady time that surpassed any previous academic learning experience.

Noah Gerhardt Weber joined us on November 19, 1980. He had spent his first five months in utero among mountain gorillas in Rwanda, where Amy succeeded in avoiding malaria and other debilitating diseases that Africa places in the path of expecting mothers and their frail charges. He emerged healthy and quickly adapted to the routine of his grad student parents, who staggered their courses and responsibilities to be with him. When necessary he went to class and, on more than one occasion, he was a well-behaved prop in courses taught by his mother. As soon as he could walk or ride a scooter, he joined the horde of children who called university family housing home. Nearly one thousand families from more than 110 countries lived in Eagle Heights, their children oblivious to the ethnic and national differences of their parents.

Meanwhile, we were reabsorbed into an American lifestyle that was almost completely unrelated to our Rwandan existence. We could drink water from the tap and skate on frozen water in winter. We could wash and dry our clothes in machines, regardless of the weather outside. Grocery stores burst with goods promising the illusion of diversity, backed by the reassurance of uniformity. A newspaper appeared on our doorstep every afternoon. Day length varied throughout the year and seasons brought changes unimaginable in Africa.

Noah was the central focus of our lives, followed by classes and course work. But two of Amy's three sisters, their husbands, and children also lived in Madison, providing us with an extended family. What free time we had we spent playing soccer, volleyball, or tennis with good friends and other graduate students. The Madison lifestyle was seductive, but we always kept one eye on Rwanda.

BILL RETURNED TO RWANDA in 1982. As a consultant to the U.S. National Park Service and USAID, he helped to produce an inventory of wildlife and protected areas across the East African region defined by Kenya, Tanzania, Uganda, Rwanda, and Burundi. In Kigali, he found the USAID director, Gene Chiavaroli, excited about an idea we had discussed before our departure in 1980. Gene wanted to know if Bill would help to design a project that addressed a comprehensive set of human needs in the region surrounding the Parc des Volcans. The original idea had been Bill's, but Gene couldn't resist the admonition that *you can't just go there and work with gorillas. This is for people.* Bill said he was still very interested and available.

The management of natural areas, especially forests, emerged as a hot issue for development agencies in the early 1980s. On the same trip, the

USAID director in Burundi asked Bill if he could help the government write a management plan for a small mountain forest in the southern part of the country. Bururi was the southernmost of an archipelago of seven forest islands that stretched north to the Virungas along the Congo–Nile Divide, a region of growing interest to us. We agreed to a contract and returned to Burundi in mid-1983. With more financial and logistical support than we had had for the entire MGP, we completed three months of biological inventories and a parallel set of socioeconomic surveys. Noah spent each day with one of us, or stayed with fellow grad student Bette Loiselle. Bette is now a distinguished professor of conservation biology, but in 1983 we awarded her our first "Vedder-Weber doctoral nanny scholarship"—paying her way to do some research in Africa in return for limited child care when we were both occupied with work. The system functioned well for all concerned, especially Noah, who spent most of his time walking with one of us in the forest or the surrounding hills or playing in the dirt with local Burundian boys. In Bururi we saw chimpanzees, serval and civet cats, and many new kinds of birds. We were also exposed to new pressures on the land from the region's Tutsi cattle herders—and we witnessed a degree of ethnic and political tension between Tutsi and Hutu that had not been apparent in Rwanda. Hutu co-workers made vague references to a series of massacres by the Tutsi ten years earlier in a neighboring province, yet both Hutu and Tutsi worked side by side on our project activities. By the end of our contract, we produced a comprehensive plan for the management of the Bururi Forest Reserve. To our satisfaction, the plan was approved by the government and implemented with support from USAID over the next several years.

From Burundi, we traveled overland north to Rwanda. Our objective was the Nyungwe Forest, which Bill had briefly explored the previous year. Nyungwe was believed to be the crown jewel of Rwandan forests. This four-hundred-square-mile forest boasted no gorillas, but it did have chimps and perhaps a dozen other kinds of primates. Its bird fauna was thought to be exceptionally rich. Yet remarkably little was known about the forest. Most of what we had learned came from a friend, Jean Pierre Vande weghe. Jean Pierre was a medical doctor based in Kigali and an avid ornithologist who constantly teased us about wasting our time in the Virungas, *when Nyungwe is so much richer and more beautiful.* In October of 1983, Jean Pierre personally showed us the wonders of his favorite forest. He was joined by Thérèse Abandibakobwa, his striking Rwandan wife, and their son, Gael. For nearly a week, Jean Pierre and Amy trekked into remote parts of Nyungwe while Bill and Thérèse took shorter hikes with Noah and Gael. After years among the northern Hutu, this offered Bill an opportunity to hear a different perspective on past and current events in Rwanda from Thérèse, a Tutsi. Thérèse's views were somewhat exceptional in that her father and uncle—both members of the deposed royal family—were captured in the 1962 Bugesera uprising and later died in prison under the Kayibanda regime. Less exceptional was the fact that her mother lived in Bujumbura, several sisters lived in Belgium, and her brothers were refugees in Uganda. They were a microcosm of the Tutsi diaspora that spread across central Africa and Europe in the wake of the Hutu revolution in Rwanda. Thérèse had complete disdain for the Habyarimana regime. If she were not married to Jean Pierre, and somewhat protected by her Belgian citizenship, she would not have returned to Rwanda. At the same time, she could joke that the members of the older generation in her family were *so convinced that they were born to rule that they were still trying to figure out how the pudgy little* abazungu *in pith helmets*—Germans and Belgians—*got the better of the Tutsi monarchy.*

Thérèse had not forgotten her roots. One evening she recalled for us some of her youthful instruction in Tutsi rituals. Her dance of the crowned crane was a sensuous re-creation of the beautiful bird's mating rituals, a flowing series of arm and head movements intended for young women to attract the attention of young men. Less alluring was Thérèse's performance of the "cow walk"—an ungainly procession in which older royal Tutsi women walked extremely slowly, with exaggerated movements of their hips and shoulders, in imitation of their beloved Ankole cattle. Thérèse was proud of her traditions, if somewhat embarrassed by her performances. She was much more embarrassed one day when we both accompanied Jean Pierre for a one-hour walk into Kamiranzovu swamp, leaving Thérèse alone in the car. Several Hutu walked by and stopped to stare at her in the car. *It looks like a woman, but it's wearing pants. It can't be a woman, it doesn't have any hips or breasts. But wait, maybe it's a Tutsi woman?* The Hutu men seized full advantage of the opportunity to mock her for the more delicate features widely believed to make Tutsi women more attractive to European men than the more buxom, fuller-bodied Hutu. This attraction was apparently a sore point for many Hutu, but Thérèse took their comments in stride, dismissing the men as *typically rude Hutu.*

The 1983 visit only raised our desire to return to Rwanda. Amy's explorations confirmed that Nyungwe was a truly special place, and she began to see herself splitting time among Nyungwe, the Virungas, and the lesser known Gishwati Forest. If the USAID project around the Virungas was approved, Bill would take that job to follow through on his earlier interests—and earn our first real salary to support our growing family. In discussions with Chuck Carr and Bill Conway of the Wildlife Conservation Society, we received initial support for our plans. George Schaller preferred that we spend the next two years repeating his 1959 gorilla and forest surveys of eastern Congo, but agreed that we could conduct some of those surveys from our base in Rwanda. George also begrudgingly accepted that Amy's annual salary could rise from $15,000 in years one and two of the project to a whopping $16,000 in year three. We had a deal. Now we just needed to finish our final courses and pick the right time to return.

In early 1984, Bill accepted a ten-week consultancy in West Africa while Amy and Noah remained in Madison. The experience gave Bill more familiarity with park and forest management issues. He was especially intrigued by agroforestry projects in the Sahel region designed to combine the need for fuel wood with the need to maintain the land's agricultural productivity. At the end of the consultancy, he returned to Rwanda to complete some surveys around the Virungas and Nyungwe for his Ph.D. He also advised USAID on the "integrated resource management project" for the Virunga region that he had helped design and was scheduled to lead sometime in early 1985. On Bill's return to Madison, we had a tight deadline and some important decisions to make. Our most weighty decision was to have a second child. After an almost four-month separation, though, we had no problem proceeding to conception. Ethan Heller Vedder was born on March 1, 1985, six weeks before Bill was due to return to Rwanda.

Viewed from the slopes of Karisimbi, the dormant volcanoes Visoke, Sabyinyo, Gahinga, and Muhavura stretch from west to east, marking the boundaries between Rwanda, Congo, and Uganda.

Sabyinyo's rugged ridges, ravines, and summits were a challenge, especially during Bill's gorilla census.

Bidele and his first wife were our nearest neighbors, living in a thatch house and farming potatoes on the edge of the Parc des Volcans.

Big Nemeye was an expert tracker who later became one of the first Mountain Gorilla Project tourist guides; shown here with Chuck and Marion Vedder, Amy's parents.

We spent our first New Year's Eve in the Virungas camping on the rim of Visoke's crater lake, surrounded by spikes of giant *Lobelia* (foreground), sunbirds, and ground orchids.

Elephant pathways create a cathedral effect through overarching bamboo.

Rural population pressure is greater in Rwanda than in any other nation in Africa, but most farmers take good care of their small family holdings.

Bill and Léopold Rwamu planting trees on Rwanda's National Arbor Day.

Students eagerly responded to Bill's talks and films as he visited schools across Rwanda.

Tutsi herdsmen with Ankole cattle during the Belgian colonial period. (Photograph by E. Everaerts)

Our younger son, Ethan, was lovingly cared for by Clementine Uwimana, who became a close family friend. She disappeared after fleeing to Congo during the genocide.

Development of the ecotourism program convinced the Rwandan government to abandon its plans for cattle-raising in the Parc des Volcans.

Ethan and Noah were welcomed everywhere.

Bagogwe herders like this man were slaughtered in a prelude to the genocide.

Fleeing Rwanda through the Nyungwe Forest, these Hutu were part of the most intense flood of refugees that the world has ever experienced. (Photograph by Rob Fimbel)

The mummified remains of Tutsi genocide victims leave an indelible memory at the Murambi memorial.

Along the Congo– Nile Divide

Chapter Twenty-three

Where Have All the Cattle Gone?

As AMY DROVE AROUND a sharp curve along the eastern edge of the Gishwati Forest, the sky ahead turned angry. Blackened clouds billowed from the hillside where the once green forest was first cut, then put to the torch. Farther on, a formerly clear mountain stream flooded the road with its thick gruel of mud and charred debris. The caravan of Land Rovers, led by Amy, picked its way through the constricted passage. From inside the vehicles, grim faces stared at the carnage.

Back in Gisenyi, along the scenic northern shore of Lake Kivu, the reluctant visitors retired to their rooms to clean off the day's collection of soot and mud, then gathered in the restaurant of the Hôtel Méridien. Some were shocked, others were resolutely upbeat as they reviewed the day's experiences. These were members of a World Bank mission, sent to evaluate the Bank's Agro-Sylvo-Pastoral Integrated Management Project in Rwanda's Gishwati Forest Reserve. Amy was belatedly added to the team as its ecologist, a few months after we moved back to Rwanda in the summer of 1985.

Working out of our new home in the nearby northern town of Ruhengeri, Amy had first set off to explore Gishwati on her own. With newborn Ethan on her back and the former gorilla tracker Rwelekana as her guide, she spent several days traveling and hiking in and around the forest. As she drove, however, she was shocked by the fires, flooding, and general chaos that characterized much of the central "management" zone. Later, she traveled to Kigali to join the rest of the World Bank team reviewing documents and interviewing government officials about the project. The "mud and fire" mission resulted from her efforts to convince the others to leave their reports and meetings to see what was really happening in the field.

THE GISHWATI INITIATIVE was supposed to be a model project. Foresters, livestock managers, anthropologists, and economists had all contributed to its design. Although no ecologists were part of the design team, the forest was correctly identified as "degraded." Centuries of traditional cattle herding by the forest-dwelling Bagogwe people, uncontrolled pit-sawing of the most valuable hardwoods, and steady encroachment by farmers along the forest fringe had all taken their toll. In response, the Bank proposed a scheme combining cattle raising, farming, and forestry in an integrated program of controlled management. In the minds of the project's designers, this approach would allow the most efficient use of the forest's various resources. But if it was so efficient, how did the Gishwati Agro-Sylvo-Pastoral Project end up being cited as one of the World Bank's biggest disasters in tropical rain forests?

Clearing the forest by fire and hoe was not part of the original project document. Yet managers in the field quickly seized on this approach as the most cost-effective means of creating improved pastures for cattle. Land-hungry local farmers would cut and burn the natural forest for the right to cultivate potatoes on the new clearings. The Rwandan passion for potatoes was again evident, not only for subsistence needs but as an increasingly valuable cash crop for export to southern Rwanda and surrounding countries. Even though local farmers benefited from this system of labor for land, they had no guaranteed tenure on the land. Eventually, the project would take their fields and convert them to cattle pastures or pine plantations. With no long-term security to reinforce the careful stewardship that most Rwandans practiced on their private holdings, the farmers mined the land to maximize short-term returns. Raised potato beds ran downhill rather than along contours and even the steepest slopes were cultivated. Farmers planted at least two crops per year on the same fields. No hedgerows, berms, or other structures broke the flow of water. With the inevitable heavy rains, the exposed soils of Gishwati—inherently poor, but enriched with recent inputs of forest nutrients—rushed down the steep slopes creating deep gullies, slumps, and massive slides. Local streams choked on their excessive loads of soil, shrubs, and entire trees, flooding roads and lower-lying fields. Without the natural forest cover to act as a giant sponge, water ran off more quickly and surrounding rivers stopped flowing during the dry season for the first time in local memory. At one point, the Sebeya River ran so low that it threatened the operation of the power plant that served the BRALIRWA brewery which supplied beer to all of northern Rwanda.

Not all of Gishwati looked like its ravaged eastern front. To the west, a network of project roads led through manicured pasturelands. Hybrid, Brown Swiss-Ankole cattle grazed on emerald fields of imported Kikuyu grass—installed at a cost of $1,000 per hectare. Lines of neatly planted exotic trees defined the large grazing blocks. Here and there, solitary *umushwati* trees— *Carapa grandiflora*—occasionally dotted the landscape in silent memory of the forest to which they gave their name—*gishwati,* place of the *umushwati.*

Two relic tracts covering no more than fifty square miles, or 20 percent of the original forest, were all that remained of the Gishwati Forest in 1985. In the north, the twenty-five-square-mile Domaine Militaire was managed by the army for its training needs and remained off-limits to all visitors, including the World Bank. In the southwest, an equal-sized block had yet to be cut, offering Amy a glimpse of the forest's former state. Gishwati's more varied vegetation supported a much greater diversity of birds than we had known in the volcanoes, though these included several species that preferred degraded or secondarized habitats. Cape buffalo persisted in this corner of the forest and Rwelekana identified the tracks of a mid-sized antelope that was possibly a yellow-backed duiker. But if Twa pygmies still lived in the forest, as some claimed, their hunting culture and economy were severely compromised by the lack of large mammals. Chimpanzees were present, but highly reclusive. Of greatest interest was Amy's confirmation of a remnant population of golden monkeys. These colorful guenons, discovered by the German colonial administrator Kandt, were known only from Rwanda and border areas of the Virunga volcanoes. Their overall status was uncertain across their range, though, and there were concerns about interbreeding with closely related blue monkeys. If the number of golden monkey pelts decorating the houses of European project staff was any indication, quick action would be needed to preserve the Gishwati population.

Mostly, Amy and Rwelekana saw domestic cattle and their Bagogwe herders. Centuries earlier, the Bagogwe had split from the central Tutsi kingdom and driven their long-horned Ankole cattle up into the mountain forest of Gishwati. There they found refuge among the northern Hutu, with whom they shared an antipathy toward the Tutsi. Over time they developed a distinctive forest-based system of cattle herding. In early colonial maps and reports, Gishwati was referred to as the Bigogwe Forest. Now, in a few short years, the World Bank project had driven the Bagogwe and their cattle from most of their former range and concentrated both at unnaturally high densities in the southwest corner of the reserve. The Bagogwe guide who accompanied Amy and Rwelekana seemed to accept his fate. Striking in his tall crocheted derby, traditional cloth skirt, and sturdy walking stick, he gave no indication that his people or cattle were suffering. But then it was not a Rwandan trait to open one's heart and mind to strangers. However sustainable their traditional practices might have been, the Bagogwe in 1985 could no longer rotate their grazing patterns throughout a much larger forest. Instead, their cattle crowded into the remaining forest, where they consumed whatever plants they could reach. Ground cover was sparse, so future tree regeneration was sacrificed to current demands for fodder. The rich Kikuyu grass planted by the project was off-limits to the Bagogwe.

Behind this wasteful loss of forest, soil, wildlife, and a way of life was the drive to convert Gishwati into Rwanda's dairy land. For the Bank, this objective could be easily rationalized. Rwanda had one of the world's lowest per capita income levels along with one of the lowest per capita animal protein consumption rates. Milk was a mainstay of the traditional Tutsi and Hutu diets. A dairy industry would generate jobs and revenues, perhaps even some regional export earnings. But Rwandan authorities made other calculations. Like the Tutsi rulers before them, the Hutu leaders of northwestern Rwanda had embraced the culture of the cow.

RWANDA WAS DIVIDED into political units called prefectures. Following the overthrow of the First Republic in 1973, political power was concentrated in the northwestern prefectures of Ruhengeri and Gisenyi. To the east of the Gishwati Forest in Gisenyi prefecture lay the Vallée des Rois. This Valley of the Kings did not refer to former Tutsi lords, but the aristocracy of the current Hutu overlords. There, away from major roads and towns, stood the huge estates of the northwest political elite. President Habyarimana, though a native of Ruhengeri, had an expansive retreat in the valley. So, too, did several members of his wife Agathe's immediate family, including her brother, the powerful *préfet* of Ruhengeri. The Catholic bishop of Gisenyi was a neighbor, as were several dozen others linked by blood, politics, and business.

It was instructive that the Vallée des Rois lay off the beaten path. From Kenya to Congo, the elites of surrounding countries were not reluctant to show off their wealth. Their almost universal preference for shiny new Mercedes-Benz cars led local wags to identify a wealthy new tribe in the region: the *wabenzi*. Rwandans, however, were much more restrained. One day, Bill was leaving the Ruhengeri branch of the Banque de Kigali with a colleague, Vincent Nyamulinda, when he asked about an elderly man who often stood outside the bank's entrance. The man looked very poor, with old traditional clothing wrapped around his body and a worn pair of sandals, yet he never seemed to beg for money from those leaving the bank. Vincent laughed. *Him! He's one of the richest men in Rwanda. He stands outside the bank because that's where his money is. He's a Tutsi, and in the old days he would have been standing out in a field watching his cattle.*

It was culturally appropriate that the Hutu would invest in their homes. The *rugo*, or household complex, was the center of Hutu life across Rwanda's thousands of dispersed hills. Maintaining the *rugo* was a matter of family pride regardless of socioeconomic status. Larger complexes were expected of wealthier families. The residents of the Vallée des Rois simply carried this tradition several steps further. In a new twist, however, the current Hutu leaders also adopted their Tutsi predecessors' fondness for cattle. Now, thanks to the World Bank project in Gishwati, they could travel a short distance from their homes to view their wealth in cattle. A few buildings within the project area were converted from their original purposes to semiprivate cabins, appropriate for meetings, family outings, or secluded trysts with a favored mistress.

The wealthy Hutu of Gisenyi and Ruhengeri were not the Bank's intended beneficiaries. The project originally aimed to improve the cattle-raising practices of the Bagogwe so that they could contribute to and benefit from the new dairy operation. But the Hutu leaders could not allow the status of the Brown Swiss cattle and fancy pastures—not to mention the increased revenue—to accrue to the Bagogwe. The Hutu regarded the Bagogwe as a barely tolerated minority in their midst. They could be driven away, squeezed into the remnant natural forest, or forced across the border into Congo for that matter. Families loyal to the Hutu leaders' clans would take their place. These families were also favored over more local groups for the right to clear and farm Gishwati's eastern slopes. Trucks and utility vehicles from various government ministries transported the potatoes to distant markets, where their sale generated high profits for the officials who misappropriated the vehicles.

In early 1986, a French veterinarian assigned to the Gishwati project wrote a report to his superiors detailing the altered land and cattle ownerships on the newly created pasturelands, as well as other abuses of project plans and resources. A few weeks later he was fired and forced to leave the country. The northwestern Hutu knew how to play hardball when it served their interests. In neighboring Congo, powerful politicians routinely stole monies intended for development projects. In Rwanda, overt corruption was still rare. This all changed in Gishwati, where the local elite hijacked the project to serve their ends, with the World Bank as a silent, if not willing, partner.

THERE WAS LITTLE AMY could do. The project should have been halted on both ecological and cultural grounds, but no one cared about the Bagogwe and there was no organized conservation presence in Gishwati. We looked back with renewed appreciation at the constellation of resources arrayed in support of the Parc des Volcans and its mountain gorillas. With no comparable global interest in Gishwati, the politicians could ignore—or expel—anyone opposed to their plans. In addition, the government agency responsible for Gishwati was the Ministry of Agriculture, not ORTPN, since the issues were ranching, farming, and forestry. MINAGRI officials showed no interest in conservation; some surely saw Gishwati as payback for their loss of the proposed Virunga cattle-raising project. You got your gorillas, now we get our cows.

Amy's final report led to a few changes, including the declaration of a fully protected nature reserve in the remaining southwest corner of Gishwati. A new World Bank project manager tried to reduce some of the worst abuses associated with clearing the eastern flank of the forest. Eventually, the Bank insisted on more fundamental changes. When those changes were not forthcoming, the Bank withdrew its funding entirely. The government then simply tapped into other funding streams to continue the Gishwati project and further enrich the local elite. The Bank may have pulled out, but the damage was done. In 1986, the Sierra Club published *Bankrolling Disasters,* citing the Gishwati project as one of the World Bank's ten worst projects in a survey of tropical rain forest destruction around the world.

Within Rwanda, Amy began to make pointed references to the fact that the country now had five forest islands—not four—thanks to the project that destroyed most of Gishwati, leaving two small remnants on its northern and southern ends. In Gishwati alone, Rwanda lost almost seventy-five square miles of forest—an area larger than the entire Parc des Volcans—representing 13 percent of the country's montane rain forest. This was a striking example of the power of the government and international agencies to quickly clear large natural areas. It reinforced our fear of further fragmentation and highlighted the urgent need to protect Rwanda's remaining forests.

With Gishwati beyond hope and the Mountain Gorilla Project working well, Amy set her sights on the Nyungwe Forest Reserve in southern Rwanda. Nyungwe's almost four-hundred-square-mile area and diverse wildlife made it the highest priority of the remaining Congo–Nile Divide forests. But another matter snapped Amy's attention back to the Virungas at the end of 1985.

Chapter Twenty-four

Another Virunga Death

THE LONG BODY OF DIAN FOSSEY stretched out beside her bed, face-up on a raffia mat that covered the cold wooden floor. It could have been a heart attack that dropped her where she fell in her nightclothes. But two machete wounds to her face and head—surprisingly bloodless—testified to a more violent end. Amy stood silently, taking in the grisly scene. Years before she had been in this cabin so many times, under circumstances good and bad. She had stood beside this same bed, talking with the very complicated and difficult woman who now lay dead at her feet. The flood of memories and feelings drowned out much of the discussion around her.

A few hours earlier, on the morning of December 27, 1985, we were in Kigali for a series of meetings. During a break, we stopped at ORTPN to make reservations for a vice president of the Wildlife Conservation Society who was coming to see the gorillas. The receptionist asked if we had heard the news. What news? *La Mademoiselle est morte.* Fossey? *Oui.* Athanase Nya-macumu, a senior ORTPN official whom we had known for years, joined us. He confirmed receipt of a radio message from the Parc des Volcans saying that Dian Fossey was found dead in her cabin, but had no further details. We knew that Dian was not in good physical condition, but because of her mental state we wondered if she might have committed suicide.

Kathleen Austen arrived at ORTPN as we were speaking. The young administrative officer was the ranking U.S. authority in Kigali at the time because the ambassador and other more senior officials were on Christmas vacation or home leave. Despite her junior standing in the foreign service, Kathleen was very capable. She probably anticipated little activity during the quiet week between Christmas and New Year, when almost all official work in Rwanda came to a halt. But now she had a major incident on her hands. A very famous American was dead on her watch and Kathleen had received a report indicating foul play. We added our own concern that it might be a suicide, especially if a gun or pills were involved. A Rwandan team on its way up the mountain to investigate the death called to ask if the embassy wished to participate. Kathleen told them to leave the body and any evidence as they found it until she could get to Karisoke. She then asked if we would help, especially with the multilanguage translations that would be required with Karisoke staff. Bill was locked into meetings, so Amy reluctantly accepted the charge. In the back of her mind she was also concerned for the gorillas: If this was a revenge killing, had they been attacked, too? Or might they still be at risk?

Amy and Kathleen arrived at the base of Visoke in late afternoon, after a brief stop at our house in Ruhengeri, where Amy changed into field clothes. They climbed quickly, then met Philippe Bertrand, a young French doctor heading in the other direction. He reported that he had certified the death and noted what he felt was a lack of professionalism among the Rwandan investigators. Amy and Kathleen continued their climb, arriving at Dian's cabin just as darkness settled over the forest. Inside, the chief inspector from Ruhengeri stood alongside Dian's dining table, interrogating the housekeeper, Kanyarogano. Kanyarogano always seemed to look guilty of something under the best of circumstances. Now he appeared absolutely terrified, as he sat alone at the end of the long table. The light of a single pressurized kerosene lamp illuminated the inspector's face, large body, and trench coat, projecting a huge shadow on the wall behind him. An assistant took notes on a portable Olivetti typewriter. Unidentified others milled around in the dim light of the main room. Within a few minutes, the inspector stopped his questioning and turned to the two white women who had just arrived. He had taken Kathleen's request to wait to mean that the American embassy intended to coordinate the investigation and asked if she wished to proceed. She assured him that he was in charge and that she was only there to observe and confirm the identity of the deceased before notifying the next of kin. With that, the inspector led Amy and Kathleen into Dian's bedroom.

The body still lay where Kanyarogano had found it more than ten hours earlier, when he let himself in to prepare her breakfast. He found the outside door was locked, as usual, but a gaping hole in the outer wall of her bedroom foretold trouble. Now inside the room, the inspector pulled back a mat to reveal the opening. It seemed raw and sinister, yet the corrugated tin was cleanly cut. The imagined sound of metal cutting metal sent chills up Amy's spine. How could Dian not have heard what was happening, or failed to react in time? She obviously made it to her feet before she was struck: a pistol lay on the floor near her hand, its loaded cartridge nearby. A machete lay a few feet away, blood and hairs stuck to the blade. Kanyarogano confirmed it was a machete that Dian kept in her room.

Inside the bedroom, everything was in disarray. Dresser drawers were pulled open, their contents spilling onto the floor. Books and papers were thrown about. The bed appeared to have been slept in, but was otherwise untouched. The incongruous sight of a full-size white bathtub against one wall jolted Amy's memories. The tub was a gift from Ambassador Crigler and his wife, Betty, while we were at Karisoke in 1978. Inspired by Dian's stated longing for a hot bath, they decided to grant her wish. Porters obligingly carried the tub up the mountain to her cabin, where Dian invited us for dinner, a viewing of the empty tub, and a good laugh. Karisoke had no running water, though, and the tub was never connected to a drain; Dian never used it. Outside, in the cabin's main room, more papers and other objects littered the floor, though several bookshelves seemed untouched. A decorated Christmas tree stood against one wall. Amy recalled how Dian loved throwing an annual Christmas party for her staff and their families, personally handing out gifts from America to every last child. It was one of her favorite events and a rare break from her reclusive existence. That year's party had been delayed for some reason, and the presents remained carefully wrapped beneath the tree.

The Rwandans seemed dismayed that the U.S. embassy had not sent its own team to investigate the crime, but soon set about their work. With rudimentary equipment, they tried to pick up fingerprints from the top of Dian's dresser. One investigator used a flash camera to take pictures of some footprints on the path outside the cabin. When it was suggested that he place a pen next to the print as a reference scale, he replied that he already knew the length of a pen. Kathleen's suggestion that someone collect the hairs she noticed still clutched in Dian's grip received a more favorable response.

That night, Amy and Kathleen were directed to share the guest house near Dian's cabin. Emotionally drained, Amy lay awake thinking of Dian. Sounds from the surrounding darkness reminded her that the killer or killers were still at large and perhaps nearby. Could it really have been a disgruntled staff member or researcher? Might it have been poachers? Were the gorillas in Groups 4 or 5 attacked? Would the military guards inside Dian's cabin wake up if Amy's cabin were attacked? Amy spent a sleepless, fitful night.

D IAN LIKED TO JOKE that if she were reincarnated, she wanted to come back as "five-foot-two, blond, and built like this," cupping her hands several inches beyond her modest breasts. She cursed like a sailor and could be very funny. She also didn't mind turning the joke on herself, especially if it had to do with her physical appearance or condition. She would have appreciated the choice of Sigourney Weaver, the actress who played her in the movie made after Dian's death. Though less striking than Weaver, early pictures of Dian show a tall young woman with an appealing face, her hair pulled back in a girlish braid. She loved birthday and Christmas parties, though dark moods often followed in stunning swiftness. Popular music from the 1940s and 1950s lifted her spirits, and Dian was ecstatic to find anyone who could sing along with her tape of Danny Kaye's "Bongo, Bongo, Bongo (We're in Love in the Congo)." It was even better if you danced.

But happiness was not a common state for Dian. Though raised in comfort in California, she had a difficult relationship with her stepfather and felt distant from her mother. This distance was reinforced by her parents' requirement that, as a child, she eat with the servants, except on Sundays, holidays, and other special occasions. This practice alone might explain her later affection for birthdays and Christmas. Dian also felt constrained by a strict religious upbringing that she resisted, reclaimed, then finally rejected during her later adult years. When we knew her, expressions of "Thank God" for this or that were often preceded by the disclaimer, "Though I spit on the Bible."

Dian left home to begin a career as a physical therapist in the early 1960s. She saved her money and even took out a loan to pursue a personal dream: a safari to East Africa. She loved the region's exotic wildlife, but wanted to see and experience more. She arranged a fateful side trip to observe gorillas in Congo. Arriving unannounced at Schaller's old cabin in Kabara, she met Alan and Joan Root, who were filming the gorillas. Dian accompanied them for several days and was captivated. Her spirits soared after her trip, and she wanted desperately to return to Africa. She later attended a lecture in the U.S. by Louis Leakey on early human evolution in Africa. She was seized by his call for more research on our closest animal relations: the chimpanzee, gorilla, and orangutan. Jane Goodall had recently taken on the cause of the chimps; Dian strode up to Leakey after his talk and informed the noted paleontologist that she would study the gorillas. Impressed by her drive and conviction, Leakey helped secure support for Dian's initial venture in Congo. There she set out "to out-Schaller Schaller," operating from his former base at Kabara. But Dian's work in Congo did not last long. Caught in the middle of the cruel civil war that was raging in 1967, she was captured by one of the many warring factions and held for several days. Possibly abused but not defeated, she fled Congo and reestablished her base camp at Karisoke, only a few miles across the border in Rwanda.

Dian's greatest source of unhappiness was a series of failed relationships with men and regrets about the family she would never have. In her own words—which tended naturally toward hyperbole—there were "two babies in jars somewhere" that resulted from her own aborted pregnancies. During her time at Karisoke, she had two serious relationships. Bob Campbell, the National Geographic cameraman who made the first two films of her work, has acknowledged his romance with Dian while at Karisoke. It was an asymmetric affair in which Campbell's reserve was seemingly mismatched with Dian's volatile passion—a passion that extended to flying barbs and bottles. Yet their relationship lasted for more than two years before the married Campbell brought it to an end in 1972.

Not long after the end of her affair with Campbell, Dian fell into another relationship with Pierre Weiss, an older French doctor who lived in Ruhengeri. Weiss had helped Dian recover from complications of her second abortion, as well as other medical problems. At some point the two become romantically involved and even announced their intention to marry. But serious problems subverted the relationship. Weiss would not live in the park, insisting instead that Dian live with him in Ruhengeri. Yet Weiss's household included a Rwandan woman named Fina, whom he described as a former mistress turned housekeeper. Dian suspected the true nature of the relationship, and Fina correctly assessed the potential threat from Dian. When Dian returned from a visit *chez* Weiss to find the windshield of her Volkswagen van smashed in with a hammer, she blamed Fina. The relationship ended in 1974. Weiss and Fina moved to France the following year. When we first arrived in early 1978, Dian asked Bill to walk arm in arm with her down the main street of Ruhengeri, "so that a certain someone will be sure to hear about us." It was a strange request that only later made sense to us.

By her own accounts, there were other men in Dian's life. Though she had an active imagination, she also had an active libido. But there would be no lasting love in her life, no partner to share her joys and sorrows, no children. She liked the Kinyarwandan name *Nyiramacibili,* which she took on in her early years. She claimed, with considerable liberty, that it meant "the woman who lives alone on the mountain." Dian's interpretation is rejected out of hand by those who know Kinyarwanda, however. The most consistent translation we have heard is "a short and feisty woman." Calling a very tall woman short would fit the Rwandan sense of humor. It is also said that Dian simply borrowed the name from Alyette de Munck, an early friend from

Gisenyi, who did fit the definition. But Dian's "lone woman" version reflected her own perception of reality.

Dian spent the vast majority of her time at Karisoke alone. Combined with other events that threatened to overwhelm her, the isolation took its toll. Accounts of her drinking in earlier years vary remarkably. But by the late 1970s alcohol became her steady companion, often in large quantities. Each week, the camp porters brought up three liters of Johnnie Walker Red and two cases of Primus, a hearty Rwandan beer. On rare occasions when others were invited to dine with her, a few of these beers might be shared. But all the scotch and most of the beer served to drown, if not salve, Dian's personal pains. On bad days she could be drunk by noon. The cumulative effect was increasingly debilitating. Dian was a chronic smoker, too, regularly consuming more than two packs of cigarettes a day. The effect on her lungs was obvious. The walk down the mountain from Karisoke that took us forty-five minutes took her a couple of hours. The return climb was a labored battle too painful to witness. Whatever else she might do to improve her health, Dian's diet did not help. Betty Crigler often asked Bill to carry large blocks of cheese and other food up to Dian when he returned to camp from Kigali. Dian, however, would feed the cheese to her dog and pet monkey, while feasting on any Fritos or other junk food that Betty might have included in her package.

A S HER HEALTH DECLINED through the 1970s, Dian's behavior became increasingly aberrant, according to those who knew and worked with her. Cattle and human incursions were a recurrent problem in the Parc des Volcans during Dian's early years. Her first responses to illegal cattle grazing near Karisoke were crudely creative. Sneaking up on one group of herders, she jumped out of the bushes wearing a grotesque Halloween mask that set the grown men to flight. In another incident, one can only wonder what local herdsmen made of the foreign words "FUCK YOU" painted on the sides of their cows when the animals were retrieved from temporary captivity at Karisoke. Later she shot at least one cow and held dozens of cattle for ransom. She kidnapped at least one herdboy. Presumed poachers received more draconian treatment. On one occasion she organized a vigilante raid to burn down the hut of a suspected poacher's family. This disturbed her Rwandan staff, who felt that homes should be inviolate, so Dian called off the raid. Yet Karisoke trackers helped capture suspects within the park, who were then brought to Dian's cabin for personal treatment. In one such incident a man was found in the forest wearing a bloodstained T-shirt. The blood could have come from a duiker or bushbuck, but this occurred not long after Digit's death, and Dian was in no mood to quibble over details of relative guilt. According to a researcher who was present, the man was bound, stripped, subjected to stinging nettles rubbed on his testicles and penis, and pistol-whipped. He was then given enough Valium to send him into a state of extreme disorientation. When he came to, Dian told him that she had taken his mind away, then given it back. Next time she would keep it. The American researcher quit Karisoke after witnessing this ordeal, telling us that he could not be a party to such torture. Other Karisoke staff have confirmed this and other incidents. But nothing comparable happened to test our tolerance for vigilante justice during our time at Karisoke.

Our tolerance for an array of lesser abuses was constantly tested, however. The research station itself was chronically mismanaged. Most of the equipment was in disrepair, leaving researchers to buy their own lamps and stoves. Base maps were redrawn by hand by each new researcher or student, leading to significant cumulative errors that could have been avoided by simple photocopying. The thin collection of books and papers euphemistically called a library was seriously out of date and accessible only during Dian's good moods. Tapes of gorilla vocalizations were not made available to us or other new researchers, resulting in confusion over the appropriate names for certain sounds. Most damaging was the lack of veterinary supplies or equipment at camp and the absence of a professional system to treat ill or injured gorillas. Mweza, Kweli, Lee, and Quince might all have been saved by straightforward medical intervention during the time that we were at Karisoke. Yet despite offers of assistance from the staff at the Ruhengeri hospital and the nearby presence of trained Rwandan vets, no program of medical intervention was devised until after Dian's death.

Karisoke was grossly understaffed. There were rarely more than three men at camp at any one time, each paid one dollar per day for a total of $90 per month. Yet Dian collected a fee of $200 per month from us (double what she had told us to budget and almost one third of our total grant for air travel, work, and living expenses) plus another $100 per month from fellow researcher David Watts. We preferred to think that our limited money did not go toward the Johnnie Walker Fund, but we were at a loss to account for its full use. No one wanted another housekeeper or woodcutter for the money: what was really needed were trackers. Amy and David monitored Groups 4 and 5 almost daily; Nunkie could have been added to the research groups if he was followed regularly; Groups 6 and Susa were often within tracking range and each had recurrent interactions with the study groups; and unknown groups and solitary males appeared from nowhere several times each year. The one tracker stationed at Karisoke could not begin to provide the most basic set of research and monitoring services, let alone do double duty on anti-poaching patrols, all of which were Dian's proclaimed priorities.

Then there was Liza. When Liza left Pablo with Group 5 and took off with another family, Dian refused to authorize the use of the camp tracker to follow the unknown group. Only after two days of unexplained refusals, and no other apparent use of the tracker, was he released to go with Bill to confirm Liza's presence in Group 6. Even more frustrating was the time that Kima, Dian's pet monkey, lost her doll somewhere on the grounds of Karisoke. The monkey usually stayed inside Dian's cabin, sharing an occasional beer with her and threatening any African who walked in. Kima liked to break free at times, though, and cavort in the surrounding *Hagenia* trees, often with her favorite stuffed monkey doll in hand. On the day when Kima lost her beloved doll, however, the tracker Vatiri was not allowed to help follow an unknown group of gorillas that passed through Group 5's range. Instead he joined the houseworker, Basira, and the woodcutter, Rukera, in a full mobilization until the doll was found not far from Dian's cabin.

Dian frequently withheld use of the single tracker from Amy and David, who became exceptional trackers by necessity and almost never lost their study groups. Amy went more than ten months in a row toward the end of our stay with almost no use of a tracker. On most days the tracking was straightforward and the opportunity to walk alone in the forest greatly appreciated. On one occasion, though, she was caught in a fight between two gorillas and limped into camp after dark. On another she risked crossing a raging torrent. Another time, she was looking over the edge of a small ravine for Group 5's trail when the ground caved in beneath her. She fell at least ten feet and landed on her back, hard enough to drive the lens of our camera into its body inside her knapsack. Once she assured herself that nothing was broken besides the camera, she realized that it would have been a long time before anyone found her if she had been seriously injured. Both David and Amy faced a variety of threatening circumstances by themselves, most notably a daily gauntlet of Cape buffalo encounters. David was also alone on that July morning when he discovered the freshly decapitated corpse of Uncle Bert and almost certainly scared off the poachers before they could finish their grisly task. The lack of trackers cost much valuable work, all of which could have contributed to Karisoke research records. It also put lives in jeopardy—those of both gorillas and researchers.

She hires us because we're too stupid to get better work anywhere else. So said the Karisoke camp staff. Stupid they were not, but their lack of French and formal education did limit their work options in a rural area with chronically high unemployment. And Dian took full advantage of their situation. Verbal abuse and accusations were common. Many of her outbursts were rendered incomprehensible by Dian's unique blend of pidgin English and gorilla Swahili. One time in Bill's presence she told Rukera, *My kichwa says you shouldn't teremuka kesho*—my head says that you shouldn't go down (off the mountain) tomorrow. The woodcutter nodded agreement, at least with the Swahili words that he understood, and left Dian's cabin. Bill later made sure that Rukera got the full translation, including the negative command hidden in the English contraction "shouldn't." Basira told David Watts that he dealt with this problem by waiting for Dian to repeat her commands until she produced a version that he could more or less understand.

Staff pay was withheld for the most minor—or imagined—transgressions. Men were terminated for little or no cause, sometimes after watching their meager earnings tossed into Dian's fireplace. Kelly Stewart witnessed the humiliating scene where a tracker was forced to get on his knees and beg for his pay. Though firings were frequent, most of the men were rehired just as capriciously. Christmas was Dian's favorite holiday, a time when she was often extravagant with her gifts to the men and their families. But she was a difficult and nasty overlord for much of the rest of the year. Her preferred term for Africans was "wogs," a borrowed colonial slur (for "worthy Oriental gentleman") that the British imported from Asia to Africa. A cheap kerosene lantern was a "wog lamp" (as opposed to the pressurized lamps that we whites imported from worthy Orientals in China). Talking to Bill about our plans for the Mountain Gorilla Project, she said we wouldn't succeed because "you don't know how to manage Africans. You have to treat them like children." Interestingly, the Rwandan camp staff's most constant refrain when complaining about Dian was *ana kichwa sawa toto,* she has a head like a child. It is a dismal part of Dian's legacy that not a single Rwandan researcher or student ever worked at Karisoke during her tenure as director from 1967 to 1981.

In the presence of Rwandan officials, Dian was generally not abusive. In private, she was completely dismissive of their power over her. She had no positive relationships of the kind built on mutual trust and respect. She assumed that all officials were corrupt, even though far fewer Rwandans were corrupt than their counterparts in surrounding countries. In the six years that we lived in the country—and another sixteen years of regular extended visits—neither of us ever paid a penny in bribes. Yet Dian assumed corruption and, by paying bribes, as was rumored in the case of Paulin Nkubili, the Ruhengeri prosecutor, she succeeded in making officials corrupt. If she couldn't corrupt, she could bully. Dian offered private visits to Karisoke and the gorillas to cement her relationships with a series of U.S. ambassadors to Rwanda. Frank and Betty Crigler were frequent visitors while we were there, and both seemed to truly appreciate the wonder of the gorillas. Their experiences also gave them bragging rights and a social standing well above the norm for an American ambassador in the Eurocentric diplomatic environment of Kigali. Frank Crigler was ready to help when needed. On certain occasions when Dian landed in serious trouble or needed to browbeat an official, she was driven to her meeting in the ambassador's official car, with the embassy chauffeur at the wheel. If that show of strength wasn't enough, Dian would insist that the meeting be conducted in English, even though she had been working in a francophone country for more than ten years. Not surprisingly, Dian usually felt that she succeeded in making her points in these one-sided encounters, while many a Rwandan official was left shaking his head in linguistic bewilderment as she was driven away.

On other issues, Dian routinely lied to ORTPN about incidents at Karisoke. If unwanted tourists appeared at one of the research groups, she would write that one of the female gorillas aborted from the stress. The refrain became a joke among ORTPN staff. Ian Redmond returned to Karisoke in 1980, after the MGP was established, only to find that Dian had not obtained the necessary authorization for him to work in the park. This wasn't Ian's fault, but the government would only allow him to visit Karisoke as a "tourist" until his authorization came through. Ian nevertheless went out on poacher patrol on Visoke's western flank one day and became involved in a dangerous altercation. When Ian surprised a group of likely poachers, one of them cut him on the wrist with a machete before fleeing. Back at Karisoke, Dian cleaned out the wound before Ian continued on to the Ruhengeri hospital for treatment. She then wrote a letter criticizing ORTPN for allowing poachers to roam freely in the park where they attacked Ian while out on a walk as a "tourist." She never mentioned that Ian was on patrol without authorization—or that he was across the border in Congo where Rwandan guards were technically not allowed to patrol. The location and circumstances were in fact similar to those surrounding Digit's death.

Ian was a favorite of Dian's. He could be critical of her actions and behavior, but he openly expressed his "eternal gratitude to Dian for saving me from a life of drudgery in England." Few other researchers shared his unswerving gratitude, at least once the pendulum of abuse swung their way. Extremely few senior scientists ever worked at Karisoke. In fact, no experienced, postdoctoral researchers ever conducted long-term fieldwork during Dian's tenure. Only five doctorates were completed during her sixteen years as director. Yet Karisoke was the world's preeminent—in many ways the only—

site for research on one of the world's most charismatic and accessible species. Still, Dian could not work alone, especially as her physical condition worsened. Her preference was to surround herself with relative nobodies: wanderers passing through Rwanda, if possible; a few of us seeking advanced degrees, if necessary; males above all. It is ironic that Dian Fossey, an icon for a generation of young girls and women, was generally dismissive of women in the field and showed a strong preference for male workers. She told both David and Amy that "women can't do this kind of work." And while she claimed to like big strong men—"real men"—she got along best with those who were most compliant with her wishes.

Dian always referred to those who worked at Karisoke as her "students." It is an odd label. Most of us doing research at the station came with independent financial support and our own set of university advisors. But we would have greatly appreciated any instruction from Dian, who was the world's unquestioned authority on gorillas. Yet Dian was extremely reluctant to discuss our work or provide suggestions. More problematic, she almost never went out to see the gorillas. During our eighteen months at Karisoke, Dian visited the gorillas no more than six times, usually with friends or important guests. She accompanied us twice on her own to see Group 5. Though she knew few of the gorillas by name and offered little advice, she seemed truly happy to be in their company. But the experience would not be repeated. Just before we left Karisoke, she accompanied a new researcher named Peter Veit so that he could take some pictures of her with the gorillas for one of her publications. There is no question that Dian spent considerable time with the gorillas in her early years. After completing her Ph.D. at Cambridge in 1974, she returned to Karisoke and maintained regular, if not steady, contact with Group 4. At some point in the mid-1970s, however, she began to spend more and more time in her cabin and less and less in the field. The simple truth is that Dian's first-person accounts of events in the field after January 1978 are virtually baseless.

It was common for Karisoke researchers to enter a "courtship period," during which Dian would tell new arrivals that they were something special. With a few, like Kelly Stewart and Amy, she went further, saying she knew she couldn't go on forever and that she wanted "to turn Karisoke over to someone like you." That this contradicted some of her statements about women did not seem to bother Dian. Yet the courtship quickly soured and turned to steadily increasing harassment the longer one stayed. Typewritten notes were Dian's preferred form of communication. The black ribbon on her compact field Royal was for basic information. She used the red ribbon when she was angry (though she said it merely saved black ink). Our first experience with red notes came when Bill was waiting for a tracker to follow an unidentified gorilla group. He waited for almost two hours before a note arrived informing him that he couldn't go out until he returned Dian's rain gear. Bill was perplexed, but responded that he didn't have her rain gear. No answer, no action, a day wasted. The next morning, Dian again demanded—in red—that he return her rain gear. Bill wrote back—in red—that "while you are very tall, I am even taller. I have two sets of my own rain gear. I don't need, or have, yours. I do need the tracker." Two hours later, the tracker was released to go with Bill, with no explanation. With time, however, Dian's actions went beyond bizarre to confrontational.

First, Amy's tracker was removed. Then our firewood ration was eliminated, then reinstated on an irregular basis. So, too, was our daily twenty-liter jerry can of warm water which we rationed to bathe our highest priority body parts, then soak our cold feet. With neither wood nor warm water, life at ten thousand feet could be quite uncomfortable. Meanwhile, Dian kept multiple fires going in her house and warmed her bed with hot water bottles. Toward the end of our stay at Karisoke, Amy let Dian know that she was done monitoring Group 5 and that Peter Veit could take over while she concentrated on vegetation sampling. The next day, Dian told David Watts to take back the plant presses that Amy would need to help with the sampling. Playing researchers against each other was a favorite pastime, and Dian was especially dismissive of those of us doing applied ecology or conservation work, as opposed to more behavioral research. Fortunately, our group, including David, Ian, and Craig Sholley, refused to play the game. We all had too much to do and respected each other's hard work and the passion behind it. But at the same time, it was impossible to take Karisoke seriously as a research station. It was an asylum, with one very famous inmate in charge.

Paranoia was a key ingredient in Dian's madness. She seemed especially troubled that someone might take over her mantle as the world's best known gorilla expert. This was absurd, given that Dian was already known to a global audience of millions who followed the work of the "gorilla woman" in articles and films. In the scientific realm, Sandy Harcourt published far more than Dian, but he remained unknown outside the small world of primatologists. Even before she published *Gorillas in the Mist,* Dian was a big draw on the international lecture circuit. Yet anyone who conducted serious research at Karisoke and showed an interest in conservation was eventually suspected of trying to take over the research station. This presumption was first applied to Sandy and Kelly, then to us. Later, Dian's active imagination lumped us all together in a combined plot to overthrow her. Ironically, she went out of her way to discuss her possible succession with both Kelly and Amy, then accused each of coveting her job.

Spying allowed Dian to release some of her negative energy. Mail was a semiprivate affair at Karisoke, delivered biweekly by porters to Dian's cabin, where it was sorted and then redistributed to the rest of us. It was not uncommon for our *Newsweek* to be delayed, then arrive on the next porter day. Sometimes it came in Dian's *Time* magazine sleeve, sometimes with cookie crumbs between the pages. Family mail always seemed to arrive unscathed. Letters from Rwandan officials or people in conservation organizations, however, encountered more trouble making it through Dian's system. Some arrived already opened. Some never arrived. Yet Dian frequently let us know that she was aware of our correspondence. Eventually, the Vimonts and Monforts agreed to receive business mail for us at their home addresses in Ruhengeri and Kigali.

One morning, we heard a noise outside our cabin. Basira, one of the two regular houseworkers, was on his hands and knees in the wet vegetation beneath our main window.

"What are you doing?"

Si kitu.

Crawling around on the wet ground isn't nothing, Basira."

Huwezi kusema kitu kwa mademoiselli?

"Of course we won't tell her."

Reassured of the obvious, Basira went on explain that Dian had been recording our conversation with her *mashini ya sauti*—her voice machine, or parabolic recording dish—the evening before. The timing made sense, in that Bill had just returned from a trip to Kigali and she might get news of some plot. Or maybe, we joked, she just wanted to record our amorous reunion. At least then she'd realize that our bed needed new springs. On leaving, however, she forgot her flashlight, which Basira was sent back to find the next morning. The flashlight was quickly retrieved near a patch of flattened grass and Basira was relieved to return to his housework. Later that afternoon, the expensive recording equipment that National Geographic had purchased for the research station was on display in her cabin. A few months later, we joined Craig and David Watts for dinner at David's. After dinner, we talked about the usual topics—gorillas, world events, Dian—as well as our pending transition to begin work on the Mountain Gorilla Project. At one point our discussion was interrupted by a loud thump on the outside of the cabin. Perhaps a rain-soaked *Hagenia* limb had fallen? Craig headed out to look and shone his flashlight on both sides of the roof before retreating from the rain. Nothing. The next morning, however, a tip from one of the trackers revealed another flattened spot behind a tree beside the cabin. Basira confirmed that Dian was again recording our discussions. He then took further delight in showing us the spot where Dian fell into the bushes as she returned to her cabin while walking under the influence.

In at least one instance, we gave Dian some cause for paranoia. Before leaving Karisoke, in mid-1979—and concerned with Dian's increasingly erratic behavior, including threats to burn down the station—we drafted a letter with Craig and David, detailing major problems with the management of Karisoke. We wrote it to the National Geographic Society, since they were the principal sponsors of Dian's work and the research station. A few weeks later, a friend told us of a confidential meeting at the U.S. embassy in which our letter was discussed, apparently so that staff could be prepared for any repercussions. We weren't surprised that affairs at Karisoke were given such attention in the diplomatic fish bowl of Kigali. However, we were very surprised that anyone knew of a letter that we never sent. After first drafting the letter, we learned that National Geographic had already decided to cut back its support for Dian, so we dropped our initiative. We don't know how Dian learned of our letter, but it would appear that on at least one occasion she was rewarded for her many wet nights of secret surveillance.

Secondhand spying was another favorite activity. In these instances, Dian would order camp staff to tell her something bad that we or someone else had done. If they couldn't come up with something convincing, she would offer to pay them. Sometimes she would even tell them what to tell her. The camp staff quickly learned and liked this game, which could generate another dollar or two toward their meager earnings. At the same time, some of them felt a bit guilty. This would lead to surreal discussions in which Basira might approach us saying, *Nili mwambia mademoiselli bongo leo.*

"Ah, and what lie did you tell her?" Bill asked.

I told her that you were playing your guitar and not working.

"But you saw that I was typing field notes all morning?"

Yes, but she told me to say that.

"Okay, thanks."

Sorry, bwana.

"*Hakuna matata,* Basira. You're just doing your job."

Much of what Dian did was laughable. Yet she also told and wrote very hurtful lies about people. Friends reported her comments to the effect that "Amy is very nice, but you know she never really works. Many days she just goes out in the fields above camp and has sex with the trackers." For Bill, the locale was changed to Kigali and diplomatic wives were his partners, but otherwise the focus stayed predictably on sex. When Dian later learned of Amy's pregnancy, she told camp staff that she felt sorry for "poor Bill, who thinks he's the father." More than one single male was subjected to completely unfounded references to his imagined homosexuality. More damning were her comments to current and potential employers in which she invented a host of personal and professional shortcomings. Dian saved her biggest whoppers for her own diary, however. There we would learn that on the day of the July massacre of Group 4, Bill did not really finish an eleven-hour census, hike around and up Visoke in the dark, meet with Dian, and go out to recover the body of Macho at dawn the next morning. According to Dian's diary, he was in Kigali neglecting his duties and pursuing loose women.

Harold Hayes documented Dian's chronic tendency to fabricate stories and distort the truth in his 1990 biography, *The Dark Romance of Dian Fossey*. But others were less discriminating in their rush to make a quick buck by publishing the twisted fantasies that she conjured up in the dark and lonely recesses of her alcoholic existence. Unfortunately, some of these malicious tales were the first accounts read by our families and friends. They were some of the last stories that Bill's parents read before his mother died of cancer and his father died of a heart attack two weeks later, in June 1988.

There is no doubt that the death of Mweza was a watershed event in our relationship with Dian. Though badly injured and maltreated during his captivity in Congo, Mweza could have survived. Dian's distrust of Alain Monfort's information delayed treatment for weeks; her refusal to permit any outside medical care allowed Mweza's condition to worsen; and her own drunken intervention contributed directly to the young gorilla's death. In the predawn hours after Mweza died, we sat in our cabin and debated leaving. How could we stay and work with Dian after what just happened? We remembered that friends had warned us not to go to Karisoke in the first place because Dian had a reputation for driving researchers out before they could finish their work. So we had decided before we left Madison that we would stay for our full eighteen months, no matter what happened, no matter what she did or said, because understanding and conserving the gorillas was too important. We couldn't have imagined an event like Mweza's death. But in the end, it strengthened our resolve. At dawn, Bill walked off the mountain to report what had happened to Ambassador Crigler and to seek the advice Crigler had offered when we first arrived. The ambassador was understanding if not immediately helpful, though later he would intervene more directly. With Bill in Kigali, Amy stayed at Karisoke, finding solace in Group 5's company during the day. At night, she locked the cabin doors and seriously wondered if a bullet might shatter the loosely shuttered windows.

We remained at Karisoke for more than a year after this event, but our relationship with Dian was strained, at best. Bill spoke with Dian about work-related matters. He reported suspected poaching activity, monitored gorilla movements away from Karisoke, and tried to bring her into discussions of the MGP, with mixed success. He avoided any personal or social contact, however. Amy took a different tack. A few days after Mweza's death, she went to talk to Dian and said she held her personally responsible for the death. She also told Dian that she was the only one who could mobilize the support needed to save the gorillas, but that she was an alcoholic who needed first to cure her own illness. This seemed to open a crack to Dian's well-protected inner world. Dian recounted her failed relationships, the babies she didn't carry to term, and a host of other sorrows. She said she knew she was an alcoholic, but she also said that she was dying of cancer and didn't have long to live. This revelation put matters in a new perspective—at least until Amy's next discussion when Dian retracted her cancer story.

Still, Amy tried to break through. Besides her tough-love message about her drinking, smoking, and generally destructive lifestyle, she tried to bring Dian out of her isolated existence. Holidays were one opportunity to provide a surprise, and perhaps a few laughs. The week before Halloween, Amy left the mountain to visit the big Saturday market in Ruhengeri. There she found a suitably round gourd and brought it back to camp. After testing her techniques on a second gourd, she proceeded to carve an animated face. On Halloween, Bill was away so Amy invited Ian along to watch Dian's reaction. Approaching her cabin in the dark, Amy placed the candle-illuminated jack-o'-lantern on Dian's doorstep, knocked on the door, then retreated to some nearby bushes. No reaction. Another knock elicited an odd, singsong "Who is it?" As Amy wondered if she should knock again, the door flew open and Dian stood there waving a pistol at the air. Looking down at her feet, she lowered the gun, then began to giggle as she lifted the ersatz pumpkin and carried it into her cabin. When Dian closed the door, Amy let out her breath and she and Ian returned to their cabins—thankful they had not been standing at the door when Dian opened it.

The following Easter, Amy repeated the experience with brightly painted eggs in a nest of *Galium,* or bedstraw. This time she didn't stick around to see the response. In the morning, however, she learned that Dian had stopped by a new student's cabin and thanked him profusely for his thoughtfulness. Beyond these efforts to lighten her spirit, Amy tried to provide some degree of social contact whenever Dian seemed open to discussion. Those times were few and far between, however, as Dian retreated more and more into the life of a recluse on her own private mountaintop. On Christmas Day 1978, Amy left at daybreak to spend some time with Group 5. The gorillas were outside their normal range, less than an hour from camp. As the morning progressed they moved closer and closer, until Amy knew they were less than ten easy minutes from Dian's kitchen door. Leaving the family, she hurried to tell Dian that the gorillas had "brought her a Christmas present and were waiting only two hundred yards away." Dian looked up the gentle slope where Amy had indicated, but said, "No, I can't." There were barriers that no one else could see.

SOON AFTER DAWN the day after Dian's murder, Amy went to wake Kathleen, who confirmed that she, too, had passed a mostly sleepless night. Amy first sought to obtain permission from the chief investigator to organize a search for Groups 4 and 5. Surprisingly, he allowed her to use the resident camp staff, although he noted pointedly that all were still suspects. The one foreign researcher at camp was rather new and spoke no Swahili, Kinyarwanda, or French, so Amy worked with him to send out separate patrols to both Group 4 and Group 5. Each patrol counted nests, while the researcher tried to identify visually as many individuals as possible in his study group.

Staying behind to help if needed, Amy walked around the field station. Everything looked pretty much the same as when we'd lived there six years earlier: a half-dozen cabins of corrugated tin, dwarfed by towering *Hagenia* trees; narrow paths of stone leading through the seeping marsh; the open meadow across the stream. Karisoke blended in like a natural feature in the Karisimbi–Visoke saddle. This was Dian's creation and part of her legacy. The fire pit still marked the site of her first encampment. Did she wonder if she would stay on those first cold, wet nights? If she could really be accepted by the gorillas? Could she transform herself into a researcher? Whatever doubts Dian might have had, she overcame them in the following years. It would be easy to say that if Dian hadn't come to the Virungas, someone else would have taken her place. But we can never know that.

Long-term monitoring of the gorillas was Dian's greatest scientific contribution. George Schaller had provided a remarkably accurate portrait of gorilla social organization and basic ecology, but his eighteen-month study provided a detailed snapshot; Dian produced a full-length movie over more than a dozen years. Characters came and went. Family dynasties formed and just as quickly dissolved. Youngsters grew up and moved out: females to other families, males to an uncertain period of bachelorhood. A few stayed to breed within their own bloodlines. Matrilines emerged as a powerfully cohesive force in gorilla social structures. Dominant females demonstrated a capacity for leadership in many situations. Patterns of succession between silverbacks slowly appeared as more cases unfolded with the passing years. Birth and death rates and other demographic trends could be calculated from the expanding set of baseline data. Despite the comings and goings of individuals, and group conflicts punctuated by spectacular displays, the gorillas earned the description of gentle giants over the years. Unlike chimpanzees, which revealed a violent side, mountain gorillas generally kept the peace—though several infanticides were documented over the years. Even studies of gorilla ecology—which Dian dismissed as boring—benefited completely from Dian's decision not to feed the gorillas, unlike Jane Goodall who had at first provisioned chimps at Gombe Stream. The result was uncontaminated feeding behavior and a reliable information base on gorilla habitat and food requirements for both scientific and management purposes. As we enter the twenty-first century, dedicated researchers continue to monitor some of Dian's original study gorillas and a host of their offspring. All that we know about this population was made possible by her single-minded dedication to establishing and maintaining a long-term study of the mountain gorilla.

Dian went to Africa to study gorillas, not to save them. But as she encountered problems, she was just as determined to fight them. In her early years, with almost no resources besides her own personality, she did battle with illegal cattle herders who had overrun the park. She didn't succeed, but she brought the issue to a head and her idiosyncratic ransoms were later formalized by Alain Monfort into an effective system of official fines. She couldn't stop the massive land conversion that occurred two years after her arrival, but her mere presence perhaps kept more of the park from being cleared. Nor could she stop the poaching, even when it wiped out Digit, Uncle Bert, and the heart of her beloved Group 4. It is even possible that Dian's early reaction to poaching and her personal role in meting out punishments helped to fuel the range war that broke out in 1978. It is impossible to know. What seems clear is that Dian's passion and tunnel vision prevented her from seeing new opportunities for expanded support and partnerships when conditions improved. As newcomers, we could see these opportunities to do more for the park and the mountain gorillas we all wanted to save. Dian could see only enemies whom she had sworn to battle, no matter how small the field or outdated the issue. And fight she did, until the battle moved on and left her behind.

Entering Dian's cabin, Amy studied the gallery of black-and-white photos that lined one wall. Some were ghosts from the past, many from Group 4. Most were still alive and roaming nearby. Dian loved the gorillas as individuals, perhaps more than as a population. She gave them names and shared their personalities and life stories with a worldwide audience that couldn't hear enough about Uncle Bert and Flossie and Simba and Tiger. Digit was as close to a household name as any wild animal would ever be. To the frame of facts from field observations, Dian would add layers of interpretation and embellishment about the gorillas from the fertile depths of her own imagination. While most scientists shy away from projecting thoughts and emotions onto animal subjects, Dian embraced anthropomorphism wholeheartedly. This personalization of the gorillas was central to perhaps her greatest success: the creation of a global constituency of millions of people who shared her passion for mountain gorillas. Many lived vicariously through her exotic life, as they perceived it. Arriving on the world stage at the time of the burgeoning environmental and women's movements of the early 1970s, Dian inspired an entire generation of girls and young women who wanted to grow up to be like her. If her personal life was messier, her personality more prickly, her research less rigorous, and her attitudes toward Africans more controversial than her admirers could ever know, so be it. There are many famous men in this world whose personal behavior and beliefs also leave much to be desired.

Throughout the cabin, antelope skin rugs and chairs and ivory statues testified to a more ambivalent view of species other than gorillas. Dian dismissed them as gifts from a friend in her early years and, in fact, such trophy furniture was not unusual among expatriates in Africa. But skins and ivory were much more common within the hunting wing of conservation than among the animal protectionists with whom Dian was more closely allied. Besides this odd impression, though, Dian's furnishings highlighted another important fact about her life: she lived a frugal existence. Even after the commercial success of her autobiography, *Gorillas in the Mist,* she had no luxuries in her cabin and few in her life. She owned two Volkswagen vans, but neither was new. Off the mountain she wore tasteful, yet unassuming clothes. Around camp, she lived in flannel and denim. Her luxuries consisted of a fireplace, two wood-burning stoves, and the water bottles that preheated her bed at night.

Dian's greatest wealth was her fame. And it may have been her greatest curse. Her happiest days, by all accounts, were during her early years at Karisoke. Free of her family, free of conventions about what she should be and how she should act, free to follow gorillas wherever they might lead and to learn from their example. Others could follow her, but only through the films and articles that packaged her for public consumption. Yet ultimately, this attention raised her profile and again exposed her to the outside world.

Critics questioned her lack of scientific training. This stung Dian, but also raised her hackles and defenses. Far more difficult for Dian, in our opinion, was the overwhelming and uncritical acclaim heaped on her by an adoring public. For someone whose outward bravado appeared to mask an underlying feeling of profound insecurity, this was a heavy burden. Her great success in understanding the gorillas seemed undermined by a seriously flawed self-image that didn't allow her to appreciate her own accomplishments. And she had seen too many gorillas die to feel comfortable in the mantle of "gorilla savior." Yet whatever her own personal discomfort with her public persona, audiences cheered and her fans and friends exhorted her to continue.

Founders are often poor managers. Special skills and attributes are required to pursue a dream, to create something where nothing existed before. Persistence, determination, and a single-minded focus on final goals are the keys to overcoming endless hurdles and pitfalls along the way. Yet once the goal is in sight, a very different set of attributes is often required to consolidate and build on that achievement. Many pioneers and founders lack the ability to make this transition, or even to see that it is necessary. Dian Fossey was the ultimate founder and a failed builder.

In 1981, Frank Crigler succeeded in convincing Dian that it would be best if she left Rwanda. He couched the move in terms of her need to finish her book, but he also knew that it was best for Karisoke if she moved on. For the next two years, she maintained a base at Cornell University while writing and lecturing on campus and around the country. Sandy Harcourt and Kelly Stewart took her place at Karisoke, followed by Richard Barnes in 1983. With publication of *Gorillas in the Mist,* however, Dian concluded a book tour and then set off straight for Rwanda. Appearing at Karisoke, she announced that she was again in charge, leading Richard to resign in protest. The Rwandans took this as a sign of her superior standing and issued Dian her first ever two-year work permit. With little to do, she quickly returned to the life of a recluse. She had scant interest in research, and five years after the start of the Mountain Gorilla Project she was sadly irrelevant to conservation efforts on the ground. Which only brought everyone back to the question: Why would anyone kill Dian Fossey?

THE PATROLS RETURNED by late morning with an accounting for all gorillas in both groups. Amy was greatly relieved to hear the news. Jean-Pierre von der Becke had radioed to say that he was coming up later that day to help coordinate patrols, allowing Amy to return to her home and family. One detail remained, however. Dian was now thirty hours dead and Kathleen had to have her body carried off the mountain. With no coffin on hand, two sleeping bags were retrieved from storage. Lifting Dian's body so that she could slide into the bag, Amy imagined Dian suddenly jumping up and laughing at all the fuss over her demise. Amy quickly slipped the second sleeping bag over Dian's head. It was a long walk to the base of Visoke, accompanying the body and the unhappy porters who carried the awkward load.

EVEN IN DEATH, Dian's life was complicated. Once notified of her death by the embassy, Dian's stepfather challenged her written request to be buried in the gorilla graveyard behind her cabin. After an exchange of communications regarding the cost of preserving her body and transporting her coffin to the United States, he ultimately acceded to her wishes. This was fortunate, because the manager of the Primus brewery warehouse in Kigali was growing frantic about the famous corpse he was secretly keeping cool on ice. For her part, Dian would have loved the irony of her temporary quarters.

Arrangements were hastily made for Dian's funeral, leaving some who might have come unable to attend. Amy felt that she should pay her final respects and represent the cadre of researchers who had known Dian through good and bad times. On the morning of December 31, she joined the small funeral procession at the base of Visoke and followed the simple casket as it made its final journey up the familiar Camp Trail. Outside Dian's cabin, the mourners were joined by the Rwandan staff of Karisoke. Amy hugged several of the men. Then, under a clear sky, the group of twenty or thirty moved to the makeshift graveyard. The Reverend Elton Wallace, a Seventh-Day Adventist preacher recruited for the occasion, presided over the ceremony. Amy thought his script of generic homilies oddly removed from the dynamic life and dramatic death of the woman he was talking about. She wondered what Dian would have thought of an evangelist having the last word over her body and soul. She wondered, too, about the thoughts of Philippe Bertrand, the young French doctor who stood silently apart from the others. He had recently befriended Dian; Amy hoped he had brought some joy to her final days. Farther to her right around the circle stood the person who had known Dian the longest. Ros Carr had already lived more than twenty years in Rwanda—most of that time as a single woman—when Dian first arrived in the Virungas. She, too, was American: an expatriate who fell in love with the land and eventually settled on the southern slopes of Mt. Karisimbi, where she raised flowers for export. Now seventy-four, Ros had witnessed the entire span of Dian's experience in Rwanda: the joys and crises, the loves and losses. Though Dian's behavior would often frustrate the older woman, and their friendship was periodically strained, Ros was always there to offer Dian a refuge at her Karisimbi plantation. She later wrote a very touching memoir of her life in Rwanda, *Land of a Thousand Hills,* with a short but respectful section on Dian.

After a final prayer by Reverend Wallace, Philippe Bertrand stepped forward to place a single red rose on Dian's plain wooden casket. Ros Carr followed with a beautiful assortment of cut flowers from her plantation. Amy then tossed a bouquet that she had gathered of the gorillas' favorite foods—wild celery, *Galium,* blackberry, and thistles—as the casket was lowered into the dark volcanic soil. Eight years to the day after Digit's death, Dian Fossey was laid to rest by Digit's side. She was at home among the mountain gorillas who she liked to claim were her only true friends and family.

Chapter Twenty-five

Rwelekana

WITHIN DAYS OF DIAN'S DEATH, journalists and would-be biographers began to swarm over Rwanda. None had any close familiarity with the country, the person, or the issues. Most seemed eager to turn up quotes or facts to make the case that Dian was either saint or sinner. For us, talking was difficult. The real Dian Fossey was a most complicated individual: tormented by an unknowable mix of experiences and fears, sadly insecure despite her great achievements, possibly burdened by the heavy mantle of fame, an alcoholic with signs of mental illness. Many of her comments and actions were personally disturbing, but then sick people aren't entirely accountable for their behavior. Dian was a flawed human being and a flawed role model for conservation. Hundreds of gorillas died and much parkland was lost during her tenure. But she was an extremely important figure for long-term research, an inspiration for millions, an effective public advocate, and a fascinating character. She was an original of a kind we will never see again in our lives. We learned the hard way, however, that any effort to provide visiting journalists with depth and perspective on Dian's life was lost in the never-ending search for juicy quotes.

The big questions in Dian's death were who and why. The answer to either would clarify the other, yet each defied easy explanation. The most popular theory was that Dian died because she was defending the gorillas, with the corollary suggestion that poachers or their powerful backers must have killed her. However, this scenario requires almost complete ignorance of the situation in Rwanda at the time of her death. Up to the death of Digit and the July massacre in 1978, Dian was the undisputed defender of the gorillas' interests. Whatever one thinks of her methods, she was actively involved in their defense. Poaching reached its all-time peak during this period of her most active opposition, however. With the creation of the Mountain Gorilla Project in 1979 and her decision not to participate, the lead role in gorilla protection passed to the MGP consortium and the government. With her departure to Cornell in 1981, even the task of patrolling the Karisoke region was taken over by official park guards, though they often worked with Karisoke trackers. By the time of her death in late 1985, gorilla numbers were rising dramatically, poaching had sharply declined, local residents were seeing the direct benefits of tourism, and government support for the park was at an all-time high. Yet Karisoke played no role in this evolution, even after Dian's return in 1983. Instead, the focus of gorilla and park protection had shifted to three new centers along the park boundary where Jean-Pierre von der Becke, Mark Condiotti, and a succession of MGP tourism advisors who came after us—the Avelings, Roger Wilson, and Jeff Towner—lived and worked. If anyone wanted to undermine the gorillas' protection, these were the people most responsible for the MGP's success in 1985, and each of their unguarded residences offered an easy point of attack. Dian's cabin was far less exposed, almost always surrounded by other Karisoke staff, and she was personally irrelevant to on-the-ground conservation action at that time.

It is still possible that poachers, angry over past treatment at Dian's hands, were responsible for her death. Her rough beatings and personalized form of justice could provide a ready motive. Several prime candidates from the 1970s, though, were dead or in prison at the time of her death; the suspected poacher she subjected to a mock hanging in her cabin two years earlier had a solid alibi.

It is possible that a Karisoke staff person killed Dian Fossey. If true, she would not be the first expatriate killed by her employees. Joy Adamson of *Born Free* fame met her fate at the hands of disgruntled workers who were apparently tired of abusive treatment on her ranch in Kenya. Dian, too, could grossly mistreat her Rwandan staff. The workers themselves would joke that *Mademoiselli would have been killed long ago in Congo* because of her capricious treatment of staff. Yet over the years, no one ever took advantage of the ample opportunities to settle scores when she was on her worst behavior. Why kill her in 1985, when she was a much subdued version of her former self? In fact, her staff remained extremely loyal and almost all had worked for Dian for years. They even seemed to have a certain respect for her mental instability, in a way that appears more common among African societies than within our own. Over time, several of her workers developed a sense of protectiveness as her powers and personal condition declined. It is extremely difficult to imagine that any of Dian's employees was responsible for her death.

Yet dead she was and the U.S. embassy quickly applied pressure on the Rwandan government to determine who was responsible for the death of the country's most famous American resident. The presence of journalists and writers added to the pressure on both the embassy and the government, as neither was accustomed to the spotlight of international attention. The flawed investigation at Karisoke revealed no useful information. Even key items like the hairs clutched in Dian's hand reportedly produced contradictory results from labs in France and the U.S. Still, the U.S. embassy wanted action. Every week, it requested an official progress report. After several months, the government acted. The Rwandan Ministry of Justice informed the embassy that the sole American researcher at Karisoke would be charged with murder. But he was not arrested. Nor was any action taken the following week as he packed his belongings, openly requested porters to carry his bags off the mountain, stayed at a friend's house for several days, and then passed through immigration control at the main Kigali airport to catch an Air France flight out of Rwanda. When that flight was canceled, he repeated the process the following day before leaving Rwanda and its legal reach forever. The government clearly didn't relish the publicity of a trial. The easy way out was to allow him to leave, then announce his guilt and sentence him to death in absentia. The government's treatment of the American's purported accomplice was another story, however.

TRACKING GORILLAS IS EASY when a large family has plowed through a hillside of thick herbaceous vegetation. It is more difficult to follow smaller groups or single individuals, and any tracking is infinitely more complicated in the bamboo zone or in high subalpine meadows. Emanuel Rwelekana was an exceptional tracker. He could follow a solitary silverback for miles through the bamboo. He could read the most subtle signs that a group had turned back on its own trail in the double bend of a single celery stalk. He could detect a distant feeding site in the bleeding white sap of a broken *Lobelia* across a hundred yards of open mountain moorland. He enjoyed the thrill of tracking poachers and raiding their camps with nothing but his ever-present curved *umuhoro* in hand, and a big smile on his face. The trackers were the fighter pilots of Karisoke and he was the top gun.

Rwelekana's confidence extended to his curiosity. On the trail or around the census campfire, he always asked questions about other places, people, and ways of life. When he stopped laughing about the idea of being served hot coffee or a sandwich from coin-operated food vending machines, he seriously wanted to know how they worked. The Apollo space program and moon landings fascinated him. If gravity was a concept that we had difficulty explaining in Swahili, he would patiently wait until we could find a dictionary to learn new terms. He wanted to know why American women were so different from Rwandan women. In a culture with one of the world's highest recorded birth rates, he saw the merits of family planning. More than any other tracker or gorilla guide with whom we've worked, Rwelekana's curiosity extended to the gorillas themselves. Karisoke policy under Dian did not allow the trackers to see the gorillas for fear that such contact would make the gorillas less wary of poachers, but Rwelekana could identify every individual in Group 4 or 5 from their nests. Walking back to camp at the end of the day, he wanted to know the day's news of the family and its members. On census work with Bill, he liked to discuss why the various groups reacted differently to our presence. One day toward the end of our stay at Karisoke, Amy asked him to join her in sitting with Group 5. He never stopped smiling as he finally was able to place faces with names known only from nests and piles of dung.

Rwelekana was a very hard worker, but not a slave to wage employment. He was a farmer and family man who would turn down work opportunities at Karisoke or on the census if they conflicted with critical harvesting periods. At the same time, he used his income from camp and census work to send all of his children to school and otherwise help his family. He borrowed money from several of us who worked at Karisoke to buy land or enhance his holdings and always repaid his loans on time. While others might spend their meager earnings on high-status commercial cigarettes or bottled Primus beer, Rwelekana always smoked rolled-up leaves of tobacco from his own farm and drank local sorghum beer, if anything alcoholic at all.

Rwelekana was one of the first to help Dian establish her camp at Karisoke. He can be seen with her in early National Geographic magazine photos, tending the fire or playing a bamboo flute. He was also probably fired by Dian more than anyone but Big Nemeye. This was partly the result of his long tenure, partly the result of his pride. Where others might apologize and grovel, Rwelekana would not back down over an issue when he felt he was right. And so he would be dismissed. Not long afterward, he would be rehired. While Sandy Harcourt and Richard Barnes directed Karisoke, Rwelekana's employment was steady. Then he went back on the employment roller coaster after Dian's return in 1983. In early 1985, he was dismissed yet again, but David Watts hired him as a personal tracker for his postgraduate work at Karisoke. When we moved back to Rwanda later that year, Amy was pleased to learn that Rwelekana was available to work part-time with her in Gishwati, Nyungwe, and the next Virunga census.

Beyond work, we were both happy to have Rwelekana again as a friend. Between stints with David or on trips to Ruhengeri, he often stopped at our house to visit. Both our boys were born in the U.S., so this was the first he had seen them. He was amused by five-year-old Noah's height—he towered over Rwandan children almost twice his age—and Ethan's carefree habit of careening around our house on a wheeled baby scooter before he could walk. Rwelekana remained as curious as ever about the world at large and brought us closer to life in Rwanda. But mostly he was a good solid person and the best friend we would ever have within the closed society of rural northern Hutu.

WITHIN WEEKS OF DIAN'S MURDER, all of the men working at Karisoke at the time of her death were arrested, along with several others who were not present on that night. Rwelekana was among the latter group. Most were released after a few days. But Rwelekana and several other trackers were held for several months of captivity and interrogation in the miserable confines of the Ruhengeri prison. After learning of their arrest, and doubting their guilt, Amy visited the prosecutor's office to express concern about the men's health. She also asked specifically about Rwelekana, since he had been working directly with her. We both thought it might help if the authorities knew that at least some foreigners were paying attention to their treatment. Amy was immediately put on the defensive and asked why an *umuzungu* would care about a bunch of unimportant *abanyarwanda* men. The discussion was neither pleasant nor encouraging. Several days later, Léopold Rwamu, one of Bill's assistants, stood in the same office getting a document processed for one of his project vehicles. Afterward, Léopold came straight to the house, where he nervously recounted a discussion he had overheard. He knew Rwelekana from visits to our house and had recognized his voice from behind the closed door. Staying as long as possible, Léopold was shocked when the interrogation turned to Amy. *Didn't she dislike Dian? Didn't she hire you?* Léopold said that Rwelekana sounded *ogopa sana*—very frightened. Amy began to worry that her visit had brought undue attention to all of the men, but especially Rwelekana, at a time when the Rwandan authorities were increasingly frustrated by their lack of success and the steady pressure from the American embassy to produce results. We bit our tongues and refrained from further direct contact. Instead, we asked if the embassy could monitor the treatment and condition of the men. Some effort was made in this regard by a helpful junior consular officer, but the official U.S. emphasis remained on the identification of a guilty party.

After seven months, everyone was released except Rwelekana. He and the American researcher at Karisoke were officially charged with Dian's murder, but the American had already fled the country. It was an absurd connection. The researcher was new to Karisoke and spoke no Swahili or Kinyarwanda, not even French. With Rwelekana's complete lack of English, it is impossible to imagine how the two might have concocted a lethal plot in mutually incomprehensible languages. More to the point, Rwelekana had no motive. His loss of a Karisoke post had landed him higher paid employment with David and Amy. He thoroughly enjoyed the opportunity to explore Rwanda's other forests. Finally, anyone who knew Dian as well as Rwelekana would know that she was never far from her guns, even in bed. Cutting though her metal bedroom wall was a sure way to get shot, and Rwelekana was far from stupid.

A few days before Dian's murder, Rwelekana seemed completely content with life. Early in the investigation, however, another Karisoke worker had recounted a tale from the distant past that gave the authorities a hook. In this retelling, by someone who may have borne a serious grudge, Dian was angry with Rwelekana about some forgotten incident. This was nothing new. But before firing him, Dian held a loaded pistol to his head and said she could kill him anytime. Rwelekana agreed, according to the informant, adding that if she didn't, he could kill her, too. Not long afterward, all was forgotten and he was rehired. It was more than a decade after that event and dozens more firings and rehirings before someone finally did kill Dian Fossey. We're certain that it wasn't Rwelekana.

A few months later, it was announced that Rwelekana had hanged himself in his cell. In fact, we later learned that he had been dead for many weeks, killed by his captors. We could imagine him stubbornly refusing to confess to something that he didn't do, despite the beatings of his interrogators. In an odd way, we were relieved to know that he was no longer suffering. But our guts were kicked in by the news. We couldn't escape self-doubts and second-guessing about our own actions taken and not taken in his defense. The reality of brutality in Rwanda—long a subject of cocktail party rumors—hit very close to home. So, too, did the realization that pressure from our own embassy had pushed the Rwandans to solve the problem in their own manner. A foreigner had been forced to flee, leaving his guilty verdict a facile formality. An innocent but "expendable" Rwandan had been framed, killed, and delivered. Case closed—though never solved.

Chapter Twenty-six

Nyungwe

T HE NYUNGWE FOREST is an exceptional refuge. In 1986, its rugged mountains, vast forest, diverse wildlife, and deep silence provided a personal retreat for Amy from the tragically linked murders of Dian and Rwelekana. While we maintained our home in Ruhengeri and Bill continued to work around the Parc des Volcans, Amy's project required that she spend much more of her time in the southern forest of Nyungwe. There she was free to explore a remarkable new world, where the *Hagenia* and *Hypericum* of the Virungas were dwarfed in size and number by more than 250 other species of trees and shrubs; a forest where no gorilla had ever roamed, but which was bursting with chimpanzees and a dozen other kinds of primates. Its health offered hope that the fragmentation and destruction of Gishwati need not be the fate of every Rwandan forest. Free of the dark clouds and conflicts that often hovered over Karisoke, and bolstered by her experiences with the Mountain Gorilla Project, Amy now faced a new set of challenges and opportunities in Nyungwe.

Throughout its history, Nyungwe has served as a refuge for the plants and animals of the Congo Basin. In parallel with past Ice Ages in the Northern Hemisphere, central Africa has experienced recurrent periods of drier weather, the most recent ending only twelve thousand years ago. During these extended cycles, the lowland rain forest retreated to a few core refuges where more humid conditions persisted. Key lowland refuges included the Niger delta region and the confluence of the Congo and Ubangui Rivers. The primary highland refuge covered the mountains encircling the Albertine Rift in what is now eastern Congo, Uganda, and Rwanda. During the most recent Pleistocene climate change, most animals and many plants from eastern Congo were forced upward by the retreating forest, into the high mountains that include today's Nyungwe Forest. There, many species survived the ecological diaspora and later recolonized the surrounding lowland forest when favorable conditions returned. Some could not tolerate high-altitude existence and succumbed to extinction. But others thrived, ultimately evolving into new species adapted to mountain life. The granitic bedrock of the uplifted complex added a degree of stability unknown in the more dynamic volcanoes to the north. The legacy of constant rain forest conditions combined with recurrent influxes of new species can be seen in a series of montane forest islands, of which Nyungwe is the largest, recognized for its great richness of species and high percentage of endemic, or regionally unique, life forms.

I N 1985, THE PAVED ROAD from Kigali to Ruhengeri was completed, permitting all-season passage from Tanzania and Burundi in the south to Congo and Uganda in the north. That same year, Chinese road crews began construction of an east–west axis from Butare to Cyangugu, working simultaneously toward the forest from both towns. As our Volkswagen van climbed past Gikongoro in June of that year, we left the fresh blacktop and entered the forest on the old road first cut by the Belgians in the 1930s. The narrow dirt track followed ridges and escarpments through alternating thick forest cover and open vistas that revealed folded mountains and almost endless forest in all directions. Bill had little opportunity to appreciate the view while driving the van, though, as recent rains had turned some sections of the road's clay surface into slippery soap. Amy usually made this drive alone in her yellow four-wheel-drive Suzuki, but this was a family outing with Noah and Ethan in the backseat.

In the center of the forest, we turned south on a smaller track toward the commercial outpost of Bweyeye on the Burundi border. It was the only other road we would encounter along the entire traverse of the forest. This path followed a series of long valley bottoms, offering fewer vistas. Yet it revealed a fascinating view of the complexities of life in a forest that lacked the full protection of a national park. Along the road, men and women carried huge bundles of goods on their heads. Some were headed to the market center at Bweyeye, but most were carrying supplies to the gold mining camps that dotted the valley floor. Stretches of tall, dense forest were broken by "war zones" that looked as though they had been subjected to weeks of constant bombardment by heavy artillery. Massive pits emptied one into the other as they descended from feeder streams to the main river embankment. All large trees within hundreds of feet were removed to build support structures for the mines. Smaller trees and shrubs were used to make primitive houses, while remaining deadwood fueled fires that cooked the miners' beans and bushmeat and kept them warm at night. It took a while to realize that each active mine pit contained as many as a dozen men, perfectly camouflaged by the slick brown mud that covered their bodies. They were barefoot and what little clothing they wore was more ragged than any we had ever seen in Africa.

We stopped below one camp to eat lunch on a bridge. Beneath its wooden planks, a clear mountain stream rushed forward to join a larger waterway whose light brown color reflected the heavy sediment load it carried from the mines upstream. Where the two waters merged like a mini-meeting of the Rio Negro and Rio Blanco, the clear dark water disappeared like tea after milk was added. These gold miners inflicted great gashing wounds on the body of the forest, from which the brown blood coursed downstream. Yet we had to admire their prodigious labor, even as we wondered what benefits they reaped for their efforts.

Returning to the main road, we pulled into a cabin at a site called Uwinka, where we spent the next several days. The Belgians had built this wooden structure, and another just below it, toward the end of their colonial reign in the 1950s. Each cabin consisted of two small bedrooms with bunkbeds, a sitting room with a large stone fireplace, and a kitchen pantry. Most remarkably, each had a bathroom with a rain-fed water supply heated by the fireplace. Outside, the site offered a commanding view of Lake Kivu to the west and the Virunga volcanoes to the north. The cabins were now owned by the Rwandan Department of Forestry and the National Institute for Agricultural Research, which shared the costs of an elderly guide named Paul Ngayabahiga, who, with his son Jean-Bosco, doubled as watchman and housekeeper.

With Noah on foot and Ethan in a backpack, we began to explore the forest around Uwinka. Giant trees of all kinds and shapes loomed above and around us. Familiar *Hagenia* and *Hypericum* were only secondary vegetation here. Even their one-hundred-foot forms in the Parc des Volcans would be dwarfed by Nyungwe's two-hundred-foot tall *Entandrophragma*—the mahoganies of Africa. Other towering emergents included the flat-topped *Newtonia,* which looked like a giant cousin of the *Acacia,* or thorn tree, so common to the drier savanna. The young red leaves of *Carapa* and the white trunks of *Macaranga* broke with the forest's uniform browns and greens. More colorful still were the great numbers of birds that flitted around each tree. While green and white wood hoopoes used their scimitar beaks to glean bugs from the bark, bright green, red, and yellow sunbirds drank nectar from bountiful canopy flowers. Large turacos—Livingstone's, Ruwenzori, and Great Blue—hopped awkwardly from branch to branch along the trees' interior architecture, until they took flight and flashed their brilliant red underwings for all to see. Above the canopy, fearsome crowned eagles glided silently, awaiting the opportunity to sink their lethal talons into an unsuspecting monkey who had ventured too far out on an exposed branch in search of a fig. And monkeys there were.

Mountain monkeys needn't worry about the eagles: their mostly terrestrial existence didn't expose them to such risks. Around Uwinka, their plush white beards and question mark tails were common sights, and they quickly became favorites of the boys. Black-and-white colobus monkeys were much more exposed as they fed on leaves in the high canopy. Yet their habit of traveling in large groups provided many more eyes and ears to detect the eagles whose wings tore the air as they swept down from above. We had already seen groups of as many as sixty colobus in a single tree, the long white fur of their tails and mantles draped like tinsel over the canopy. Even larger groups were reported to exist, and one of Amy's goals was to document this singular phenomenon and determine how such associations could endure.

Blue monkeys, white-nosed monkeys, gray-cheeked mangabeys, and chimpanzees added to the menagerie. Jean Pierre Vande weghe called Nyungwe the crown jewel of Rwanda's forests, the home of its greatest diversity, and the primary challenge for conservation. Even a short visit could confirm this fact. Yet Nyungwe's much greater biological riches failed to attract the interest generated by the single species star quality of the mountain gorilla, or the savanna wildlife spectacles of Rwanda's vast Akagera Park to the east. In the mid-1980s, Nyungwe remained shockingly unknown and precariously protected.

IN 1907, THE NYUNGWE FOREST'S boundary extended twelve miles to the east of its current location. In the west, the forest stretched even farther to reach the shores of Lake Kivu. The German administrator and dedicated naturalist Richard Kandt was concerned by reports of steady clearing along the forest fringe, however. Kandt issued a decree claiming Nyungwe, Gishwati, and Virunga as colonial Crown land with restrictions on all clearing and certain forest uses. The Belgians continued many of the German forest policies after acquiring Ruanda-Urundi, yet with little actual enforcement of restrictions. By the 1930s, the eastern border of Nyungwe was receding under pressure from hoe and axe at the astounding rate of almost one kilometer per year. In 1933, the Belgians declared Nyungwe a national forest reserve and acted to first slow, then halt the clearing. With independence in 1962, another wave of clearing began, driven by demand for land and a populist desire to re-assert control over the nation's land and resources. The natural infertility of the soils in eastern Nyungwe eventually ended settlement, though illegal wood-cutting was widespread. In the west, some of the richest forest continued to be cleared.

In 1973, the First Republic was overthrown in the coup led by Juvénal Habyarimana. The new government led by President Habyarimana was more supportive of conservation measures across the board, from parks and wildlife to soil and forest conservation. One of the government's earliest and most steadfast partners in this enterprise was the Swiss technical assistance program in forestry. The Swiss placed a primary emphasis on Nyungwe. Their first action was to plant a buffer of fast-growing pine, *Eucalyptus,* and black wattle around the northwest border of the forest. These exotics provided an alternative to native wood sources within the forest. More important, they marked and fixed the boundary of the forest reserve in a nearly permanent, easily recognizable manner, ending decades of steady encroachment. Once completed with help from other foreign assistance programs in the mid-1980s, this buffer fully surrounded a block of natural forest covering almost four hundred square miles—five hundred if one counted the contiguous Kibira Forest Reserve across the border in Burundi. This made Nyungwe one of the largest and richest tracts of African forest with a full range of montane vegetation zones between five thousand and ten thousand feet in elevation. To consolidate this holding, the Swiss then worked with the Forestry Department to develop a long-term management plan. The first Five-Year Plan was produced in 1984. It foresaw three different management regimes for Nyungwe: a core protected area, a surrounding zone of multiple use—including commercial forestry and subsistence harvesting—and the outer band of buffer plantations. The extent and exact location of the first two zones were left intentionally vague, to be determined by further surveys and inventories. The project designed by Amy to look at the conservation needs of Rwanda's remaining montane forests fit perfectly with the government's stated need for more information.

The Nyungwe management plan was to be implemented by four different donor organizations. The Swiss took responsibility for Sector One in the northwest, the European Development Fund took the northeast, the World Bank the southeast, and the French the southwest. Some partners were more equal than others. The Swiss enjoyed the closest relationship with the

Forestry Department and played a central coordinating role in Nyungwe; the World Bank was the country's largest donor and had its own relationship with senior government authorities—cemented by the runaway cattle-raising scheme still operating in Gishwati.

The same team that Amy had joined earlier in 1985 to evaluate conditions in the Gishwati Forest also helped plan the World Bank's activities in Nyungwe. Soon after their depressing drive through the fire and mud of Gishwati, the team moved south to assess the Bank's potential role in Nyungwe. Ethan joined the team this time, flying over the forest in a single-prop Cessna, hiking the trails on Amy's back, sitting in on meetings, and discreetly breast-feeding whenever he needed. The Rwandans on the team took this in stride, accustomed to women and children engaged in this most natural of functions across all levels of their own society. The European men had difficulty dealing with a nursing mother. They seemed to have an even harder time listening to her. Remarkably, some of the team members proposed that the Gishwati model be applied to the Bank's sector in Nyungwe. The disastrous failings of the Gishwati project were dismissed as easily corrected errors in application and management: the core design was still valid. Amy listened in shock as the team leaders put forward this point of view, with little criticism from others. She would not allow Nyungwe to be drawn, quartered, and dismembered like the bygone forest of Gishwati.

Development agencies in the 1980s were beginning to assemble multidisciplinary teams to design and assess their projects. This often meant adding an anthropologist to work with the standard economists and technical specialists, who might be foresters, veterinarians, engineers, or agricultural advisors, depending on the nature of the project. Ecologists were rarely included. Amy joined the World Bank team in Rwanda only because Jacques Daniel Stebler, chief of the Swiss forestry program, knew of her background and interests and asked that she be added. The leader of the Bank team was an economist from the Food and Agriculture Organization, a Rome-based bureaucracy operated under the auspices of the United Nations, and a frequent partner of the World Bank. He was responsible for the final mission report, which meant that he could also filter and edit the subsection reports of other team members. This potential weakness of multidisciplinary teamwork was partially balanced by allowing individuals to control the content of their own sections. Amy dutifully submitted her report, with a condemnation of the Gishwati project and a forceful call for more effective conservation in Nyungwe. She did not demand a radical reorientation of the Nyungwe Five-Year Plan. Instead she called for appropriate surveys to identify the most important areas for full protection and realistic systems to monitor and evaluate the effectiveness of pilot forestry and subsistence activities in the multiple-use zone. It was a reasonable and realistic plan, but it was rejected. More to the point, it was rewritten. Whole sections were deleted, strong language was softened, and in more than one instance, the entire sense of her writing was reversed. When Amy saw the draft, she protested to the team leader, then asked that her name be dropped from the document when he would not change his edits. She then sent copies of her original environmental section to the Rwanda program officer at the World Bank, Kathleen McNamara, and her immediate superior, Jane Pratt. She also delivered her original report and its revised version to the director of forestry, Phénéas Biroli, with whom she reviewed the two texts sentence by sentence. Biroli seemed genuinely shocked. So, too, were McNamara, who had accompanied the original field mission to both Gishwati and Nyungwe, and Pratt. The final report was not rewritten and the overall direction of the Gishwati project was not altered. But in Nyungwe, top authorities from the Bank and the government made certain that many of Amy's recommendations were followed.

FORESTS REVEAL THEIR SECRETS SLOWLY. They lend themselves to rumor and myth. Persistent reports from Nyungwe's rare visitors spoke of vast troops of *Colobus angolensis ruwenzorii,* the race of black-and-white colobus endemic to Nyungwe and perhaps one or two other Albertine forests. We had seen associations of fifty to sixty colobus during our brief 1983 reconnaissance. Michael Storz, a biologist hired by the Swiss to carry out preliminary wildlife surveys in 1984, reported more than one hundred in a single group. Jean Pierre Vande weghe said that he had personally seen double that number. Jean Pierre's enthusiasm left him prone to hyperbole at times, but his years of birding expeditions in Nyungwe also gave him unequaled firsthand knowledge of the forest. Amy's almost immediate immersion in conservation issues allowed her little time for research. But the mystery of the colobus became her part-time passion.

There are many advantages of group living. From a security standpoint, a large group offers greater opportunities for collective defense. The presence of hundreds of eyes and ears in a colobus troop permits early detection of crowned eagles above, leopards below, and snakes in their midst. Large groups can also harvest the food resources of a forest more efficiently than many smaller groups by grazing wide swaths of the canopy in a rotational manner, like shifting cultivators. This is especially attractive to folivores— leaf eaters—like the colobus that are adapted to feed on the younger leaves and seeds of certain tree species. Thus red colobus from Uganda have been noted in groups as large as fifty to sixty individuals. Other black-and-white colobus were once observed in an association of up to one hundred members near Uganda's Sango Bay. The known record for tree-dwelling primates was held by another folivore, the Chinese golden monkey, which formed groups of as many as two hundred individuals. Yet in the latter two cases, the large groups were believed to be temporary: combinations of smaller bands that united for a limited time, presumably to take advantage of a localized resource abundance or to meet certain social needs of the species.

For fourteen months, from late 1986 to early 1988, Amy allocated five days per month to the colobus. On each of those days, she would locate her main study group near Uwinka in the early morning, then follow the group until sundown. Throughout that time she recorded their numbers, movements, feeding behavior, and key social interactions. Early on, it was easy to get lost in pure appreciation of the monkeys' physical beauty and sublime adaptation to their arboreal existence. All subspecies of *Colobus angolensis* combine characteristic patterns of black and white hair, but no others look like the adult Nyungwe colobus. The monkey's jet black coat of moderate length is offset by epaulets of four- to six-inch-long white hairs, which drape in a distinctive fringe from its shoulders. Two thirds of the way down the tail, another patch of white begins, ending in a bushy plume of nearly eight-inch hairs. Colobus faces are a rich ebony, like the gorillas', though their constantly darting eyes betray a much more wary nature. This seriousness, in turn, is offset by large shocks of brilliant white hair that grow outward from each cheek, creating a clownlike appearance. At birth, colobus babies are pure puffs of white. But all adult colobus, male and female, share the same striking mix of short black and long white hairs. One notable exception in Amy's study group was a male who had brown hair instead of black. He cut an attractive figure in his brown-and-white suit and allowed easy identification of the group.

From a distance, a tree full of white-fringed colobus looks as if it is draped with confetti or Christmas tinsel. Yet this decoration comes alive as the monkeys fling themselves through the forest canopy. They do not swing from branch to branch like brachiating gibbons, but instead leap from perch to perch more than one hundred feet above the forest floor. Bracing against a branch, they uncoil and explode through the air, white fringes flying behind, then grasp the next branch with both hands and feet as they land. Entire groups moving quickly in this fashion create a dizzying blizzard of motion. Twice in her first year, Amy witnessed aerial blunders. Each time, she happened to be watching an individual as he or she leapt through space, only to miss or slip off the intended landing perch. She then watched helplessly as the colobus plummeted to earth, with only the ground vegetation to break its fall. Both times she hurried to the spot, but never found a body, mangled or otherwise. Studies of most arboreal primates reveal a high percentage of broken bones, but these lithe acrobats were especially tough.

Michael Storz named his primary transect path though Nyungwe the Sado-Maso Trail. Amy was surprised to learn that Nyungwe was in fact more rugged than most of the Virungas, with no large saddle areas to break up its succession of steep ridges and ravines. Yet topography also helped her work. Often, she was able to stand on a ridge and observe monkeys or birds in a tree's canopy almost at eye level. This reduced much of the neck stress known to lowland rain forest biologists who must constantly stare up at their subjects. It also made counts much easier and more reliable. The only precise counts, however, came on days when the colobus passed through sites where they were funneled into narrow columns or single file. On a January day in 1987, Amy's study group began to cross a small ravine. She hurried ahead to find a prime spot from which to count. Over the next hour and a half, more than 260 individuals crossed the ravine. This was a world record for aboreal primates. It was also a minimal count, since Amy knew that she had missed some of the group. Several months later, she anticipated that the group would soon cross the main road through the forest and again ran ahead to secure an observation point. Over the next several hours, she counted every individual as the group descended to the ground to cross the road in full view. Three hundred fifty-three colobus crossed that day. Amy was numb and elated. Over the course of her study, she and a team of assistants confirmed a consistent number of more than three hundred monkeys in several counts. Only once did the group split, and then for only one day. Something special was happening in Nyungwe.

Many arboreal folivores are hind gut fermenters: a slower form of digestion that allows them to process the lower-quality leaves on which they largely depend. The Nyungwe colobus were no different in this respect, although they did seem to be eating more mature leaves than had been reported in the literature on other folivores. Amy noticed they also spent considerable time and energy eating the lichens that covered a high percentage of branches in the forest. This unusual feeding behavior could definitely help account for the Nyungwe colobus's ability to sustain such large group sizes. Amy added another piece to the puzzle when she recorded extensive colobus feeding on *Sericostachys*. This kudzulike creeper sometimes blanketed areas half the size of a football field and seemed to be taking over large parts of the forest. It was especially common in areas of disturbance, such as landslides or logged slopes. By adding lichens and creepers to their menu, the

Nyungwe colobus transformed the forest into a world of almost boundless food abundance, a world in which resources placed few limits on group size. An expanded diet at least explained the "how" of the colobus's exceptionally large group size, if not the "why." It still was interesting to note that sub-groups of a half-dozen males sometimes competed aggressively over certain trees. Were these six-on-six contests a sign of some unseen constraint on food availability? Were they sublineages within the group competing over status, leadership, or females? So many questions. So little time. So many colobus.

Whatever advantages large group size conferred on the colobus, the monkeys were not averse to sharing the benefits with other species. Within her first few weeks of observation, Amy realized that other species often traveled along with her main colobus troop. She commonly saw blue monkeys, large guenon-type monkeys of the genus *Cercopithecus,* embedded within the group, quietly eating fruit while the colobus grazed on nearby leaves and lichens. Later, Amy realized that another guenon was also a fellow traveler. Smaller than the blue, the mona monkey was much less common. Its presence in Nyungwe was reported but not officially confirmed until Storz's survey in 1984. Amy detected its presence for the first time when her binoculars picked up a cream-colored belly among the colobus. Thinking at first that it was the brown-suited colobus male, she then detected the brilliant black fur that ran along the outside of the monkey's arms, legs, and midsection, separating its light underside and reddish brown back. This contrasting pattern on the body, the golden diadem on its forehead, and its elf-like yellow ear tufts make the mona monkey one of the most attractive primates in all of Africa.

Besides their intrinsic beauty, Amy's monkeys also proved good teachers about the forest environment. In her fifty hours each month following the colobus, she came to know a great number of the tree species and their individual characteristics. Giant *Symphonia* grew tall straight trunks with candelabra-like elbowed branches. Their red flowers provided fruit and nectar for blue monkeys, and their bright yellow sap streaks brightened the forest as they turned orange-red with oxidation. *Balthasaria*'s three-inch orange trumpet flowers provided added color, at least until the colobus broke them off to eat the bottom of the stem. *Parinari* trees formed tight broccoli tops, from which the colobus ate seeds while the blues ate their fruits. *Syzigium* produced a sweet purple fruit that some of the monkeys ate, but which chimpanzees adored. The chimps were secretive in Nyungwe, but their presence was easily detected by the piles of masticated *Syzigium* fruit they left along forest trails. Peter Trenchard, who was working on a grant from the Wildlife

Conservation Society in Burundi's contiguous Kibira forest at this time, coined the term "chimp chewies" to describe these distinctive purple droppings. Many of these fruits and flowers were also shared with the 270 species of birds that called Nyungwe home. Mostly, Amy's time with the colobus allowed her to gain a much deeper appreciation for the richness and ecological complexity of the forest. Not to mention its magnificence.

SOLVING THE MYSTERY of the colobus group size provided an important research contribution. But Amy's project revealed much more about the forest, its wildlife, and the threats to both. Just outside of Nyungwe's official borders, her surveys found a few isolated populations of white-nosed monkeys *(Cercopithecus ascanius)* surviving in relict patches of natural forest. Some of these populations were already interbreeding with similarly isolated mona monkeys. In southeastern Nyungwe, Amy and an assistant, Katy Offutt, provided the first observational evidence since the colonial period that golden monkeys, *Cercopithecus mitis kandti,* persisted in Nyungwe. Even more exciting, Katy later found and photographed the owl-faced monkey, *C. hamlyni,* in the same region. This was the first record of the sturdy species with the striking white nose stripe in Rwanda and the only known population east of the Great Lakes Rift. Both monkey populations used the bamboo zone that distinguishes Nyungwe's southeastern sector, where the World Bank was the lead donor organization. This zone was also used by local people who harvested the bamboo and peeled it into slats to sustain a major market in handmade baskets.

Bamboo was one of many products that local residents had cited in Bill's earlier surveys of forest uses in Nyungwe. Honey, medicinal plants, vines, and certain trees were also listed. Even more than in Bill's Virunga surveys, the people around Nyungwe had difficulty acknowledging forest values beyond those connected with use. If you could hunt an animal, it had value. If you could cut a tree, it had value. Local people saw almost no value in protecting animals and trees for their own sake, however. A small minority mentioned the potential of tourism, but Nyungwe was not a tourist destination in Rwanda at that time. Another study funded by the project looked at the economic and environmental impact of gold mining within the forest reserve. Kurt Kristensen and Shaban Turykunkiko confirmed that mining was almost completely destructive of all life within affected streams and for a considerable distance downstream. Miners also cut virtually all trees and other vegetation within a few hundred yards of their mines and adjoining camps. Trap hunting was a common practice and almost certainly explained the lack of any significant terrestrial wildlife within mining zones. Only Rwandan taboos on eating monkey meat protected Nyungwe's primates. Most important, the Kristensen-Turykunkiko study indicated what many had long suspected: that the thousands of miners trying to pry gold from Nyungwe's grasp received precious little reward for their prodigious labors. If profits were being made, they stayed in the hands of the powerful few who funded the mining operations and recruited desperate workers from the growing ranks of Rwanda's landless poor.

There was much that Amy's project could not do. The entire issue of sustainable use was left ill-defined in the government's Five-Year Plan. Donor organizations seemed uncertain of what sustainable use meant, and the global conservation community was just beginning to recite the mantra of sustainable use without much thought to its implications. Local communities obviously needed to be included in any discussions of controlled local forest resource use, but democratic inclusion was not a strength of either the government or the major donors working around Nyungwe. The Wildlife Conservation Society had neither the money nor the standing to take unilateral action. Amy could advocate, but she needed partners to act.

The government was a strange and sometimes difficult partner in Nyungwe. In the Virungas, we worked in the land of the favorite sons, the northern Hutu. After we overcame initial resistance, our work with the Mountain Gorilla Project brought tourism jobs and revenues to the Ruhengeri and Gisenyi prefectures, the epicenter of Rwandan political power. Criticizing the Gishwati cattle project in 1985 placed Amy in the uncomfortable position of opposing the interests of this region's powerful elite. Yet this was no worse than a fly biting a cow, and her critiques were just as easily brushed away. Nyungwe was in southern Rwanda, though. This was not a favored area for development, and, in many ways, the southern prefectures had fallen behind since the Habyarimana government took power. More to the point, the Forestry Department was dominated by southerners, who saw in Nyungwe and its related donor projects a rare opportunity to exploit their power. The arrival of an independent voice for conservation was not widely welcomed by senior Forestry staff. Maybe this was simply a matter of retaining control over an area where they held both geographic and institutional authority. Certainly they had reason to be suspicious of their ability to control Amy after she had bypassed the World Bank chain of command over her Gishwati and Nyungwe report. Perhaps there were darker reasons, as indicated by growing rumors of corruption within the department and at least some of the donor projects within Nyungwe. Whatever the reason, it took much longer and greater effort to work with the Rwandan forest service and its multiple partners than we had experienced with the Mountain Gorilla Project in the north.

IN ITS FIRST FEW YEARS, the Nyungwe project achieved some remarkable successes. It increased the number of primate types known to inhabit the forest from eleven to thirteen—one of the highest totals for any African forest. This fact alone, in combination with an already high total of bird species, placed Nyungwe for the first time on several lists of key forests for conservation in Africa and around the world. Within Rwanda, the project also raised the forest's profile among government agencies and officials. When the director of forests dragged his feet on certain issues, Amy encouraged ORTPN to extend the park service's legal mandate for wildlife protection to Nyungwe. She also offered the carrot of potential tourism development in the forest. Whatever the reason, ORTPN responded favorably and sent its first-ever mission to Nyungwe—seventeen guards under the command of chief warden Shaban Turykunkiko. A growing number of expatriate residents, including many who worked for development agencies, came to Nyungwe to hike the research trails and view the forest's birds and primates. Jim Graham, director of the U.S. Agency for International Development, often drove down for weekends to camp in the forest with his wife and two daughters. He was impressed with Nyungwe's tourism potential even as he was challenged by the sustainable use issues that confronted the forest's future. He and Amy began to discuss a possible project through which USAID could help the Wildlife Conservation Society amplify its impact with the government and other donors. Finally, the Nyungwe project succeeded in hiring almost entirely local staff. This reflected Amy's commitment to identify, recruit, and retain people who would increasingly see their future intertwined with that of the forest.

Chapter Twenty-seven

In the Shadow of the Virungas

W HEN WE LEFT THE Mountain Gorilla Project we knew that even the most successful tourism program could not meet the needs of more than a small percentage of those who lived in the shadow of the Virungas. Without more secure livelihoods for the majority of others, it seemed clear that local people would continue to exert pressure on the forest and its wildlife. Before leaving Rwanda in 1980, Bill had proposed that USAID-Rwanda design a project to address the subsistence interests of those living beyond the boundary of the Parc des Volcans. The suggestion was politely dismissed as irrelevant to the recently opened mission's priorities. In 1982, USAID director Gene Chiavaroli had more resources at his disposal and a well-developed network of Rwandan contacts that helped him shape a more expansive program. He invited Bill to discuss his ideas further. As a lifetime development specialist, Gene's interests lay with people and their basic needs. He was indifferent to wildlife protection for its own sake. Yet he recognized the economic benefits of the MGP tourism program and agreed with our plan to buffer the park and gorillas by addressing the land use pressures in the surrounding agricultural landscape. Over the next two years he supported a series of consultancies and design missions to craft a project that met our common interests. At the same time, the USAID regional program for Africa was developing field expertise in what was called integrated natural resource management. Across East Africa, USAID had teamed with the U.S. National Park Service and a consortium of American universities to explore ways to improve land, forest, and water management practices in a manner that would increase productivity while reducing pressures on remaining natural areas and wildlife. By the mid-1980s, the consortium was looking for a model project to demonstrate all that they had learned about integrated resource management. The objectives of the regional and national programs intersected in Rwanda.

In 1984, a team was formed to design the final project. Members included an economist, an anthropologist, an ecologist, a management specialist, a women-in-development specialist, and Bill as a conservationist. It was an impressive group, with a lot of experience and energy, but the director of forestry, Phénéas Biroli, suggested that the team's focus on the Virunga region was all wrong. He reasoned that the park itself was fully protected, yet surrounded by a densely populated agricultural landscape, neither of which offered many opportunities for flexible resource management. Instead, he argued that the project move to Nyungwe, where a broad array of multiple and sustainable use issues lent themselves almost perfectly to the concept of integrated management. He didn't have to add that Nyungwe was under his administrative purview, as opposed to the ORTPN fiefdom in the north. Still, his arguments had merit. Bill was left in a particular quandary. Knowing that Amy's project was already approved by the Wildlife Conservation Society with a major focus on Nyungwe, he could now design a project that would allow us to work—and live—together. Yet he was also committed to build on the foundation of our work with the MGP.

To consider Biroli's suggestion, the team made an unexpected detour to Nyungwe, where it spent several days before returning to brief Gene Chiavaroli and other USAID staff. The idea of a project in the south made technical sense. So did the original plan for the north. A combination project was briefly considered, but rejected on the grounds of management complications and insufficient resources. Some of Chiavaroli's Rwandan sources also made it clear that the government would look favorably on a decision to locate the project in the politically correct north. The design team remained split, but USAID chose the northern option. In a further compromise, USAID agreed to work only within the administrative structure of the Ruhengeri prefecture. Bill had proposed the entire natural area formed by the Virunga watershed, but politics prevailed over ecosystem integrity. In any event, Ruhengeri covered roughly 90 percent of the watershed and included the vast majority of the Parc des Volcans.

Bill made one more trip to Rwanda to fine-tune the proposal. In April 1985, he returned as director of the Ruhengeri Resource Analysis and Management, or RRAM, project. The name was selected by the design team and approved by USAID, even though it was almost impossible to translate into comprehensible French. Its only redeeming quality was that it did reflect the project's goals of first analyzing the region's problems, then moving on to a second phase of action.

～～～～～～

WHEN WE HAD FIRST ARRIVED in Ruhengeri in early 1978, the town looked like something out of the Wild West. Single-story buildings of nondescript color lined either side of a broad street. The main thoroughfare was a deeply rutted mix of mud and standing water for most of the year, until the dry season transformed it into a dust bowl. A large open market on the south side of the street was the focus of most economic activity in this town of no more than eight thousand residents. If there was any need to underscore the political primacy of Gisenyi, a resort town on the shores of Lake Kivu forty miles to the west, it could be seen in the fact that the paved road to Gisenyi started on the western outskirts of Ruhengeri.

By the time of our return in 1985, the newly paved road from Kigali to Gisenyi ran through downtown Ruhengeri. The Muhavura Hotel had been renovated, as had several other commercial establishments, to accommodate the growing gorilla tourism traffic. The Coca-Cola company paid to repaint most of the buildings in town as part of its contract to promote Coke and related Fanta products. Ruhengeri began to look like a modern African town. Beyond the commercial center, a few dozen large homes housed a small expatriate population and a few senior Rwandan officials. Slightly farther out, along the main western and northern axes, the more modest residences of *commerçants* and mid-level officials lined the roadsides. Next appeared the huddled structures of rough clay and corrugated tin that housed the middle class and working poor. These looked exactly liked the thousands of one- and two-room dwellings that had recently swelled the population of Kigali, except that in Ruhengeri there was still space to plant bananas and perhaps a few beans behind the houses.

Outside the commercial center of Ruhengeri there were no other towns in the prefecture worthy of the name. Instead, more than 600,000 Rwandans lived in traditional *rugos,* or individual household complexes, in a dispersed pattern across an area of almost 700 square miles. This overall population density of 921 people per square mile was higher than that of any of Rwanda's other nine prefectures, but the difference was not immediately noticeable to the eye. The relatively flat and fertile lava plain between the town of Ruhengeri and the Parc des Volcans could support a higher than average population density, and more wetlands were being drained for agriculture in 1985. Yet the park itself and the two large lakes of Bulera and Ruhondo lowered the total area available for settlement within the prefecture. Elsewhere, the Ruhengeri landscape looked like the rest of Rwanda. Most households were situated on the lower talus of hills, preferably with a woodlot and banana stand located just uphill. A fortunate family would have use rights to the valley bottomland below, where they could grow crops on naturally irrigated raised beds throughout the year. Another six to eight small fields might be dispersed at various elevations across the surrounding hillsides. There women would grow beans, peas, sorghum, and white potatoes, if at all possible. Still, a detailed 1984 survey found that the total area of all these fields averaged barely one acre per family. Rwandans were gardeners more than farmers, and their fate was inextricably tied to that of the land on which 95 percent of them lived and worked.

B ILL HAD ABUNDANT RESOURCES to tackle the initial problem identification phase of the RRAM project. Whereas Amy had less than $15,000 per year of discretionary funding for use in Nyungwe, Bill had nearly $500,000 for his eighteen-month assessment phase alone. The difference in financing—and salaries—available for private conservation versus publicly funded development initiatives was made painfully clear within our family. At the same time, the Ruhengeri funding was not excessive for the task at hand.

Bill's earlier 1979 survey of local attitudes and knowledge about the park and the gorillas included a series of questions about land use. Even though a slight majority wanted to open the park for agriculture at that time, he also found that a large majority of local farmers felt they could meet their subsistence needs from their own landholdings. Among those who felt a need for more land, emigration was the most commonly suggested solution. In 1984, Bill repeated his survey and found much greater support for the park, especially because of tourism. Yet he also found fewer people able to sustain their families from agriculture and fewer still who saw emigration as an outlet. In part this reflected the failure of a World Bank resettlement project in southeastern Rwanda, to which many residents of Ruhengeri had been attracted in the 1970s. After several years of converting the natural vegetation of the Bugesera region to charcoal, however, many of these people had returned to their original homes in Ruhengeri because of the poor quality of the underlying soils. Alternative solutions to the land shortage suggested by local farmers included modernized agriculture and, surprisingly, birth control. Among the most striking of the 1984 findings was that a large majority of farmers felt that they could no longer subdivide their land with even their oldest son, as Hutu inheritance tradition required.

In 1986, Bill was able to use the expanded funding base for the RRAM project to survey a larger population, this time targeting more than six hundred families across the prefecture. In addition, he left many of his questions more open-ended than before. This meant he didn't ask specific questions about problems related to the park or land, but rather asked people to list their personal concerns in priority order. The park and gorillas did not appear on a single response. Land was the top problem, followed by poverty, soil erosion, and shortages of water, wood, and food. Health came next. Overpopulation appeared in eleventh place. When asked what problems would confront their community in ten years, however, overpopulation jumped to second, behind only the lack of land and just ahead of food shortages. Rwandans in the mid-1980s—at least those in Ruhengeri—were keenly aware of the land crisis confronting them. Most people also knew that the rising population was part of the problem, if not for themselves, then for their community. It was not a coincidence that family planning services were expanding at that time, despite opposition from the Catholic church hierarchy, and that Rwandan birth rates were starting to decline for the first time in history.

At its most fundamental level, overpopulation can only be determined in relation to the resources available to a given population. We can debate the relative merits of fewer people for crowding, noise, and other quality-of-life considerations. We can note with dismay or condemnation that the United States consumes almost 40 percent of the Earth's resources, even though its population represents only 6 percent of the world total. But the rural people of an impoverished nation like Rwanda, which lacks the economic and geopolitical clout to reach beyond its own borders, must survive on resources within walking distance of their homes. The RRAM project set out to quantify the availability of key resources within the Ruhengeri prefecture, then set those figures against known and projected human population levels.

Wood is a basic resource in Rwanda. It is used to build houses, cook meals, boil water, and provide heat in a cold, damp environment. In the early 1980s, the Forestry Department conducted a survey that indicated the average Rwandan consumed slightly less than a cubic meter of wood each year, 9 percent of it for firewood. Yet no one knew how much wood was available to the population. In Ruhengeri, the only trees outside the Parc des Volcans grew in either commercial, communal, or private woodlots. Commercial woodlots were usually maintained for use in tea plantations. Since Rwanda produced extremely high-quality tea, which earned valuable foreign revenues, these plantations were well documented, well managed, and off-limits to the general population. Each of the sixteen communes within the prefecture also maintained its own tree plantations, and most families planted woodlots on their own lands. Communal foresters dutifully provided dubious area figures each year to the government; private holdings required even greater guesswork. The RRAM project devised a sampling methodology for both kinds of woodlots, as well as isolated trees of particular value, and proceeded to survey the entire prefecture. The results showed that communal plantations had been slightly overestimated, whereas private lots contained far more wood than anyone imagined. A high percentage of households also protected certain native tree species for medicinal products, boat construction, and even spiritual purposes. Unfortunately, the annual increment from all of this wood—the amount that would permit a sustainable harvest over time—met less than one third of the estimated demand for the Ruhengeri population in the mid-1980s. Projecting from current rates of tree planting and population growth, Bill's project calculated that the shortfall would only increase with time.

Water was plentiful in Ruhengeri, as one would expect in an area where rainfall averages five feet per year. Yet this critical resource was unequally distributed and of highly uneven quality. The best water came from the park, where the Virunga forest served as both filter and water storage tower. An old Belgian delivery system fed park water by pipe to dozens of points along the boundary. Dozens more outlets were broken, however, and long daily walks were a fact of life for most of the region's women and girls. In eastern Ruhengeri, Lakes Ruhondo and Bulera provided ample water. But entering the lakes to fill their jerry cans, local women were exposed to bilharzia, a debilitating disease carried by freshwater snails. In a society where women performed the vast majority of all domestic work, the collective hours spent gathering water and recovering from waterborne diseases represented an annual loss of more than 100 million hours that could be spent on farming, education, or other more productive tasks.

Soil productivity was reported to be on the decline by almost three fourths of Ruhengeri's farmers, primarily due to erosion. But hard figures on this serious problem were difficult to find. For this, RRAM pioneered the use of a geographic information system in Rwanda. A GIS is a computerized program that permits the combination of detailed data sets for analysis and subsequent mapping. As we enter the twenty-first century, GIS is a highly advanced technology used by resource managers to help understand complex problems around the world. In 1985, it was a much simpler, yet still very effective tool. With information on basic soil types, slope, rainfall, and population density as a general measure of land use intensity, we were able to map areas of highest potential erosion risk. These maps proved of instant interest to Rwandan agricultural planners responsible for combating erosion at the regional and national levels. Where tables and figures had a mind-numbing effect on these same officials, visual forms of information, such as maps, proved more effective. Ultimately, we entered all of the data that lent itself to mapping into the GIS system.

The final RRAM survey moved out of the agricultural landscape, up into the Virungas. The Mountain Gorilla Project seemed to be doing extremely well at that time and required little assistance from Bill's program. But the bottom line in any wildlife conservation project is how your critters are doing and no one had counted the gorillas since 1981. So in 1986, Amy left her work in Nyungwe to organize another census. This time, however, the financial resources of the RRAM project helped to greatly improve the process. First, we paid for representatives from Congo, Uganda, and Rwanda, as well as from private conservation groups, to meet and plan the census together. The result was a greater degree of advance communication and planning than ever before. The planning also facilitated obtaining permits necessary to work across all borders—a formality that had never been brokered before. Once we established standard methods, mixed teams from the three countries blanketed the Virungas to begin work in a coordinated manner. Because of limited resources, previous censuses had moved across the Virungas in an irregular, noncontinuous manner, with the risk that certain gorilla groups might move or otherwise avoid detection. With near simultaneous coverage of the entire range, this was much less likely to occur. RRAM financing also allowed Amy to hire more staff, greatly reducing the amount of time each team spent in the forest. This in turn helped with maintaining high morale during long, grueling days.

The first results from Mt. Mikeno in Congo brought devastating news. George Schaller had counted 250 gorillas on Mikeno alone in 1960, and the 1978 census had revealed a shocking decline to eighty-four gorillas. In 1986 Amy's teams found only thirty. One team member wanted to halt the rest of the census, fearing the effect of adverse publicity on government and international support for his project. But better news soon followed. More gorillas were living in areas not far from Mikeno, perhaps indicating that they had merely shifted their ranges. The Visoke count came in next at a record high, with many young gorillas. Even the eastern sector of Sabyinyo, Gahinga, and Muhavura showed some improvement. In the end, the census found a total of 290 gorillas: the first recorded increase in the otherwise dismal history of Virunga counts. The increase of almost 12 percent since Bill's 1978 census was unprecedented for a large mammal that produced only one infant every four years. And now there were more infants than at any time since Schaller's first survey. For us it was the best news since we came to Rwanda—and a strong validation of the Mountain Gorilla Project.

EXCEPT FOR BILL, the full-time RRAM staff was entirely Rwandan. All were asked to perform work beyond anything that they had ever done before, and all performed with considerable competence. A secretary who had learned to use a typewriter just the year before moved seamlessly into the world of computerized word processing. Only the wonder of a French language spell checker continued to amaze her. A geography graduate who had made a few maps by hand in college became a GIS specialist using digital technologies unknown at the University of Rwanda. In his spare time, he also mastered the use of Excel spreadsheets and served double duty as the project accountant. The RRAM survey workers proved highly capable, whether conducting questionnaires or sampling trees. Our watchman even learned a new trade, tending a nursery of native and exotic tree species that we thought might improve soil quality. On National Tree Day, the entire office staff planted these seedlings in the Ruhengeri countryside, rather than the standard *Eucalyptus* and black wattle. It was a good team.

Vincent Nyamulinda was the RRAM assistant director. He had just finished his master's degree at the Ruhengeri campus of the national university when Bill hired him. He was good-looking, with a great smile, and far more socially outgoing than most Rwandans. He was also very bright and a fast learner. He enjoyed spending time with the half-dozen or so Western consultants who helped jump-start the RRAM inventory process in 1985 and he was a sponge for whatever information they might pass on. One night over dinner at our house, Vincent offered another advantage of working with so many different foreigners: *I've always had trouble telling white people apart; but now that I've worked with so many of you, it's much easier.* Vincent was also proud. Both of his parents were uneducated farmers, and he knew he had come a long way from his native hills of Byumba, just to the east of Ruhengeri. The fact that Byumba was not as favored, politically or economically, as Ruhengeri or Gisenyi was a sore point. *We're northern Hutu, too. We're part of the northern coalition. We support the government, but we don't get respect.* Vincent was offended when he saw government officials from Ruhengeri claim illegal parcels from a wetlands conversion project in the northeastern corner of the prefecture. He wasn't sure what he thought about the environmental implications of the project, but he didn't like the fact that wealthy politicians were taking land intended for poor local farmers. When the powerful *préfet,* or governor, of Ruhengeri asked to use the RRAM project vehicles to transport potatoes for sale across the border in Congo, Vincent recognized the illegal personal enrichment scheme for what it was. He was pleased when Bill said no. His critical observations about powerful elites were surprising, and refreshing.

Over time, Vincent's pride grew and his perceptions of power changed. He wanted to be recognized as important, too. First, he bought a more expensive home in the right part of town. Then he needed a project vehicle after hours, with the driver. Once a month, the vehicle would drive him to his home in Byumba. His wedding was a wonderful affair, but it cost almost half a year's salary. In all instances, Vincent was clear about his need to maintain a certain standing, which, he claimed, would also help the project. Bill doubted this claim, but tried to meet some of Vincent's desires in accordance with USAID policies that required him to pay for any personal vehicle use. The use—and abuse—of the driver, however, became an issue.

Léopold Rwamu grew up in eastern Congo. His Tutsi parents were among the many refugees who left Rwanda in 1959 when the Hutu took control. He was able to attend Congolese schools and obtain an education, but ethnic politics and limited family finances kept him out of college. Instead, he learned to drive and maintain cars, and eventually returned to work in Rwanda. He was the first person Bill hired for the RRAM project, and the only one who became a true family friend. This was all the more remarkable after he sideswiped a wall with our personal vehicle and ripped off part of a door. But Léopold was extremely humble, painfully honest, and completely engaging. He would drop by on weekends and spend hours talking with us and playing with the boys. He became close friends with our nanny, Clementine—a relationship as remarkable for the fact that it was between a Rwandan man and a Rwandan woman as for the fact that it was between a Tutsi and a Hutu. Driving around the countryside with Bill, Léopold moved easily from joking and telling stories to serious discussions of life in Rwanda. He openly talked about Hutu-Tutsi relations and even politics—to a point. At work, Léopold lectured others for wasting their time and money on drinking and womanizing. Rwanda was the epicenter of the AIDS crisis in Africa at that time, and Léopold felt that it was his duty to point out the dangers to anyone who would listen—including one American consultant who entertained a succession of Rwandan men in her quarters. When Léo married in 1986, he, too, borrowed money from the project. But it was typical that he used the advance to purchase a plot of good land with a woodlot, leaving a modest amount to pay for his wedding.

Léopold's strong moral compass made him a trusted assistant. As the project expanded, he was responsible for carrying monthly bankrolls that surpassed his annual salary. He performed a variety of other duties far beyond the role of chauffeur. He even did volunteer work for Amy's project. There was no question that Vincent was his superior within the project hierarchy, but the relationship between the two was tense from the beginning. When Vincent began to use Léopold as his personal chauffeur outside of work hours—a necessity since Vincent, like most Rwandans, didn't know how to drive—matters steadily deteriorated. In Léopold's account, Vincent was rude and overbearing, treating him worse than a servant. Vincent saw Léo as insubordinate. Bill suspected an ethnic element in the conflict and suspended all personal use of project vehicles in the interest of project unity and performance.

IN LATE 1986, the RRAM project was ready to present its findings. A large conference was organized at which we outlined our major conclusions and released the two-hundred-page report detailing our inventory and analysis phase. More than four hundred resource managers and local leaders were invited; Protais Zigiranyirazo, the *préfet* of Ruhengeri and brother-in-law of President Habyarimana, agreed to chair the initial session. Opening remarks by the *préfet* and Bill were followed by a requisite series of tributes to the project and political leadership that bordered on adulation. In the midst of one of these talks by a local communal leader, the speaker gasped, clutched his chest, and dropped to the floor. Bill shot up from his chair on the dais, but the *préfet* grabbed his arm and gestured that he sit. As others gathered around the man and carried him out, the *préfet* gave every indication that this was an underling, nothing for real leaders to be concerned about. The show must go on. At the noon break, it was confirmed that the man had suffered a heart attack, and was recovering in the Ruhengeri hospital.

Conference participants endorsed the results of our surveys on priority problems facing the prefecture, though the question of where to rank population growth provoked much discussion. Bill presented a calculation that if the annual population growth rate of 3.7 percent continued unchanged, then in 225 years the population density of Ruhengeri would be one person per square meter. This fact was greeted first with silence, then nervous laughter. But ultimately the point about limits hit home. Bill then added that if the entire Parc des Volcans were cleared and settled at the average Ruhengeri density of almost one thousand people per square mile, the park would provide enough land for only one third of one year's population growth in Rwanda. Set against this background of a population-land imbalance, issues of wood, water, and other resource shortages took on added clarity. Yet just when everything seemed hopeless, the GIS maps helped to identify certain areas for priority attention and some immediate approaches to mitigate some of the more serious problems. The conference ended with an endorsement of a set of second-phase activities for RRAM.

R RAM II WAS DESIGNED to address three problems confronting the people of Ruhengeri. The problems were the need to improve agricultural productivity, control erosion, and produce wood. Agroforestry provided a logical integrated solution to all three. In agroforestry, trees and shrubs are selected that improve soil quality by fixing nitrogen with their roots and adding organic matter through natural leaf fall. Their roots and trunks add a physical barrier against erosion and, at maturity, they provide material for firewood and construction. It is an ideal package for Rwanda. But no one had tested agroforestry systems in the northern highlands. In fact, no one had ever tested the erosion control systems already in place for their relative effectiveness. RRAM II set out to develop agroforestry systems adapted to the conditions and needs of Ruhengeri and to monitor their effectiveness in comparison with other methods.

Working in three communes with the highest erosion risk and severe wood shortage, the project set up model agroforestry plots in late 1986. Selected trees and shrubs were planted along contour lines on steep slopes made available to the project. Alongside were equal-sized areas in which grass hedgerows were planted or anti-erosion ditches cut at equal intervals across the slope. A final plot had no erosion control structures at all. Local women then planted the same crops in all of the plots. Wooden boards enclosed each two-hundred-square-meter plot that drained into a cement pit at the bottom of the slope. After each rainfall, a monitor measured the amount of liquid and solid runoff from each enclosure. The result was a direct measure of the effectiveness of each method to control erosion. Local farmers began to appear after each storm to check out the erosion traps, too. Soon they were asking for the trees and shrubs that seemed most effective to plant on their own lands. Meanwhile, Bill worked with the Ministry of Agriculture to see if a stratified approach to erosion control could be developed based on the concept of relative risk. Current practice was to require every farmer, under threat of penalties, to plant hedgerows or cut ditches on every field, at the same intervals, regardless of slope. With limited resources, it made more sense to focus on those steeper hills and fields at greatest risk.

Soil erosion and agroforestry took Bill far from his initial focus on gorillas almost ten years earlier. In some ways, he enjoyed the new challenge. The director of soil conservation, Aloys Habimana, also proved to be a pleasant colleague. He had just finished a master's program at the University of Minnesota and seemed excited to be back in Rwanda with major responsibilities before him. He joined a cadre of well-trained young professionals committed to building a better nation. Within a few months, though, Habimana seemed depressed. This was not unusual for Rwandans who returned from well-equipped offices and labs at U.S. universities, only to find that their exalted new position came with a dank cement office with a single bare lightbulb, no computer, no journals, and few funds. RRAM could help provide some support, but that wasn't enough. One day, Aloys asked if Bill could help him obtain a scholarship to return to the U.S. When Bill asked why, the young Rwandan was surprisingly candid. Yes, his job was overwhelming, but the real problem was political. Every weekend he was required to attend meetings called by the political wing of the Mouvement Révolutionnaire National pour le Développement, the single party that ruled Rwanda. These meetings were time-consuming, but Aloys implied that they were also disturbing. *They want us all to believe in the same things, to think the same way. Some of those things I don't believe in.* That was as far as he went, leaving Bill to imagine what was happening in the hidden realm of Rwandan politics.

Meanwhile, the internal politics of the RRAM project itself turned ugly. Bill had already agreed to help Vincent gain entry into an American graduate school, once the second phase was under way. But Vincent was increasingly arrogant and demanding with other office staff. Bill learned that he had purchased plots in a recent wetland conversion scheme—the same action that Vincent had condemned the previous year. He also began making disturbing comments about Tutsi. One day, Bill stopped to talk with a lone *umugogwe* herder along the road to Gisenyi. When he got back in the project Suzuki, Vincent smirked and asked if *it didn't smell too bad.* Bill was struck by the use of "it," but shrugged it off as a reference to the smell of cattle. Vincent then added, *All Bagogwe smell bad, they're Tutsi.* A few days later, Vincent entered Bill's office and asked to close the door. *You know the new person you hired to work on the erosion monitoring is a Tutsi?* Bill said he had no idea if the person was a Tutsi and that he didn't care. Vincent replied that it would be *very bad for the project to be known for hiring Tutsi. We already have too many,* clearly referring to Léopold. Bill grew angry, telling Vincent that race and ethnicity were not factors in hiring and shouldn't be matters for discussion. He told him to drop the subject. Vincent responded pointedly that the subject of ethnicity was of great interest to the *préfet* and other higher authorities, then turned and left the office.

Bill had pushed the RRAM project as far as he could. Now it was entering new areas of technical expertise in soils, farming, and forestry far from his original conservation focus. He didn't like being a government advisor. He didn't like the politics, especially ethnic politics, that had entered into his working environment. He began to doubt whether he liked Rwanda.

Chapter Twenty-eight

Living in Rwanda

IT WOULD BE ROMANTIC to report that we raised our boys in the remote wilds of Rwanda, living in a primitive cabin as we had before. Amy did spend most of her time in the mountains and our entire family joined her as much as Bill's work permitted. But, in fact, our base accommodations were not very different from housing in America. Our lifestyle, on the other hand, *was* rather different.

Our Ruhengeri house was a large one-story relic from the colonial era. Its exterior walls were white, with electric blue trim. A stone path lined with rocks painted the same shocking color code led from the house to a crushed rock driveway. Indoors, the floors were dark red cement. Ethan, our youngest, could hit surprisingly high speeds over these floors, tearing around in a four-wheeled scooter as fast as his first-year legs could fly. He was especially skilled at knowing when the outer door was left open, so he could dash out onto the terrace. A crash landing lurked at the bottom of three cement stairs at the far end of the terrace if he pressed his freedom too far. In Kigali, where we moved for the RRAM second phase, the stylized metal *grillage* that covered the windows for security purposes provided Ethan, now eighteen months, with a ready-made latticework jungle gym, on which he liked to climb all the way to the ceiling. Noah, older by four and one half years, watched Ethan's escapades as he would a monkey in the forest.

Both houses had running water most of the time, and a limited supply of hot water much of the time. Shortages were more a function of citywide shutdowns than our own plumbing problems. If we wanted heat, we could light a fire to offset the cool night air in Ruhengeri at eight thousand feet and in mile-high Kigali. Unfortunately neither elevation was high enough to be completely free of malaria, so we always slept under mosquito nets. Still, Noah suffered for a week from a nasty case of malaria toward the end of our first year. It was gut-wrenching to hear our five-year-old in such pain that he said he wished he were dead. It was no easier to bring Ethan through a debilitating bout of amoebic dysentery before he was old enough to talk.

The best part of the Ruhengeri house was its lawn and gardens. USAID required all American contractors to have a twenty-four-hour guard at their houses in response to fears of Libyan-sponsored terrorism (a laughable proposition at our Kigali residence, which was directly across the street from the Libyan embassy). We hired a gardener to fill the day guard position. Isaac had worked at our house for previous occupants and had personally planted every flower and shrub in the yard. He cared deeply for his plants and kept them healthy and strong, despite the steady damage inflicted by our family's errant Frisbees, Wiffle balls, soccer balls, and volleyballs. We also tended our own vegetable garden, in which we grew broccoli, carrots, and tomatoes, while failing dismally in our efforts to grow sweet corn. Lemon, papaya, and pomegranate trees grew around the yard, as well as flowering jacaranda and flamboyant trees. Royal sunbirds and paradise flycatchers resided year-round; hadada ibises roosted outside our bedroom window, waking us with the raucous calls from which they get their name. From time to time, a solitary crowned crane would land on the lawn, as strikingly beautiful as it was improbably large.

Our night watchman was Radio, or Ladio as many Rwandans pronounced it in their habit of reversing Rs and Ls. He apparently earned his name from a habit of nonstop talking in his younger years, which someone likened to listening to Radio Rwanda. In 1985 he could still talk a streak in excited Swahili, but mostly he was a peaceable old man with a dull machete. Fortunately, he never had to defend us against anything more than a noisy cat. Sembabuka was a reasonably good cook and a proud man. He could whip up a mean crêpe, soufflé, or tarte, but his main courses tended toward the ordinary. Whenever possible, we contrived to have him leave the dinner cooking to us. The cook would ordinarily be the head of the household staff, but in our case that role fell early to Clementine Uwimana. Clementine was the first and only nanny our boys ever had. The day she came to interview for the job, she picked up Ethan and held him close throughout. He returned her hugs and the job was hers. During the long hours when both of us were at work, she was also responsible for Noah, and she cared for him deeply. But playing games with him as he got older did not come as naturally to Clementine as loving an infant.

Noah attended the French kindergarten in Ruhengeri and the Belgian first grade in Kigali. He quickly became fluent in French and developed a keen ear and mocking tongue for the flat accents of many visiting Americans. Both schools attracted a mix of European, African, and Asian children, providing a great opportunity for Noah to see life outside the cultural cocoon in which most American students learn. The downside came with a brutally ambitious Belgian curriculum that required six days of school each week and one or two hours of homework every night. His first-grade Belgian math class covered subjects that Noah would not see in the U.S. until fifth grade. Fortunately, Noah's real school was outside the classroom. He had friends from Rwanda, Congo, Sri Lanka, and Europe, as well as a few Americans his age. His best friend was Gael Vande weghe, the son of Jean Pierre and Thérèse. Both boys loved the wild, and we developed an exchange in which Noah would stay with Gael in the Akagera Park, where Jean Pierre was then working, and Gael would come to Nyungwe with our family. In Akagera, the boys had free reign of the savanna around Jean Pierre and Thérèse's simple house, where browsing impala, topi, and zebra were a common sight. They learned to be wary of the Cape buffalo and *phacochère,* or warthog. The warthog became Noah's favorite. Gael loved the English name, which he pronounced "war dog." We all thought it a much better name for the sturdy tusked wild pig. In Nyungwe, the boys hiked the forest with Amy on days when their running and talking wouldn't interrupt her work. They took full advantage of their freedom, but Amy also used the forest as a classroom. One day, they were engaged in a running battle with wooden swords when they came upon a strangler fig. Amy explained how the fig had grown up around a live tree, gradually choking out its light, then its life. Eventually, the tree died and rotted away, leaving only the intricate exoskeleton of its killer standing like any other tree in the forest—except that it had no interior substance. The boys absorbed the tale of intrigue and murder with wide eyes, then resumed their running sword battle.

During our first year in Rwanda, we visited the Akagera Park on several occasions when we could both find a weekend free. The park covered the eastern tenth of Rwanda and, in many ways, was more interesting than more famous reserves that we had visited in East Africa. Its rich landscape mosaic included open grasslands, wooded thickets, rocky ridges, and an extensive wetlands complex through which the Akagera River slowly wound its way toward Lake Victoria. Yet Akagera was off the beaten tourist circuit. Their loss, our gain. Lions, leopards, zebra, elephants, hippos, topi, sitatunga, and Cape buffalo could all be seen. Lilac-breasted rollers, multicolored bee-eaters, shoe-billed storks, and abundant fish eagles were among the phenomenal total of almost seven hundred bird species recorded in the park. Noah absorbed the wildlife spectacles and can still recall certain sights and experiences from those early years. Ethan was equally attentive, but remembers only the day that an elephant stuck its trunk in the window of our Volkswagen van.

One weekend, Bill and Noah drove to Akagera with a visiting friend from Madison, Kate Noonan, while Amy and Ethan stayed in Nyungwe. Normally we camped in tents in Akagera, but this time we rented rooms at the Gabiro Lodge after a full day in the park. Gabiro was an attractive stone structure in the northern part of the park, where we had often stopped for a meal during previous visits. A captive group of chimpanzees always attracted Noah. Seized by Rwandan authorities from smugglers and tourists, ORTPN decided to put them on display in Akagera, despite the park's historical lack of chimps. On that day, the chimps displayed considerable agitation when we arrived and Bill told Noah to stay away from them. Then, while checking in, he noticed that Noah was missing. Bill bolted toward the chimp cage, just in time to see Noah holding out a handful of leaves to the still raucous chimps. In an instant, a long black arm reached out past the proffered food to grab Noah and pull him toward the cage. The bars kept his upper body outside, but several of the chimps began biting his arm as the others screamed and hooted. Bill arrived in time to hit two of the chimps and pull Noah out of their reach. Noah's arm was covered in blood; an exposed artery in the crook of his arm pulsed rapidly, but seemed to be intact. Racing inside to see if there was a doctor visiting the park, Bill settled for a washcloth and ice from the restaurant's kitchen.

The lodge manager led Bill to a U.N. field hospital only ten miles to the north. The hospital was tending to a small number of Tutsi refugees from Uganda, but the one staff doctor had little medicine at his disposal. While Bill held down the screaming Noah, the Sudanese doctor thoroughly cleansed more than twenty bite wounds with alcohol-soaked gauze. He used tongs to probe and clean the deep gash around the exposed artery. Bill raced back to Kigali that evening, with Kate holding Noah in the back of the van. They arrived at the home of Noah's Belgian pediatrician around 10:00 P.M. Working in his pajamas and robe, Dr. de Clerc examined the wounds, complimented the U.N. doctor's work, told Noah he was a brave boy, and gave him a powerful antibiotic cocktail. Many of the wounds called for stitches, but Dr. de Clerc left them open because of the high potential for infection. Now a strapping young man, Noah need only look at his right arm for a series of unwanted reminders of his most frightening day as a child in Rwanda.

A few weeks after the Akagera attack, Noah was walking with Amy in

Nyungwe when they spotted chimps in nearby trees. The forest was home to hundreds of the apes, but it was truly special to see them so clearly. Amy moved close to Noah and asked what he thought. *Nice.* You're not afraid? *Nope.* Why not? *Those chimps in Akagera were crazy from living in cages. These chimps are happy 'cause they're in their own homes.* Nice.

Noah's experience with the gorillas was even better. We wanted to take both boys, but also knew that we couldn't trust the rambunctious Ethan to understand how to behave when surrounded by such large, powerful, curious creatures. As director of Karisoke, David Watts approved a visit to Group 5 for the two of us with Noah. It was a very kind gesture by David, who made it even better by accompanying us into the field. Amy had not seen her gorilla family in almost seven years. Beethoven had died of natural causes in the interim and Icarus had left the family. Ziz was now the lead silverback. Pablo was barely starting to silver. Puck and Tuck—the ex-brothers—had each given birth a second time. We were excited to see all the gorillas again; it was incredibly moving to sit among them with our son. It took Tuck less than a minute to approach us and stare into Amy's face. Noah stared back, wide-eyed, from a few yards away. The next thing we knew, Cantsbee—Puck's seven-year-old—tried to untie the laces of one of Noah's boots. As Bill gently shooed the young male away, Pablo made his entry with a bluff charge. It was typical Pablo, creating a ruckus and calling attention to himself. The display was a bit unnerving for Noah, who hadn't known Pablo as an endearing young juvenile delinquent. Just to be sure, Bill held Noah in his arms as the group calmed down and returned to its normal routine of resting, feeding, and play. Toward the end of our stay, Tuck approached Bill, who was lying on his stomach with Noah beside him. A video taken by David shows Tuck sitting on Bill's legs, looking around and scratching her chin. She then begins to drum on his rump, provoking a wry smile from Bill—and a look of pure wonder from Noah, only a few feet away from the gentle giant who had turned his father's butt into bongos.

WE ALSO TOOK OUR Ruhengeri house staff out to see the gorillas. By that time in 1986, thousands of foreigners had visited Rwanda's gorillas, but extremely few Rwandans had shared this experience. We gave Clementine, Sembabuka, Isaac, and Radio a day off to join this select group, paying their way to join Amy as "tourists" to see Group 11. Despite some initial nervousness, all of them became completely absorbed by the individual behavior and group dynamics playing out before them. Clementine turned to Amy to point out a strutting young gorilla, saying *Just like Ethan.* Mostly there was rapt attention with little talking. The gorillas, too, proved attentive, perhaps intrigued by the rare experience of seeing so many black faces after a steady stream of white apes. After close to an hour, the gorillas and their guests were left to ponder the day's experiences as they headed their separate ways. Radio lived up to his name, broadcasting his experiences with the gorillas for any and all who would listen. But he was not alone. Each of the Rwandans said they returned to their homes and recounted their visit to eager family, friends, and neighbors. They were great ambassadors from another world, not far away.

The spirit captured in the visit to Group 11 was broken a few months later. We had taken the boys to Gisenyi to swim in Lake Kivu on a clear Sunday, leaving Sembabuka responsible for the empty house. He apparently started drinking soon after we left and was quite drunk when Clementine arrived. She had stopped by to socialize with whoever was there on her day off, but Sembabuka became verbally abusive. He asked if she was there to see her "Tutsi friend," meaning Léopold. Clementine, a northern Hutu like Sembabuka, told him that *we're all Rwandans,* adding that her friends were her business. Sembabuka became more abusive and Clementine left quite shaken. When we returned that afternoon, the gate was open, the house unlocked, and no one in sight. Later, Clementine returned to tell us of the day's events. Sembabuka had already been warned about drinking at the house, and he was fired when he showed up late the next morning with no excuse. But underlying ethnic tensions were again rising to the surface.

After sixteen months in Ruhengeri, we moved to Kigali. Clementine came with us and lived in a bedroom annex on the house. She shared all our meals and joined us for most social occasions. We had recently acquired a video cassette player and she enjoyed our evening viewings of everything from *Star Wars* to *Sesame Street.* When Rwandan authorities issued a pamphlet warning of the dangers of AIDS, Clementine asked Amy to discuss the subject with her. Some Belgian friends suggested that Clementine was becoming too familiar with us and that we were spoiling her to ever work for anyone else. But we couldn't imagine any other way to live. Up to the age of one, Ethan often rode on Amy's back as Amy hiked through Nyungwe. Once he could walk, though, he wanted no part of being carried. So Clementine traveled to Nyungwe and accompanied Amy on her hikes, occupying Ethan whenever Amy needed to concentrate on her work. Unlike many Rwandans, who were uncomfortable in the forest, Clementine seemed to revel in it. She became an expert spotter of monkeys and birds. Amy wondered whether she wouldn't make a great guide once Ethan outgrew her services as a nanny.

Life in Kigali presented many distractions. French and Belgian friends were always available for good food and long discussions, and many a pleasant evening was spent *chez* Alain and Nicole Monfort, whose hospitality was boundless. Jean Pierre Vande weghe and his wife, Thérèse Abandibakobwa, were also close friends and confidants. The Cercle Sportif offered a high level of clay court tennis and mediocre basketball, although within a neocolonial ambiance that was broken by only a few black faces. The American Club offered swimming, videos, and hamburgers on demand, with an equally white mix of diplomats, evangelists, and the occasional odd conservationist. On Sundays we could play softball with the Québécois, who were surprisingly good at *la balle molle* and always ready for a party afterward. A few of us joined with a handful of Europeans to play the Russians in hardcore, set-and-smash volleyball, followed by shots of chilled vodka and a sincere sense of camaraderie.

There were ample distractions. The housing was excellent. We had hired staff to do our cooking and cleaning. It was a very seductive lifestyle, but we decided to leave it behind. The question was whether we would leave the city to return to the forest, or leave Rwanda altogether.

IN THE SPRING OF 1987, we returned to the U.S. on home leave, passing through New York so that Amy could update Wildlife Conservation Society staff and donors on the status of her work. After her two talks, everyone complimented Amy on her accomplishments and exceptional speaking skills. Bill, who had no speaking role, was then offered a job. In fact, he was asked to be the deputy director of the entire international program, based in New York, at a time when the director was based in Nairobi. This was not the first time that WCS had overlooked Amy, or ignored our complementary skills as a team. They would recover from their oversight later. The offer of a U.S.-based job was not the first for either of us, but it was attractive. Now we had the most dramatic choice possible: Nyungwe or New York.

By the summer of 1987, Bill had given notice to USAID and selected his replacement as director of RRAM II. In August, the entire family moved from our large Kigali house into a small cottage in Kitabi, on the eastern edge of Nyungwe. In theory, our options remained open. But the move to Kitabi prepared us to leave Rwanda, professionally and emotionally. Kitabi was the name of a tea plantation, the most productive along Nyungwe's border. All around our cottage lay the neatly cropped plantations, each row tracing the contour of the underlying hill. Conditions were almost perfect for producing a tea that often garnered top dollar on the London tea market. Interspersed with the tea grew large blocks of *Eucalyptus,* the fast-growing exotic tree from Australia that provided the prodigious amounts of firewood required to slowly cure the tender tea leaves. Beyond the uniform plantations of trees and tea, the unbridled exuberance of the natural forest dominated the western skyline.

Our five months as a family in Kitabi were the best of the three years we spent on our second tour in Rwanda. With no more commuting between Kigali, Ruhengeri, and Nyungwe, we spent much more time together. Bill finally enjoyed the luxury of spending whole days in the forest, taking the time to appreciate its plentiful offerings. We often explored the forest together with Noah and Ethan. In Kitabi, Bill and the boys developed the habit of taking daily walks and bike rides along the network of narrow plantation roads while Amy focused on her fieldwork. Noah and Ethan became best friends in Kitabi. They were almost inseparable, creating a bond that has carried them through their teenage years and into adulthood. They readily welcomed other Rwandan boys who would appear on most days to play, alternating between Wiffle ball and soccer. But mostly they amused themselves.

Without Clementine, who stayed in Kigali to marry and work with friends, the boys fixed their attention on two Rwandan men who performed multiple chores around Kitabi for the Rwandan and European staff who lived in surrounding cottages. Alexi and Justin worked hard, tending gardens, cutting wood, and maintaining the buildings. They were also extremely tolerant of the two little blond boys who followed them everywhere. In the garden, they put the boys to work hoeing and carrying buckets of fruit and vegetables to the kitchen. They also carried firewood, and Noah learned to wield the traditional Rwandan axe. The boys learned a very important lesson from watching Alexi and Justin slaughter goats and chickens for dinner: if you are going to eat meat, you should at least witness where it comes from. Not all animals were on the losing end of encounters, though. Ethan walked into the house crying one day, telling Amy, "Big sheep butt Ethan. Blood run down face. I cry, I cry, I cry. I tell Amy, big sheep butt Ethan, blood run down face . . ." The tragic tale was repeated many times for all who cared to listen. Reluctantly, Ethan accepted the idea that squaring off against big animals with horns is a bad idea.

On most weekdays after the tea factory closed at five, workers would gather for as many volleyball games as they could play before sundown. Alexi, Justin, and Bill almost always joined in, as did Amy and Katy Offutt, her assistant, when they weren't in the forest. The games were high-intensity and great fun. They also offered an opportunity to get to know another group of Rwandans. The tea workers were better educated than the gorilla guides, although less so than Vincent and other government officials connected with RRAM. They were uniformly kind to our entire family, within the restrained strictures of Rwandan culture. It was a pleasure to leave most of our accumulated household goods, bicycles, and toys to Alexi, Justin, and others from this group.

Noah was excited about returning to the U.S. He enjoyed both sets of grandparents on our home leave and he looked forward to more time with them and to attending an American school. We took turns helping Noah learn to read and write in English, skills which we had earlier set aside in favor of French. The oddities of the English language, blithely accepted at a younger age, he now challenged as illogical and painful to learn. He did the same with the role of his parents as teachers. But Noah made the transition without too much difficulty. Perhaps anticipating our return, we also placed more emphasis on American holidays. Amy collected a large, four-inch *Carapa* seed from the forest and carved a ghoulish face into its hard, brown exterior. With a small candle inside, it made a quaint jack-o'-lantern that the boys insisted on lighting every night, long after Halloween had passed. Christmas of 1987 was our last time together as a family in Rwanda before Bill and the boys left for the States. Noah and Ethan would stay with Grandma and Grandpa Vedder for three months, while Bill settled into his new job and looked for affordable housing around New York. Amy stayed to finish her work in Rwanda through March, when the family was reunited.

THE PAVED ROAD THROUGH Nyungwe was completed in 1986. We were concerned from its inception that Rwandans avoid their standard practice of planting *Eucalyptus* along the roadside. Not only would the fast-growing exotic block all views, but it would then spread into the natural forest, where it would almost certainly outcompete native species for both sunlight and space. More than a year earlier, we had argued in favor of roadside recolonization with native shrubs, trees, and other ground cover. Despite assurances that this would happen, National Tree Day in 1987 saw the arrival of an army of Rwandan citizens bearing thousands of sacks of *Eucalyptus* and black wattle to plant along the new road. We raced to Kigali the next day to tell the director of forestry, who expressed surprise and concern. He was reluctant, however, to take on the *préfet* of Gikongoro, who had orchestrated and paid for the plantings, of which he was now quite proud. There was nothing to do.

Over the following weeks, Bill took many late afternoon drives into the forest with the boys. At first, Noah watched in shock as his father pulled the newly planted trees from the ground. Then he and Ethan joined in with great gusto, uprooting more than a hundred of the *Eucalyptus* and black wattle seedlings each afternoon, throwing them into the back of the car, and dumping them in a deep ravine just outside the forest boundary We removed only a fraction of the ten thousand offending trees in this way before we left Rwanda. But the boys learned about good trees and bad trees. And they learned the more important lesson that there are good decisions by authorities and bad decisions by authorities. As they carried out their acts of eco–civil disobedience, Bill was reminded of his flagging respect for Rwandan authorities across the board. It was time to leave.

Chapter Twenty-nine

Ten Years After

IN 1989, MORE THAN six thousand tourists paid almost $200 each to see the gorillas in Rwanda's Parc National des Volcans. After a decade of steady growth in visitation—and ticket prices—since the creation of the Mountain Gorilla Project, direct entry fees earned by ORTPN surpassed $1 million in a single year for the first time. These same tourists spent another $3 to $5 million that year on food, lodging, rental cars, and souvenirs during their stay. Tourism had grown from a negligible economic activity in 1978—when the Parc des Volcans earned only a few thousand dollars—to Rwanda's third largest source of foreign revenue, behind exports of coffee and tea. Gorillas featured prominently in Western nature and wildlife magazine ads throughout the 1980s as Rwanda became an elite destination on the East African tourism circuit. Across the border, a parallel project started by Rosalind and Conrad Aveling successfully adapted the MGP formula of tourism, education, and improved security to conditions on the Congolese side of the Virungas. Visitation to Congo's Virunga Park reached three thousand per year as a result.

Between 1959 and 1978, the mountain gorilla population in the Virungas had crashed from at least 450 to no more than 260 individuals. With no evidence of an increase in natural mortality, poachers were almost entirely responsible for this decline, killing an average of at least ten gorillas per year. Yet in the ten years after the creation of the MGP, only three gorillas were known to have been killed. Even more important for the future of the population, the MGP halted all habitat loss—most notably the proposed Virunga cattle scheme. A 1989 census team coordinated by Craig Sholley with help from Amy was hopeful that the gorilla population would continue the increase first recorded in 1986. With good cooperation again from the range states of Rwanda, Congo, and Uganda, it took less than two months to find the answer. The mountain gorilla population in 1989 stood at 320, an increase of more than sixty individuals, or almost 25 percent since Bill's 1978–79 survey. The percentage of young remained high at almost 45 percent. All of us who worked with the MGP could be proud of the project's role in this dramatic turnaround.

In the meantime, the Mountain Gorilla Veterinary Program added a component to conservation missing from both the MGP and Karisoke. Our own frustrating experience at the research station had shown that many gorilla lives could be saved by the most basic veterinary intervention. Started by Jim Foster in 1987, with support from the Morris Animal Fund, the veterinary program set out to provide the best available medical care under field conditions to ill and injured gorillas. Its ultimate goal was to assure that as many of these gorillas as possible remained in the wild to contribute to the population's breeding success and survival. In his first years, Foster lived and worked out of the RRAM project guest house in Ruhengeri. Later, he oversaw construction of the Virunga Veterinary Center near the park headquarters in Kinigi. By 1989, several mountain gorillas already owed their lives to Jim, Barkely Hastings, and their Rwandan colleagues.

Following Dian's death, Karisoke returned to the more open management style begun under Sandy Harcourt and Kelly Stewart in the early 1980s. A broader set of research subjects was encouraged and Rwandan students were once again recruited to work at the station. Karisoke was still supported by the Digit Fund and operated independently of both the Mountain Gorilla Project and the veterinary program. Yet under the direction of David Watts and Diane Doren lines of communication were opened between the research center, conservation organizations, ORTPN, and the National University. It would have helped if all of the nongovernmental organizations belonged to a single management consortium, but central office politics and competitive fund-raising precluded this option. Still, the degree of cooperation among the various Virunga field projects in 1989 was better than at any time in the past.

Political support for the park and gorillas peaked in 1989. The Parc des Volcans had recovered from years of neglect to claim its place as the jewel in Rwanda's conservation crown. National authorities were elated by the economic benefits of gorilla tourism, as were many local officials. A small tract of degraded communal grazing land between Sabyinyo and Muside was even returned to parkland because tourist Group 13 commonly passed though the site. Political support, however, did not extend to revenue sharing with local communities. Of the six communes that bordered the park, only one received significant direct benefits from tourism. Kinigi was home to the park headquarters and to most of those who earned a living as guides, guards, or private sector employees. The other five communes received little or nothing from tourism. To address this inequity and reinforce general support for the park, we suggested a form of revenue sharing in which 5 to 10 percent of park entry fees would be distributed in the neighboring communities that bore the greatest burden of park protection. This total amount could be divided by each commune's length of park border, population, or effectiveness in controlling poaching in its sector. All ideas were rejected outright by ORTPN. Still, dozens of local jobs were created by the MGP tourism, security, and education programs; hundreds more positions were supported by the regional tourism economy. Gorilla revenues paid for a quadrupling of the park guard staff to seventy-two—a world record of one guard per square mile. Surplus profits from the Parc des Volcans also helped cover the costs of guards in the unprofitable Akagera Park, as well as the initial contingent of guards assigned to the Nyungwe Forest.

With success came problems. The new ORTPN director, Laurent Habyaremye, came from the hotel business. He recognized gorillas as the "golden goose" of tourism in Rwanda, but he wanted to squeeze out a few more eggs. First he ended a two-tiered system of entry fees, which allowed Rwandan citizens to pay around $5 to see gorillas during nonpeak seasons. Only a few hundred Rwandans ever took advantage of this system, but Habyaremye's decree sent the broader message that the park and gorillas were there for the *abazungu,* not for the native *abanyarwanda.* Symbolically, this was a return to the days of off-limit colonial Crown lands. Workers like Clementine, Isaac, and Radio could never afford the $200 ticket. Next, Habyaremye moved to increase the number of daily visitors allowed to see each gorilla group. Our limit of six tourists per group per day had held constant for the ten years after Bill's encounter with Brutus. In 1989, the ORTPN director decided that the standard should be doubled to twelve. Craig Sholley, who had replaced Jean-Pierre von der Becke as head of the MGP in 1988, dug in his heels in opposition. He turned to us for arguments in support of the old limit of six. Our original rationale remained valid: gorilla safety, control of tourists, and maintaining a high-quality experience that justified a high price. We added that the MGP had increased annual park revenues by 30,000 percent over a ten-year period and that the government ought to be satisfied with this windfall. Craig used these and other arguments in a protracted battle that no one really won. Daily visitation rates were increased slightly to eight tourists per group, with the caveat that only six would be taken to an undefined category of "small" gorilla groups.

The spring of 1989 brought an important reminder of why visitation limits were a critical issue. Six gorillas had died in recent months and everyone was deeply concerned. Autopsies revealed that five had died of respiratory ailments—a likely result of extremely cold, wet weather during the heavy rainy season that year. Taken alone, these results could have served as a sad reminder of the continuing impact of earlier habitat losses that forced the gorillas to live at higher—and colder—elevations. But tests on the sixth dead gorilla revealed evidence of possible exposure to measles. There was no indication when an exposure might have occurred, nor any certainty that she had actually contracted the disease. Yet measles among gorillas could have the same devastating effect already seen among nonresistant populations of humans around the world. Measles was the number two source of infant mortality in Rwanda. A Rwandan was the likely source; it was virtually impossible that an adult Westerner would carry the disease. Yet since this gorilla was from one of the tourism groups, ORTPN acted on the advice of the veterinary program and decided to inoculate all of the gorillas visited by tourists. Nicole Monfort, one of those mobilized to conduct the vaccination campaign, called us in the U.S. to ask what we thought of this approach. We didn't like the idea of widespread darting and inoculation, but we lacked any firsthand perspective from which to make a clear decision about the degree of threat. It was an uncomfortably gray and disturbing set of circumstances. Craig Sholley supported the plan, though, and worked with the veterinary program to inoculate as many Karisoke and tourism gorillas as possible. Infants and possibly pregnant females were not darted, but most other gorillas were vaccinated. No other gorilla has ever contracted or tested positive for measles in the Virungas.

Habyaremye grew to dislike the MGP and its bluntly honest director, Sholley. Habyaremye wrote a bizarre letter criticizing Craig and other whites who "were more interested in living like monkeys in trees than in helping Rwanda solve its problems." The absurdity of the letter made Habyaremye look ridiculous, but the ORTPN director was not going to be stopped. After months of a wasteful standoff, a face-saving compromise was found. The MGP announced that it had achieved its major objectives and was closing down its operations in Rwanda. The Rwandan government officially thanked the project for its contributions—then looked the other way as all major MGP functions were absorbed by a replacement project. The International Gorilla Conservation Project was managed by the same consortium of conservation groups and shared many of its objectives. The difference was that the IGCP no longer focused solely on Rwanda, but took a regional perspective to work in the three countries that shared the Virunga forest habitat and mountain gorilla population. A Dutch woman named Annette Lanjouw served as a roving advisor across the region, with backstopping from Rosalind Aveling, who had moved into a managerial position in Nairobi with the African Wildlife Foundation. Rwandans took over the day-to-day management of tourism, education, and security in the Parc des Volcans, assuming their new responsibilities with few complications. This smooth transition was a final tribute to the handful of individuals who had dedicated themselves to training competent Rwandan counterparts during the ten-year run of the Mountain Gorilla Project.

L AURENT HABYAREMYE was more supportive of conservation activities and organizations in the Nyungwe Forest in 1989. In part this reflected his businessman's view of the vast southern forest as an emerging tourism market and revenue magnet. In part, it reflected the rising political profile of Nyungwe. After a surprise visit to the forest and Nyungwe project headquarters, President Habyarimana personally ordered the removal of two illegal settlements within the forest interior. Pindura and Karamba were shantytowns along the main road that served as market centers for gold miners and various illegal activities. The closing of Pindura allowed the development of a nearby tourism center at Uwinka. Karamba, near the reserve's western border, became the headquarters of the new ORTPN presence in Nyungwe. Habyaremye, however, did not provide any operating or construction funds for the contingent of guards assigned to the forest. Rather than take the standard bureaucratic approach of not working until he received the funds, chief *conservateur* Shaban Turykunkiko ordered his men to build their own housing and office structures. While these positive steps were taking place, the Swiss were completing a road that linked their center of operations to the north with the main road through the forest near the recently evacuated town of Pindura. Many more trees were cut and much more land was cleared for this road than the former residents of Pindura could have ever imagined, yet the road proceeded unopposed.

While Swiss actions sometimes clouded their clear intentions for conservation, the World Bank in Nyungwe acted with all the passion of a recent convert to preservationism. The Bank planned no forestry operations in their southeastern sector until the results were in from a massive biological inventory. Roads, too, were on hold. The inventory project cut survey transects, though, at one-kilometer intervals across the entire Bank sector, providing

excellent access for poachers and illegal woodcutters. The transects also generated considerable new data about the forest. Rwandan biologists collected
this information, but no one was trained to process or analyze the resulting
massive data sets in any meaningful way. Nor was the forest service prepared
to base its actions on indicators of diversity and disturbance that were poorly
understood and unrelated to any of their traditional management guidelines.
Ultimately, most of the Bank's inventory effort was wasted.

A complementary research program, based at Uwinka, was started by
Tim Moermond of the University of Wisconsin. His dedicated team of
graduate students—Beth Kaplan, Chin Sun, and Kurt Kristensen—focused
on the role of birds and mammals in dispersing fruit seeds throughout the
forest. Tim used his Fulbright position at the National University—where
his infectious enthusiasm overcame his serious limits to expression in
French—to attract students and professors to the project. These researchers
and local assistants then lived and worked with Beth, Chin, and Kurt at the
Uwinka site—a hands-on training process that produced the best field biologists in Rwanda. Meanwhile, Amy's former research assistant in Nyungwe,
Samuel Kanyamibwa, finished his own Ph.D. to become Rwanda's first
doctoral-level field biologist. Amy was disappointed when Kanyamibwa took
a position at the National University and subsequently spent less time in the
field. However, she also knew that this was a common phenomenon across
Africa and many other developing countries where the attractions of job security, status, and an urban existence generally outweighed those of a career
in field biology, which include cold showers, rain gear, and muddy boots.

The biggest change in Nyungwe in 1989 came in the form of a USAID-
funded initiative called simply the Projet Conservation de la Forêt de
Nyungwe. PCFN was managed by the Wildlife Conservation Society, with
Amy as the project's principal advisor and Bill as director of the WCS Africa
Program in New York. The increased funding from USAID supported a
greatly expanded range of activities and raised the political profile of
Nyungwe to a much higher level within the government. Tourism was the
most visible new activity. Under the direction of Chris Mate, a Peace Corps
volunteer assigned to the project, a scenic network of trails was created in the
area surrounding Uwinka. Visitors could arrive at the main office overlooking Lake Kivu and pick from a series of interlinked trail circuits through the
forest below, each ranked by difficulty. There were no wrong selections.
Chris had a great eye for landscape aesthetics and the use of topography to
enhance wildlife viewing possibilities. His Rwandan crews created sinuous
trails that combined gradual climbs—not a traditional Rwandan characteristic—with eye-level opportunities to view monkeys and birds in the canopies

of valley bottom trees. The huge associations of black-and-white colobus, first identified by Amy, were now followed for tourism purposes. Almost all of the wildlife guides and trail crews came from the village of Banda, a forty-five-minute walk from Uwinka on the forest's northern border. They were a hardworking group, led by Félix Mulindahabi and Martin Sindikubwabo, whose dedication would be tested in ways unimaginable in 1989.

Four hundred and forty-one paying visitors came to Nyungwe in 1988, the first year of organized tourism. In 1989, 2,896 people visited the forest. They paid $5 each to hike Nyungwe's trails and observe its birds and monkeys. Most paid an additional $5 for a guided tour to see the colobus. The majority of visitors were expatriates who came first to see mountain gorillas, then extended their stay by two or three days to explore Nyungwe. ORTPN was ecstatic with this development of a second international tourism site, which was already generating another $1 million per year in valuable foreign revenues. Visitors were pleased, too. Not only did the colobus provide an instant attraction, but the abundance of birds and primates offered an exceptional wildlife experience. The trail network opened an unknown world of giant rain forest trees, colorful butterflies and orchids, strange sounds, and exotic smells. In addition, the guides in Nyungwe quickly gained a reputation as the most knowledgeable and personable in the country, eclipsing the gorilla guides as Rwanda's best ambassadors to the outside world. Wilderness Travel was one of the first groups to combine Virunga and Nyungwe visits, as well as specialized bird-watching tours. As 1989 came to an end, Wilderness Travel's field coordinator, Brad Goodhart, was adding capacity and reserving tickets as fast as they were offered. The combined experience of the Volcanoes Park and the Nyungwe Forest would soon be cited as successful models for the ecotourism movement that emerged in the 1990s.

Nyungwe was also primed to contribute to the interrelated fields of "sustainable development," "integrated conservation and development," "development through conservation," and "community-based conservation" that emerged and evolved through the 1990s. The government's Five-Year Plan still called for a core protected area within the forest. It also mandated controlled use of certain forest products in areas bordered by human settlements. All of the major donor projects acknowledged this need, but no one was prepared to take the first step. We stepped into this void feeling we needed to find the least damaging use options to preclude more destructive activities. We suggested several small-scale projects that might test the concepts of sustainable use without subjecting the forest ecosystem or its core protected areas to serious risk. These included gathering medicinal plants, collecting honey from man-made hives, and selective harvesting of bamboo for basket-making. Everyone agreed that these should be manageable, but we had many concerns. How do you monitor use? How do you control overuse? How do you limit illegal activities by people allowed in the forest for legal purposes? We didn't have answers, but we felt that Nyungwe might be large and resilient enough to offer a laboratory to test various systems for monitoring and careful management, from traditional enforcement to creative new systems of community control. What we needed were donors and the Rwandan forest service to open up to the idea of small-scale experimentation and careful monitoring.

In 1989, the first conference ever held on the conservation of mountain forest ecosystems in Africa was organized by the Wildlife Conservation Society, funded by USAID, and hosted by Rwanda. The selection of Rwanda was a reflection of its emergence over the preceding ten years as a model for its neighbors. It was a model in the designation of protected areas; a model in building international, national, and local support; a model in ecotourism development; and an emerging model for community involvement in a part of the world with little local democracy. The meetings took place on the western edge of Nyungwe, with field trips into the forest. More than forty participants came from Rwanda, Congo, Uganda, and Burundi. The conference was marked by a high degree of cooperation and enthusiasm. But it was the last time that the four neighbors would cooperate without constraints and recriminations.

CHANGES IN THE RWANDAN landscape in the late 1980s were visible, yet surprisingly subtle. Surprising because the Rwandan population had increased from 4.8 million to almost 7 million people over the preceding decade. Perhaps 100,000 of those additional people lived in Kigali or other towns; the other two million or so were dispersed across the countryside. We noticed that the line of fields in many places had risen higher onto the steepest slopes. Elsewhere, there were simply more active fields—and fewer fallows—than before. Wetlands offered the most conspicuous transformation. Traditionally a dry season reserve for grazing or home to a few scattered raised fields, almost every valley bottom in 1989 was marked by the same pattern of parallel raised beds. Only population pressure and the need to feed more mouths could justify the considerable expenditure of labor to create those mounds and maintain the flow of water that irrigated high-yield sweet potato crops. We still counted crowned cranes in these wetlands, as we had since our first truck rides between Kigali and Ruhengeri. But there were fewer cranes to count and the population of this long-lived species was en-

tering a period of slow death: deprived of its breeding sites in natural wetlands, the aging population would eventually disappear.

For several days in the spring of 1989, torrential rains pounded northern
Rwanda. The exposed and overworked soils of the region could not hold the
accumulated charge. Walls of mud roared down from hundreds of hillsides,
carrying millions of tons of crop debris, soil, and water into the already engorged bottomlands below. Thousands of homes were damaged and hundreds of people died. The raised beds in the valley bottoms were covered by
giant fans of silt, or carried downstream to wreak further damage. Many of
the worst slides occurred in areas identified as "high erosion risk" by the
RRAM project. Bill even received a few perverse compliments on the
project's predictive skill. In fact, the RRAM analysis was intended to predict
erosion of the kind susceptible to human controls. The catastrophic landslides seen in the spring of 1989 resulted from exceptional rainfall, and a century of stretching the land's capacity to support its growing human
population.

Those of us who had lived in Rwanda in the 1970s were struck by the network of paved roads that crisscrossed the country by 1989. From Rusumo
Falls on the southeastern border with Tanzania to Gisenyi in the northwest;
from Cyangugu on the Congolese border in the southwest to three different
points along the northern border with Uganda, an asphalt gridwork covered
the major axes across the land. Interior connections, like that across
Nyungwe, were being added on a steady basis. The newest, between Gikongoro and Gisenyi, was the most improbable—until one remembered that it
passed through the Valley of the Kings and added a second point of access to
the increasingly mobile elites of Gisenyi and Ruhengeri. The number of vehicles using these roads had increased tenfold since 1978, and the percentage
of Mercedeses and other luxury vehicles was also rising. Public transportation consisted of a fleet of new Japanese-made minibuses that replaced the
open-backed pickup trucks of our early years. A high percentage of vehicles
were labeled with the names of the projects and agencies for which they
worked. For most Rwandans, though, paved roads represented little more
than walkways free of mud and dust. Ancient footpaths etched in the hills
still carried them where they needed to go. Traditional round houses were
giving way to rectangular forms, their attractive thatched roofs replaced by
shiny caps of corrugated tin. It wasn't clear whether the trend reflected
greater prosperity or changing cultural preferences. In towns across Rwanda,
Coca-Cola was doing its best to turn the urban landscape red and white and
squeaky clean.

THE MOST DRAMATIC TRANSFORMATION in Rwanda was invisible. Across the country, from its expanding urban centers to its rural heartland, death stalked the land.

Rwanda in 1989 was at the center of the AIDS epidemic in Africa. The figures passed quickly from mouth to mouth, from factoid to factlet. Eighty percent of all prostitutes tested positive for HIV exposure. More than half of the Rwandan army was infected. Thirty percent of the blood supply was infected. Twenty-eight percent of all mothers giving birth at the Kigali hospital were HIV positive. Many African countries responded to the threat with official denials and threats against anyone who wrote that AIDS was even present in their countries. Rwanda was different. The government responded quickly and constructively to the very real threat, whatever its leaders believed or feared from the figures. Booklets about AIDS were translated into Kinyarwanda and tens of thousands distributed to people like Clementine throughout the country. Foreign researchers were welcomed, and various populations were made available for some of the world's most comprehensive AIDS studies. The fact that HIV transmission in Rwanda occurred almost entirely between non-drug-using heterosexual partners made these studies a valuable complement to parallel studies in America and Europe. Research teams came from government agencies and universities in the U.S., France, and Belgium. If AIDS researchers weren't the most numerous foreign assistance constituency in Rwanda by 1989, they were the most visible.

Shortly before we left Rwanda, Bill took Noah to see Léopold. He was still on the RRAM payroll as a chauffeur, but he had been too sick to work for the past few months. A Belgian doctor confided what Bill had feared. Léo was dying of AIDS, the victim of years of blood transfusions—a common Rwandan medical practice—from a spreading pool of infected blood. If Léopold knew of his death sentence, he had not mentioned it in earlier visits. He did note with a wry smile that none of the others brought into his four-bed suite had left alive. When Noah entered his room, Léo was alone. He was terribly thin, his cheeks so sunken that Noah didn't recognize him at first. Léo looked at Bill and smiled, then added in a gentle voice, *You can look at me, Noah, I'm not dead yet.* Noah looked up and smiled at his sometimes playmate and family friend. It was a good visit. We talked about his newborn daughter, the boys, the project. Noah told stories from Nyungwe. We filled time with warm conversation and left with a round of hugs. Léopold died a few days later.

Léopold was lucky that he died quickly and retained his mental health to the end. Thousands suffered more protracted and horrible deaths, often preceded by a complete physical breakdown and dementia in a country with few facilities for the treatment of even the most conventional diseases. Ironically, we would soon learn that Vincent Nyamulinda, Léopold's antagonist within the RRAM project, was also infected. Before leaving for graduate school in the U.S., Vincent had increasingly lived the lifestyle of the professional male in a culture with a pronounced double standard. Successful men were expected to have mistresses. Unprotected sex was the norm. Death was the result.

Léopold and Vincent took very different paths to their deaths, but the killer was the same. In following years, Laurent Habyaremye would fall to AIDS, as would one of his top assistants at ORTPN. So, too, would the director of forestry, Phénéas Biroli. Several hundred thousand Rwandans from all walks of life eventually would be infected. Cruel voices whispered that AIDS would help control rampant population growth. But the truth was that AIDS in Rwanda and across much of Africa struck first at the thin layer of professionally trained males who were responsible for planning and implementing desperately needed programs in education, agriculture, health, economic development—and conservation. As more women were infected by their husbands and lovers, the percentage of children with HIV and AIDS increased dramatically. Orphanages swelled and multiplied. The medical, economic, and human toll was incalculable as the deadly plague spread across the land.

We visited Rwanda frequently throughout the late 1980s. There was a growing sense at the end of the decade that Rwanda was facing a disaster of unprecedented proportions. It was universally believed that that disaster was AIDS. But another catastrophe would soon shake the nation to its foundations.

Section Five

In the Face of Madness

Chapter Thirty

The Cauldron Churns

ON OCTOBER 1, 1990, several hundred soldiers dressed in Ugandan army fatigues overwhelmed the small Rwandan border guard force at Kagitumba in northeast Rwanda. They were the vanguard of the Rwandan Patriotic Front, or RPF: Tutsi refugees dedicated to returning to their homeland in Rwanda—by force, if necessary. They were led by Major-General Fred Rwigyema, a charismatic veteran of the long bush war that overthrew Milton Obote and brought Yoweri Museveni to power in neighboring Uganda. Rwigyema and his men had fought with distinction in Uganda and they were prepared for another long struggle in Rwanda. But they hoped to achieve more immediate results with their surprise attack.

Within a few days, the RPF forces★ advanced forty miles to take over Gabiro, headquarters of the Akagera National Park, and rout a nearby Rwandan army garrison. On the second day, however, Fred Rwigyema was killed by a single shot to the head. Like the loss of Stonewall Jackson to the Confederacy, the death of "Commandant Fred" cost the RPF its most popular and inspiring leader in the field. And like Robert E. Lee, the RPF commander-in-chief, Paul Kagame, lost his most daring and effective military commander. The RPF offensive sputtered. The Rwandan government's army recovered from its initial shock and soon rushed superior numbers and equipment into the battle. Supplies and logistical support were openly provided by the French. The Rwandan army even staged a fake nighttime "at-

---

★ The military wing of the RPF was technically called the Rwandan Patriotic Army. Both groups are lumped under the RPF umbrella in this section to avoid confusion.

tack" on Kigali, complete with artillery and bomb bursts, that succeeded in scaring the expatriate community enough to pull French and Belgian paratroopers into the capital. These European soldiers were ostensibly there to protect their own citizens, but their presence effectively served to reinforce the Rwandan army.

The RPF attack failed. Kagame was forced to rally his scattered and dispirited troops, retreat from Akagera, and seek refuge where he could reorganize his proud but battered forces. Like Mao, Castro, and outnumbered insurgents throughout time, he found safe haven in the mountains—this time in the Virunga volcanoes. As Museveni, his mentor and colleague, had done in southwestern Uganda, he used the forest to shelter and sustain his men while plotting his next move.

In 1981, Bill wrote that the Rwandan government would always have higher priorities than conservation because, after all, "gorillas don't overthrow governments, guerrillas do." It would be satisfying to cite this quote as a prophetic warning, except that Bill was thinking of disgruntled Hutu peasants. Neither of us imagined that the guerrillas would be Tutsi refugees from Uganda. When the October attacks occurred, we were angry. *Who are these people and why are they disturbing this poor, struggling nation?* Their retreat into the Virunga forest made us worry about the gorillas. The potential for further mayhem raised concerns about the various projects we had started and still sustained. The beleaguered peasants of northern Rwanda certainly didn't need this new burden. If anyone should be revolting against the government, it was they. The aggressors weren't even "real" Rwandans in our eyes. They were young men who had grown up entirely in Uganda. They spoke English. Their leaders had names like Fred and Peter; even their organization had an English name. To us, this was a Ugandan invasion of Rwanda. But we lacked the Tutsi perspective.

In the three decades since the 1959 overthrow of their monarchic rule in Rwanda, the Tutsi had gone from colonial kingpins to members of a vast network of displaced persons. By 1990, as many as a million Tutsi refugees and their descendants were dispersed around the world. Thousands lived in New York, Montreal, Paris, and Brussels. Hundreds of thousands lived in neighboring Burundi, Congo, and Uganda. They stayed in touch through newsletters and informal networks that sustained their identity as Rwandans. But the reality of their homeland faded with the first-hand memories of their elders, and many younger Tutsi sought success in their adopted lands with the drive of second-generation immigrants any-where.

In the years after the failed 1990 invasion, Amy spoke with several RPF soldiers. They described growing up in Uganda and joining the rebel forces of Yoweri Museveni in the early 1980s. They were attracted by the appeal of Tutsi leaders like Kagame and Rwigyema, who were already trusted lieu-tenants in what was called the National Resistance Army, or NRA. They were dedicated to the overthrow of Milton Obote, who was strongly anti-Tutsi and in many ways more ruthless than his predecessor, Idi Amin. But they also joined with the clear idea that the Uganda liberation struggle would provide training for their own eventual return to Rwanda. As the NRA gained strength, Obote turned up his anti-Tutsi rhetoric in an effort to un-dermine Museveni, who came from the related Ankole ethnic group. In Oc-tober of 1982, more than thirty thousand Tutsi were driven out of Uganda into Rwanda's Akagera Park. Bill visited one of their camps, its thatch huts abandoned as quickly as they were built, leaving dead livestock and freshly dug graves behind. Vultures stood guard on termite mounds as heavy rains worked to erase all signs of human passage. Nearby, the International Red Cross and the U.N. High Commission for Refugees created a tent city sec-ond in size only to Kigali. Within a few months, the surviving refugees fil-tered back into Uganda, denied the right to remain by the Hutu government of Rwanda. The international relief agencies packed their camps, but they would return.

Museveni and the NRA succeeded in overthrowing Obote in 1986, but talk of "too many Rwandans" grew louder among Uganda's other ethnic groups. The powerful Baganda business class was especially bothered by growing economic competition from the Tutsi community. Most visible were military leaders like Kagame and Rwigyema, who now occupied senior positions in the Ugandan army. Meanwhile, the Tutsi within the Ugandan military openly created the RPF in 1987 with the express goal of promoting repatriation to Rwanda. A smaller group secretly made plans for armed ac-tion, if necessary. The following year, a World Congress of displaced Tutsi passed a strong resolution demanding the "right of return to Rwanda" for its people.

The Hutu perspective on events was also different and more complicated than most outsiders could imagine. The bubble of the government's gener-ally good performance—and excellent foreign press—burst in 1988. Up to that time, the "Habyarimana regime . . . was in general one of the least bad in Africa if one considers only its actions and not its intellectual underpinnings," according to historian Gérard Prunier.★ Through most of the 1980s, Rwanda outperformed its neighbors in terms of per capita economic growth, expanded its agricultural productivity, increased its spending on education, and created classroom openings for more than 60 percent of its primary-school-age children. Nine percent of those seats were reserved for Tutsi, in line with conservative government estimates of their representation in the larger population. This was a sharp decline from the Tutsi's favored status under the Belgians, but a marked improvement over the extreme discrimination and repression faced by ethnic minorities in many other African countries. The same 9 percent share was supposedly reserved for government posts, but access to government power was actually much more restricted, and senior military representation of Tutsi was nonexistent. If Tutsi were to succeed in Hutu Rwanda, they would have to do so in the business sphere, where fewer restrictions existed.

In 1986, the world market for coffee crashed, devastating countries like Rwanda that were overly dependent on a single export commodity. The regional drought of 1988–89 dealt a further blow to millions of Rwandan farmers dependent on subsistence production from their withered fields. Deprived of its principal source of foreign revenue from coffee sales, the government slashed its budget by 40 percent in 1989. The following year, the Rwandan franc was officially devalued by 40 percent as part of a structural adjustment loan from the International Monetary Fund. The black market devaluation was even more dramatic. International economic assistance had risen steadily under the perceived benevolence of the Habyarimana regime. Now this foreign infusion of money was essential to the most basic functions of government. Even more important, it became the lifeblood of the increasingly corrupt elite that surrounded the president.

No one was more elite than Madame Habyarimana herself. Born Agathe Kanzinga, she was a direct descendent of traditional Hutu rulers from the Gisenyi region. Her family was wealthy and powerful, unlike that of her self-made husband from a neighboring Ruhengeri prefecture. Using her husband's position, she brought three of her brothers—including Protais Zigiranyirazo, the *préfet* of Ruhengeri—and other family members and close associates into top government and advisory roles. Together they formed the *akazu*—the "little house" that represented the true power behind the throne.

★ This chapter and the next draw on multiple sources, including direct personal experience, firsthand accounts from friends and colleagues, and published reports in the international press. Both chapters owe a considerable debt to excellent works of recent history by Gérard Prunier *(The Rwanda Crisis: History of a Genocide)* and Philip Gourevitch *(We Wish to Inform You That Tomorrow We Will Be Killed with Our Families)*.

Nowhere was the greed of the *Clan de Madame* more evident than in the corruption associated with the World Bank's cattle-raising scheme in the Gishwati forest. The blatant openness of this corruption, with its land grabs and displacement of intended local beneficiaries, undermined the previously unquestioning support of even the local Hutu majority. To reinforce its hold on economic and political power, the inner circle also seemed prepared to kill. Prying journalists, critical legislators, and competing military officers were cut down in a remarkable series of murders during the late 1980s. Even President Habyarimana's choice as eventual successor appears to have been assassinated on orders from within the *akazu*.

The spreading corruption and violence undermined respect for authority, especially that of the president. Juvénal Habyarimana had been easily re-elected as the single candidate of a single-party state in 1983 and 1988. A new constitution allowed him to run at least one more time. Yet by 1990, Rwanda's foreign supporters were growing tired of the Habyarimana regime and its increasingly rotten core. With the recent end of the Cold War, tolerance for Africa's West-leaning dictators was waning. French President François Mitterrand told Habyarimana that Rwanda must open the door to competing political parties. The United States began to insist on democratization as a condition for continued economic assistance. Habyarimana heard the chorus, as did his opponents. The president announced his support for multiple parties and elections in July 1990; a call for more expansive democratic reform was issued by an anonymous group of Rwandan intellectuals in August. With tensions rising, and no guaranteed place at the newly reconfigured political table, the RPF launched its October attack.

IN EARLY APRIL 1991, George Schaller walked into Bill's office at Wildlife Conservation Society headquarters in New York. He placed a small tape recorder on the desk and walked out, adding cryptically, "I thought you'd like to hear some sounds from Rwanda." In the place of chestbeats or gorilla vocalizations, the tape projected a staccato stream of automatic weapon fire, punctuated by the bass thump of heavy artillery. In January, George and his wife, Kay, had fulfilled a dream of returning to the Virunga volcanoes and mountain gorillas that were the focus of their first field study, more than thirty years earlier. Film crews from National Geographic and IMAX accompanied them there to film the occasion. The Rwandan government was eager for good publicity, but a military guard accompanied the crews in case of problems. On the morning of January 22, just as George and Kay climbed to the top of Mt. Visoke, a radio message arrived advising them to evacuate the park. By that afternoon, they had reached in transit house in Ruhengeri. The next morning, Kay awoke at 5:30 to what sounded like firecrackers. Three months after their strategic retreat, the RPF forces were again on the attack.

Throughout that day, George recorded the sounds of the assault that raged around them. Sometime after nightfall, a convoy of French paratroopers and government soldiers reached their house and escorted the Schallers, film crews, and other expatriates south to Kigali. The RPF forces could not hold Ruhengeri in the face of a counteroffensive from the much larger Rwandan army. Yet their lightning strike accomplished many of their goals. They captured weapons and ammunition from the Ruhengeri military camp, as well as vehicles and gasoline from town. They freed nearly one thousand prisoners from the infamous Ruhengeri prison, including Théoneste Lizinde, a former close associate of President Habyarimana who was convicted of plotting to overthrow the government in 1980. Most of all, the attack shook the confidence of the Rwandan public in its government and military. After all, Ruhengeri was the birthplace of the president, second only to Gisenyi as a center of political power. As fears of renewed RPF attacks spread, several hundred thousand local residents moved out of the battle zone to camps farther south. In Uganda, local populations had greeted Museveni and Kagame as rescuers from an unpopular regime, but in Rwanda, Kagame's Tutsi troops were not welcomed by the northern Hutu.

In the wake of the first RPF attack, marauding gangs of Hutu men killed at least three hundred of their Tutsi neighbors in a wave of ethnic killing that swept over the Gisenyi prefecture. Following the attack on Ruhengeri, hundreds more Bagogwe men, women, and children were slaughtered in the highlands around the Gishwati Forest. Although the Bagogwe had lived in peace with their mountain Hutu neighbors for centuries, their distant connection with the Tutsi doomed them to be hunted down and crushed like *inyenzi:* the derisive "cockroach" label used by some Hutu to describe the Tutsi. Their killers were mostly poor, rural Hutu. Those who incited the killing were the same Hutu leaders who had cheated the Bagogwe herders— and impoverished local Hutu—out of their rightful lands in the Gishwati cattle-raising scheme.

⌒⌒⌒⌒

OVER THE NEXT TWO YEARS, the Rwandan cauldron churned. Military skirmishes were interspersed with seemingly endless political negotiations, delays, plots, counterplots, and occasional agreements that were disavowed or undermined before the ink was dry. Our firsthand experience with events was reduced to a series of punctuated views as visitors, since we were no longer permanent residents. We had planned to live in Rwanda every summer when Noah and Ethan weren't in school, but that plan was scrapped with the outbreak of hostilities. Instead, we made regular visits to Rwanda and more than a dozen other countries in our roles as senior managers for the WCS Africa Program.

When Amy arrived in Rwanda for a five-week stay in June of 1991, there were many positive signs. The government had revised its constitution to allow for multiple political parties, and four new parties had already announced their intention to contest the next round of elections. They reflected the prevailing European political spectrum from social democrats on the left to Christian democrats on the right. All claimed to represent the Hutu majority. A notable exception was the Parti Libéral, which tried to reach across ethnic lines. This reflected the thinking of one of its leaders, Landoald Ndasingwa, a Tutsi businessman married to a Canadian. We had known Lando and Hélène since the days when they ran the American Cultural Center in Kigali, then as owners of La Taverne and Chez Lando, two popular restaurants in Kigali. The politics of harmony and inclusion fit with Lando's worldview and personal life, and even some key Hutu supporters were attracted to a party where ethnic issues were secondary. But the Parti Libéral remained a minor group.

Driving through the countryside, Amy was struck by the appearance of colorful flags proclaiming adherence to the various parties. The same flags and banners were carried by enthusiastic supporters in political marches and demonstrations. These were new and heady developments in a country with no history of comparable open expression. The excitement was palpable in her discussions with some Rwandans, especially those from the National University, who tended to support the new opposition parties.

Meanwhile, thousands of Tutsi were summarily arrested in the wake of the first RPF attacks, including the Nyungwe project administrator, Marie-Paul Sebera, and a former Nyungwe project employee, Eugène Rutagarama. Eugène was working for ORTPN in the Parc des Volcans when he was arrested as a suspected collaborator. He was among those released in the Ruhengeri prison raid; then he fled to Burundi. Marie-Paul was held for five months in the Butare prison before being released to join her family in Congo. The American director of the WCS Nyungwe project, Rob Clausen, lobbied to gain Marie-Paul's freedom, then eventually resigned to assure the safety of his own Tutsi wife and their children. Such precautions were reinforced as rumors of the Bagogwe massacres filtered out of the north. Leaving Rwanda on July 23, Amy found herself seated next to Alison DesForges, a specialist on ethnic relations in Rwandan history. DesForges had just completed a review of the situation on the ground for the organization Human Rights Watch—Africa. Her conclusions were sobering, contradicting the burst of optimism that had greeted Amy's arrival barely one month before.

Political interactions were heating up by the time Bill arrived in May for the first of two visits in 1992. The creation of an extremist Hutu party—the Coalition pour la Défense de la République, or CDR—introduced physical violence as a political tool. Demonstrations by rival parties now routinely ended with attacks by young CDR thugs, eerily reminiscent of the role of the Brownshirts in Hitler's early rise to power. Deadly grenade attacks in crowded markets and taxi buses spread terror among the general public. When Bill returned in November the increased use of land mines on rural roads in southwestern Rwanda made his drives through the Nyungwe Forest disquieting. Recently burned communal woodlots along Nyungwe's western border also testified to growing resentment of central authority, especially among southern Hutu. Observing the charred remains of one woodlot, a French forester repeated a slogan from the 1968 student riots in Paris: *incendie, démocratie . . . incendie, démocratie.* Democracy to some was the freedom to burn government-owned woodlots. Later, it might mean the freedom to clear and farm the Nyungwe Forest Reserve. Part of Bill's mandate in November was to participate on an international team reviewing the government's Five-Year Plan for managing the country's forest reserves. He was responsible for a report on ecotourism opportunities and socioeconomic factors in conservation. Fighting and terrorism had reduced tourism to a trickle by late 1992, however. And it was difficult to focus on conservation concerns in the darkness of a widening civil disturbance that threatened to overwhelm the social and political order of the entire nation.

In late June of 1993, Amy arrived in Kigali for the beginning of a twenty-three-day visit. Officials at USAID were eager to discuss the Nyungwe project in which we were partners, but they tried to dissuade Amy from a field visit. A recent campaign of terror in nearby Cyangugu prefecture had included rapes and killings of Tutsi civilians. Despite the presence of government soldiers and tanks, U.S. officials felt the situation was insecure. Amy was unaware of most of the ugly details of the situation and believed that she could avoid any serious trouble. She also felt a duty to show our field staff—American and Rwandan—that WCS cared enough to share their work conditions in the field. She made the trip south, limiting her travel to daylight hours, and experienced no problems. The number of roadblocks near our Cyangugu office, though, made travel in that area uncomfortable. Most of the soldiers were very professional, but some were clearly drunk. Every checkpoint renewed the discomforting realization that the young men behind the guns held almost unbridled power over those they stopped.

Once out of Cyangugu and back in the Nyungwe Forest, Amy thought life seemed almost normal. Rob and Cheryl Fimbel were now running the Projet Conservation de la Forêt de Nyungwe—a complex program of research, monitoring, and management—with the help of a few Rwandan senior staff and dozens of dedicated local residents who worked on various projects. Rwandan staff also maintained basic tourism services for those who still came to visit the forest. Tim Moermond, Beth Kaplan, and Chin Sun from the University of Wisconsin continued their research in Nyungwe with Rwandan colleagues like Isaac Kabera. In Butare, Amy met with students and faculty at the home of Samuel Kanyamibwa, her former assistant, who was now teaching ecology at the National University. The university crowd was caught up in the enthusiasm of recent agreements among the government, a coalition of opposition parties, and, for the first time, the RPF. Ultimately approved in August of 1993—and known collectively as the Arusha Accords—these agreements called for a transition government, a multiparty assembly, and a unified military before full elections could be held. The pace of change was dizzying and the scope was daunting. The excitement at Kanyamibwa's house was real in anticipation of a blueprint for peace. But one realist noted that "agreements are easy; the hard part is making them work."

WE CONSIDERED OURSELVES careful observers, attentive to changes both dramatic and subtle in a land where we had lived for many years. We had many friends who were still year-round residents of Rwanda. But our view was limited as both *abazungu* and visitors. As white foreigners, we traveled through a surreal world in which the dark underside of life was almost always out of view, constantly rotating away from us like the far side of the moon. The Rwandan government was hiding most of its plans and deliberations for what it saw as an internal affair between Hutu and Tutsi. Our perspective was further limited as transient visitors. We saw vivid snapshots and segments of day-to-day life in motion. We joined in lengthy discussions of events and their implications with Rwandans and other residents. But still we couldn't follow the changing cast of characters and complex plot.

Kurt Kristensen was one of many foreigners who lived through the increasingly chaotic events of the early 1990s. After two years of research in Nyungwe, he completed his master's at the University of Wisconsin in late 1992, then returned to live with his family in Kigali, where his wife directed U.S. AIDS research. Kurt met with Amy in the summer of 1993 to discuss his plans for an expanded gold mining study in Nyungwe. But he also wanted to raise his concerns about the escalating violence in Rwanda. From his porch, Kurt could watch a steady stream of demonstrations along a major road through the valley below. Over time these demonstrations became more and more aggressive, commonly ending in confrontation and bloodshed. Gunshots and grenade blasts punctuated the night. Reports of political killings were widespread. Kurt didn't understand much Kinyarwanda, but fearful Rwandan colleagues recounted increasingly inflammatory anti-Tutsi broadcasts over the radio. These included frequent repetitions of a Hutu supremacist tract known as the *Ten Commandments*. This racist screed castigated Tutsi businessmen as thieves, Tutsi women as seductive secret agents, and any Hutu who helped a Tutsi as a traitor to his kind. Kurt repeated several of the "commandments" to Amy and noted growing tensions among the mixed Hutu and Tutsi staff who worked at his house. Two Dobermans and a shotgun reinforced the level of his concerns.

The blur of events and the protective cover of a secretive culture made it impossible for any foreigners to see what was really happening in Rwanda. Only after the fact could historians analyze the surface flow and underlying currents of events in Rwanda at that time. Ethnic killings were silently spreading across the countryside. Most were localized and unreported. The persecution of the Bagogwe and Tutsi in the northwest was more systematic. Several hundred Tutsi had been rounded up and executed by organized death squads in the southern Bugesera region in March of 1992. The language of day-to-day work and farming was increasingly perverted in the cause of death. The collective labor of *umuganda*—organized groups of men who ostensibly volunteered to help with public works—provided cover for killers. Communal death squads "went to work in the field." Their members were called *interahamwe*, or "those who work together." Their task was to "clear the brush." Their targets were the *inyenzi*, or Tutsi "cockroaches," and their *ibyitso* "collaborators" among the Hutu. Behind this lethal perversion of language was the *Clan de Madame*, extending its power though a clandestine network of local supporters. In May 1993, even the popular leader of a major opposition party, Emmanuel Gapyiyi, was gunned down in front of his house.

RESPONDING TO MASSACRES of Tutsi in northwestern Rwanda, and eager to bolster their negotiating position with the government, the RPF launched another major attack in early 1993. Sweeping south through the prefectures of Ruhengeri and Byumba, RPF soldiers even threatened Kigali until French forces blocked their way. Nearly one million Hutu refugees flowed south, stretching the capacity of the government and international relief agencies to provide the most basic services. The Virunga region was largely abandoned to the RPF. So was the fate of its gorillas. Karisoke was occupied and stripped of essential equipment by RPF personnel, who recorded each detail on a list of items, almost all of which they returned after the fighting ended. A veil then closed over the Virungas, preventing outside observation.

In August of 1993, the final Arusha Accord was signed in Tanzania. The Rwandan population was ready for peace. Thousands of Tutsi were already dead, but the Hutu majority had its own concerns. The massive human displacements caused by fighting in the north compounded the economic suffering caused by the devalued currency, doubling of the national debt, and near tripling of the defense budget. The novelty of political expression had worn off, replaced by a general distaste for the strange diet of political maneuvering force-fed to a largely uninterested population. Yet the final Arusha Accord was a prescription for paralysis in the guise of progress. It allocated twenty-one cabinet positions across six groups—the existing government, four opposition parties, and the RPF—yet required a two-thirds vote for any action. The transitional assembly suffered from an equally well-intentioned and dysfunctional imbalance. The president was to be a European-style figurehead. Habyarimana chafed at this prospect and pursued a policy of delay. He proved especially adept at playing opposition parties against each other and coopting key individuals. Yet time was running out.

Soon after the accords were signed, Radio et Télévision Libre des Mille Collines began transmission. Radio Mille Collines, as it was known, was the first privately licensed broadcasting network in the history of Rwanda. But its message combined general support for the Habyarimana government with strident advocacy for the Hutu cause. Increasingly, it became the voice of the Hutu Power militants within the CDR and other extremist opposition parties. In a language that few foreigners understood—but which the RPF monitored with rapt attention—Radio Mille Collines spewed forth a venomous brew of anti-Tutsi propaganda. *Rwanda's problems are not among its Hutu majority; they are caused by the "devil" Tutsi.* This ethnic bias ignored deep regional divisions within the Hutu population and widespread dissatisfaction with the government. Yet the mix of satanic imagery, ancient antagonisms, increasingly hateful language, and modern music proved attractive to the public. Radio Mille Collines was especially popular among the tens of thousands of young Hutu men who had no jobs, no access to farmland, few prospects for marriage, and a high likelihood of dying from AIDS. They were tinder for the sparks of hatred that streamed forth from their radios.

On October 21, Melchior Ndadaye was assassinated. The murder of the popularly elected Hutu president of Burundi at the hands of rebellious Tutsi military officers convulsed the region. At least fifty thousand Burundians died in the ensuing fighting; 150,000 Tutsi fled the countryside for the relative safety of their stronghold in the capital of Bujumbura; and several hundred thousand Hutu sought refuge across the border in Rwanda. Events played perfectly into the hands and mouths of the ethnic agitators at Radio Mille Collines. The radio increasingly took over the communication functions of the Habyarimana government, which it saw as all too ready to accommodate the Tutsi. Demonstrations to undermine the accords signed by the government were announced over the radio. Even individuals targeted for execution—both Tutsi *inyenzi* and Hutu *ibyitso*—were openly broadcast over the airways.

The first United Nations peacekeepers authorized under the Arusha Accords arrived in Kigali in November. French forces allied with the Habyarimana government conspicuously departed at the same time. The Canadian commander of the U.N. forces, Colonel Roméo Dallaire, was appalled by the open distribution of weapons to civilians in Kigali and the explicitly hostile calls for ethnic confrontation that had replaced any pretense of discussion. Word came of RPF preparations for battle in the north, while rumors circulated of Burundian Tutsi maneuvers in the south. In the midst of this mayhem, a high-level informant told Colonel Dallaire of detailed preparations for genocidal killings of Tutsi and Hutu collaborators. Dallaire's request for additional peacekeepers and an expanded mandate for protection was rejected by his U.N. superiors in New York. Demonstrations grew increasingly violent. On February 21, 1994, the leader of the opposition Social Democratic Party was assassinated. In retaliation, his southern supporters lynched the leader of the right-wing CDR, who had the misfortune to be visiting Butare the same day. Hutu were now killing Hutu over leadership within the majority ethnic group. Massive riots followed in Kigali.

On April 6, President Habyarimana was returning from another effort to restart the peace process in Arusha. As his Falcon 50 jet approached the Kigali airport, two rockets shot up from the ground below. They hit their mark and the jet plunged in flames, killing its French crew of three, the new president of Burundi, and President Habyarimana. Parts of the plane crashed in the yard of Habyarimana's household complex on the outskirts of Kigali. There were many suspects, from the Tutsi fighters of the RPF to the hardcore Hutu supporters of Madame Habyarimana herself. But there was no time to dwell on arcane questions of responsibility: it was time to kill.

Chapter Thirty-one

Genocide

WITHIN HOURS OF President Habyarimana's death, a premeditated campaign of genocide was launched across Rwanda. More than 900,000 Tutsi men, women, and children fled their ancestral homes, forced to seek refuge from predatory death squads that turned the land of a thousand hills into a thousand killing fields. On Nyungwe Forest's eastern edge, thousands of Tutsi gathered at the main Catholic church in Gikongoro. Hutu and Tutsi shared a strong Catholic tradition and churches had served as effective sanctuaries during past times of trouble. The bishop of Gikongoro, however, advised his frightened flock to move on to the new technical school being constructed on a hill called Murambi, just outside of town. There, he assured the Tutsi, they would be protected by government *gendarmes*. Tens of thousands of Tutsi followed this advice and assembled at Murambi, where *gendarmes* did indeed stand guard—to assure that no one escaped when government soldiers opened fire on their helpless victims with rifles and grenades on April 19, 1994. The slaughter continued unabated for three days, as masses of local Hutu joined the killing frenzy with machetes and spears.

EMMANUEL MURANGIRA pulled the key chain slowly from his pocket. As he stared down to sort through the large set of keys before him, the deep depression where he had been shot in the forehead was painfully visible. Selecting the right key, Emmanuel opened the door to the classroom, gesturing silently for Amy to enter. Inside, a half-dozen long tables were lined with what looked like grotesque plaster mannequins or perhaps the ash-embalmed remains of the victims of a volcanic blast. They were victims, but not of any natural disaster.

In August 1999, Amy visited the site of the 1994 Murambi massacre. There are seventy-eight rooms in the Murambi school complex, spread over thirteen separate buildings. Each room is filled with at least forty bodies exhumed from mass graves, then treated with a chalk-colored preservative. The ashen appearance of the corpses creates a surreal effect until one is overwhelmed by the brutal reality of their deaths. Bodies contorted, arms reaching, fingers grasping. Gaping wounds and shattered skulls. Each terrible story sealed behind a tight-lipped death grimace. In some rooms, there are only children. Almost all were killed with machetes. Some are missing heads, some are cut in half. Their small, shattered bodies lie in mute affirmation of their killers' chilling logic: *none must survive.*

Amy walked slowly through eighteen of the seventy-eight rooms. Sick from the sights and smells, and overwhelmed by the magnitude of the horror before her, she stopped to talk with Emmanuel. He was tall and thin, with a receding hairline and other classic Tutsi features. His gentle demeanor belied the fact that he was one of only a handful who survived the Murambi massacre, left for dead with a bullet wound to his head. He told Amy how he had crawled away at night and hid in a nearby woodlot until the killing stopped. Then he watched as the dead and dying—including the other forty-eight members of his extended family—were thrown into massive pits and buried. Now his mother, father, and brother were among those whose remains were on display in the newly opened memorial.

Nearly three thousand bodies can be seen at the Murambi memorial. Another 24,000 bodies have been exhumed from mass graves around the site and reburied. At least 27,000 human beings were killed at Murambi. That is only a small fraction of the 800,000 Rwandan Tutsi believed to have been killed in the Rwandan genocide. One hundred days of frenzied butchery. One murder every ten seconds, all day and through the night. Many were killed by neighbors, whose faces were the last image they saw before they died. It was the most intense and the most personal genocide of the bloody twentieth century.

A S THE CHARRED RUINS of the presidential jet still smoldered on the outskirts of Kigali, death squads were rushing to implement a highly organized campaign of mass murder. Tutsi civilians would pay the most fearsome price for this campaign, but many of the earliest victims were Hutu leaders of opposition parties. This fact underscores the political framework for the killing that went hand in hand with ethnic genocide. At some point in 1993, Hutu Power elements had decided that the extermination of Rwandan Tutsi civilians was the only way to rob the RPF of its internal support base, thereby settling which ethnic group would ultimately rule Rwanda. This Final Solution logic was promulgated in the name of Hutu salvation. Left unsaid was the self-interest of the Hutu ruling clique in maintaining political and economic power at the expense of opposition Hutu challengers. With the signing of the Arusha Accords in mid-1993, it was clear that the inner circle would lose power and that opposition Hutu groups as well as the RPF would gain political and military advantage under the proposed new arrangement. This combined threat to ethnic sovereignty and personal privilege offers the most likely explanation for the attack on the president's plane: that it was the work of an inner circle of Hutu extremists, including, perhaps, members of even the *Clan de Madame*. This cataclysmic event was then blamed on the Tutsi, providing cover for a campaign of terror against a broader set of perceived enemies.

In the months before the presidential assassination, as Radio Mille Collines and Hutu Power newspapers spewed forth a steady stream of menacing ethnic propaganda, Colonel Théoneste Bagosera, a confidant of Madame Habyarimana, took the lead in coordinating planning for the genocide. Former government soldiers were recruited to join the *interahamwe* and other local militias to carry out the murderous task at hand. Finding disillusioned young men ready to support the Hutu cause—and perhaps gain some bounty in the process—was not difficult, especially among the urban poor and recent Burundian refugees. Weapons were openly distributed to these militias, but ammunition was in short supply, so Hutu businessmen paid to import enough machetes for every third Rwandan male. There was much brush to cut.

With preestablished lists of targets in hand, death squads set to work in Kigali within an hour of Habyarimana's assassination. One of the first to be killed was Lando Ndasingwa, our old acquaintance and leader of the Liberal Party, along with his Canadian wife, Hélène, and their two children. Among the opposition Hutu killed on that first day was Agathe Uwilingimana, the transitional prime minister. She was executed in cold blood along with ten disarmed Belgian soldiers who formed her U.N. bodyguard. Journalists who didn't toe the party line and human rights activists were also early targets. In the following weeks, the killing spread to Butare, where Hutu professors at the National University and Hutu doctors at the nearby hospital were added to the death lists. But the vast majority of those killed were Tutsi. In Kigali a few thousand people were saved at the national football stadium and the principal Catholic church, Ste. Famille—though individual Hutu and Tutsi were removed from both sites for execution and the military occasionally lobbed artillery shells into the stadium. The Hotel Mille Collines provided more effective protection for several hundred Tutsi, thanks to the tireless efforts and diplomatic skills of its Hutu manager, Paul Rusegabagina. He was aided in his efforts by high-level foreign contacts, a secret fax machine, and the judicious distribution of liquor to prominent Hutu leaders—many of whom also kept their Tutsi wives and mistresses at the hotel. Still, the successes were minor. More than sixty thousand people were killed in and around Kigali during the first few weeks. Radio Mille Collines called on all available trucks to help remove the bodies, which were piling up faster than they could be buried.

THERE WERE FEW FOREIGN WITNESSES to the bloodbath in Rwanda. Kurt Kristensen, the former Nyungwe researcher, experienced the first few terrible days in Kigali. On the evening of April 3, Kurt was returning from a visit to a friend's house. Driving in his Suzuki jeep with his fourteen-month-old son, Ryan, and nanny, Sincère, he was surprised to encounter a roadblock at a traffic circle not far from his home. The airport and the headquarters of the president's MRND political party were not far away, but he had never seen a roadblock at that intersection. He was even more surprised by the agitated state of the soldiers manning the roadblock. They demanded his papers and those of Sincère. *Why are you out at night? Where are you going?* After much shouting among themselves they returned the papers and told Kurt to go home and stay there. He readily complied. Bad as conditions had been, he sensed they were taking a turn for the worse.

Around 8:30 on the night of April 6, Kurt heard a distant roar outside his house. It was a powerful explosion, followed by complete silence. After twenty minutes the silence was broken by the sound of small weapons fire. Then all hell broke loose. In contrast to the pounding sounds of machine guns, grenades, and artillery rounds, Radio Rwanda was broadcasting classical music—a first in Kurt's experience. He turned his walkie-talkie to the frequency used by the U.S. embassy to broadcast emergency information. After an extended wait, he learned that the explosion he had heard was the sound of the president's plane crashing. The gunfire was the bloody aftermath.

The following days passed in a blur of painful memories. Kurt kept Ryan in a windowless bathroom and an adjoining hallway. Sincère moved into the hallway, as did an American Peace Corps volunteer and his Tutsi wife. They slid a refrigerator into the open end of the hallway to block any stray bullets or shrapnel. Kurt kept a loaded shotgun and a Magnum .380 pistol by his side as he stayed up all night. On the morning of April 7, artillery pieces moved onto the hill above the house, which shook with the sound of their discharges. Far worse was the whistling of incoming shells above the roof when unknown opponents briefly returned the artillery fire. Later that afternoon, things calmed down and Kurt took out his spotting scope—normally used for viewing birds and monkeys—to survey the valley below. Black smoke seemed to seep from the ground, filling the air. A U.N. armored troop carrier approached a roadblock manned by dozens of machete-wielding young men. After several minutes, the armored vehicle yielded to the machetes and turned back in the direction from which it had come. It was a pathetic sight. Later that afternoon, Kurt heard screams from his neighbor's house just downhill. A Tutsi family lived in the house. Then shots rang out and the screaming stopped. Within an hour a truck appeared and a gang of unknown men began to carry away the belongings of the murdered Tutsi family. Kurt wondered how long he could defend his son and the others if his house were attacked. Later that evening he listened helplessly as his two-way radio crackled with the pleas of another American with a Tutsi wife whose house was under attack by Hutu soldiers. There were no American troops in Rwanda to protect them, and the U.N. peacekeeping force was nowhere to be seen.

On the morning of April 8, the U.S. embassy spread the word that all American citizens would be allowed safe passage out of Rwanda traveling by road to Burundi. Everyone was to gather at the U.S. ambassador's residence by 1:00 P.M. Kurt loaded Ryan, Sincère, the Peace Corps volunteer and his wife, and his two Dobermans into his jeep. Fortunately, it was less than a mile to the ambassador's. More than two hundred people crowded inside the compound: half of them Americans, the other half Tutsi who worked or lived with Americans and were desperate to escape. At one point men in Rwandan army uniforms appeared and began shooting through the gate into the crowd. Several children were killed and many others wounded. Finally the commander of the U.N. peacekeeping force, Colonel Dallaire, arrived. He gathered everyone together and explained the rules for safe passage. Only Americans, their spouses, and pets would be allowed to leave; all others must remain. Kurt looked around. It was a death sentence for dozens of Tutsi friends, lovers, and colleagues. Kurt decided he would pretend that Sincère was his wife. Dallaire then ordered everyone into their vehicles and the convoy slowly pulled away.

To reach the road to Burundi, the convoy first had to pass through central Kigali. Kurt was shaken by what he witnessed.

It was chaos. Bodies were everywhere. Stores were looted and buildings were burning. A group of Hutu women stood amid the wreckage, pointing at us and laughing as we drove by. A ten-year-old boy leaned against a building, smoking a cigarette. He had a machete cradled in one arm and a string of grenades around his waist. He stared at me with bloodshot eyes and I thought, This place is done for.

At the outskirts of Kigali, the convoy halted for a roadblock. It was the first of many stops, each an opportunity for harassment. Kurt realized for the first time that the convoy had no U.N. or other escort. Miraculously—or perhaps because of the two Dobermans in his car—no one challenged his relationship with Sincère. Just outside Butare, they were joined by Rob and Cheryl Fimbel and others who had come from the Nyungwe region. Finally, the convoy straggled across the Burundi border, arriving in Bujumbura after midnight. What was normally a four-hour drive had taken twelve very long hours. Kurt and his charges collapsed in exhaustion after sleepless days in a living nightmare.

Across Rwanda hundreds of other expatriates made their way out of the country by any means possible. Some headed north to Uganda, others slipped across the western border with Congo. All foreign embassies were evacuated. Most, like the Americans, left their Tutsi staff behind to fend for themselves. The French, however, found room for Madame Habyarimana and dozens of her entourage. Fewer than thirty foreigners remained in Rwanda through the genocide. Most important, there were no international television crews to document the carnage on the ground.

IN THE DAYS AFTER THE PRESIDENT'S plane was shot down, the genocide quickly spread beyond Kigali. Exhorted by the radio and by local political leaders, bands of *interahamwe* executed known members of Hutu opposition parties and intellectuals. But the overwhelming majority of victims were Tutsi. At schools like Murambi and Catholic churches like Ntarama, they were killed by the thousands. Elsewhere, they were dragged from their homes or hunted down like helpless prey in swamps and woodlands. Many were doomed by their ethnic identity cards—a regrettable relic of the Belgian colonial period. Individuals of uncertain ethnicity were denounced by their neighbors. Some unfortunate Hutu were killed simply because they were tall and thin, or had a receding hairline, or otherwise "looked like" Tutsi. Husbands and wives were told to inform on their spouses; parents were expected to turn in their "mongrel" children. The fortunate victims were shot, sometimes even paying for the use of a bullet and the "right" to a quick death. Most died slowly from the repeated blows of machetes and even hoes. Many were mutilated and dismembered. Unknown thousands of Tutsi women and girls were raped. Most were then killed; others were left to bear the children of their assailants.

In early May, almost one month after the killing began, the outside world saw its first evidence of the scale of human slaughter in Rwanda. Forty thousand bodies had washed down the Akagera River into Lake Victoria. Now the Western press had pictures to sell the news. American and European TV showed grisly video images of bloated bodies, floating facedown beneath a bridge along the Tanzanian border. We had been following events as closely as possible through the written media and knew that terrible things were happening in Rwanda. But we were unprepared for the visual horror of that first footage. We dissolved in tears. When the tears ran dry, we walked around for days with a hollow, kicked-in-the-stomach feeling. Some of our colleagues at the WCS office in New York had invested comparable time and energy in other cultures, and the office was an oasis of understanding and sympathy. In the Hudson Valley community where we lived, families busied themselves with the rituals of suburban existence: shopping and shuttling children to and fro. Passion was reserved for weekend soccer and lacrosse games. We followed the rituals and grieved in silence. Noah joined our discussions at home. At thirteen, he could understand the meaning of genocide, if not the reasons. One of his classes had covered ethnic killing in the Balkans. But for Noah, the news was personal. Was Gael okay? What happened to Clementine? To Isaac? He was especially concerned about Alexi and Justin, the two workers he adored from the Kitabi tea factory. We didn't know the fates of any of our former friends or co-workers. We could only wonder who was killed—and who did the killing. Was Nyungwe a refuge from the madness, or a killing ground? Was the ravine where Bill and the boys threw unwanted *Eucalyptus* trees now filled with bodies? There were no pleasant thoughts, no answers to our questions.

Sometimes salt was rubbed in our psychic wounds. The week that the first gory videos appeared, we were contacted by one of the major U.S. television networks. We were asked to talk about the fate of the mountain gorillas on the network's prime-time evening news magazine, which would also be showing the footage of bodies in the river. We replied that it would be wrong to discuss our concerns for gorillas against a background of unspeakable human suffering. We did offer to appear if we could talk about the country and its people, but the woman on the phone assured us that more people were interested in the gorillas. Sadly, she was probably right. We declined the request.

The use of the adjective "tribal" in press accounts was an added irritant. The Western news media apparently decided that the violence in Rwanda was most easily understood as a tribal conflict. Tribalism was an umbrella concept used for decades to account for most of Africa's woes. Unfortunately, no one took the time to determine whether Hutu and Tutsi—two groups that shared a common language, territory, history, culture, tradition of intermarriage, and religion—in fact represented different tribes. More to the point, no one asked why it was accepted practice to describe conflicts among Africans as "tribal" when conflicts among most non-African peoples were "ethnic" in nature. Concurrent reports of deadly violence from Bosnia and Serbia in the spring of 1994 were especially revealing in this regard. Reporters consistently used the adjective "ethnic" with reference to the various Balkan adversaries and their activities, however atrocious they might be. Croats, Bosnians, Serbs, and Kosovars were far more different from one another than were Tutsi from Hutu in most important ways—except that the West saw them as "white." Yet their differences, disputes, and deadly antagonisms all merged into an image of ethnic conflict. Even genocidal killings in the Balkans were described as "ethnic cleansing." Sadly, the best media were as guilty as the tabloids in accepting the premise that "black is tribal, white is ethnic." We wrote an op-ed piece on the subject, which we sent to the *New York Times*. They didn't publish it, but they did stop using the word "tribal."

Western press coverage was frustratingly uneven. Story lines would develop, capture our attention, then leave us hanging with no follow-up. In the first weeks, articles contained grossly inaccurate information. The histories of Rwanda and Burundi were melded and confused. A legacy of constant violence was presumed to exist, whereas the reality was extended calm periods broken by intensive bursts of bloodshed. Habyarimana's relatively benign rule for most of his reign was equated with the brutal dictatorship of Idi Amin. Reporters did the best they could with a country and culture about which they—and their viewers, listeners, and readers—were woefully ignorant. To their credit, several reporters quickly educated themselves about Rwanda and its people, writing increasingly accurate and insightful pieces. Part of the problem was that our own demand for news about the situation was insatiable and our expectations too high. We belonged to an extremely small group of Americans who had firsthand, in-depth understanding of a small, landlocked country with little geopolitical significance. But this lack of information contributed to our feelings of utter helplessness about events in the spring of 1994.

If we felt helpless, the Western world was paralyzed. The United Nations had sent nearly 1,500 peacekeepers to monitor an expected political transition in late 1993. When their commander, Colonel Roméo Dallaire, called for additional troops and a revised mandate to respond to the growing possibility of violence in early 1994, his request was turned down without debate by the Security Council. When ten Belgian soldiers were disarmed, then executed on April 7, the day after Habyarimana's jet was shot down, Belgium withdrew its remaining troops from Rwanda. Many discarded their U.N. blue berets in disgust as they boarded the plane to leave. In New York, the U.N. secretary-general, Boutros Boutros-Ghali, described the situation in Rwanda as a "two-way" genocide, using the proper word but completely failing to appreciate the one-sidedness of the slaughter being conducted by the Hutu against the Tutsi. On April 21, the U.N. threw up its hands and voted to reduce the Rwandan peacekeeping force by almost 80 percent. The remaining 270 monitors would be lightly armed and constrained from any meaningful intervention in the conflict. Two weeks later, on May 6, the Security Council reversed itself and approved a contingent of 5,500 soldiers with expanded powers. It took three months, however, for the force to deploy—too late to be of any help.

Behind the U.N.'s lethal vacillation was a U.S. government that could not say the word "genocide." Less than one year earlier, in the summer of 1993, U.S. military intervention in Somalia had gone horribly wrong. A helicopter crashed while on a mission in an urban no-man's-land, and seventeen marines were killed in the ensuing firefight. When the Somalis paraded one of the dead Americans through the streets and displayed his body on Somali television, U.S. public opinion turned against the mission and troops were quickly withdrawn. Stung by this humiliation, the Clinton administration did not favor any intervention in Rwanda. The U.S. government, however, was a signatory to the 1948 Convention on Genocide. Written with the horrors of the Nazi Holocaust fresh in everyone's mind, the convention obligated the international community to intervene whenever and wherever evidence of genocide again appeared. Faced with this clear-cut requirement—and an extreme reluctance to become embroiled in another African conflict—the U.S. played deadly word games. For more than two months, as the worst of the killing continued unabated, our government systematically refused to utter the G-word. When a spokeswoman finally used the word "genocide" on June 10—after half a million Tutsi were already dead—she tried to dilute its meaning with qualifications. Even worse, the U.S. actively lobbied the U.N. and its European allies to adopt a similar position vis-à-vis the genocide and the need for intervention. Madeleine Albright, then U.S.

delegate to the U.N., played a key role in this blocking maneuver. When the U.N. finally acted to intervene, the U.S. refused to authorize any American troops or financial support for the operation. The world's greatest super-power not only failed to act in Rwanda, it used its considerable influence to stop others from doing their humanitarian duty. It stands as a shameful episode in American foreign affairs.

Bᴜᴛ ᴛʜᴇ RPF didn't wait for outside help. The Tutsi refugee army had monitored Radio Mille Collines from its Virunga stronghold and knew that an outbreak of hostilities against Rwandan Tutsi civilians was brewing. The attack on President Habyarimana's plane and the speed of the subsequent political assassinations, however, took them by surprise. Once stirred to action, Kagame's forces moved quickly to take control of most of Byumba, eastern Ruhengeri, and northern Kigali prefectures. As they swept past the Parc des Volcans, they paused to rearm the mostly Hutu guards, instructing them to protect the park and its wildlife. Anyone turning their weapons on the RPF, they made clear, would be executed. The southeastern region of Kibungo fell next, providing a link with Burundi. As they moved forward, the RPF integrated Rwandan and Burundian Tutsi civilians into their army, which numbered only 25,000. Bypassing Kigali, Tutsi forces captured Gitarama on June 13 as they pushed toward the southwest, where the slaughter of Tutsi civilians continued. By then the Rwandan government had retreated to Gisenyi in the northwest, where virtually no Tutsi remained alive.

As the RPF continued to gain ground against the larger and better equipped Rwandan army, a strange thing happened. The French decided to intervene. Longtime patrons of the Habyarimana regime, the French had withdrawn from Rwanda when the first U.N. troops arrived in late 1993. Now, ten weeks after the killing began, the government of François Mitter-rand announced its intention to stop the bloodshed. According to historian Gérard Prunier, who was an advisor to the French at that time, the socialist government had a legitimate humanitarian interest in the mission, especially given the paralysis of the U.S. and the U.N. However, he also noted that the English-speaking RPF forces from Uganda had revived deep-seated French fears of anglophone encroachment onto what they perceived to be their own francophone African turf. These fears were only deepened by talk of a South African intervention in Rwanda led by the widely respected—and English-speaking—Nelson Mandela. Still, its past role in arming and training the Rwandan army and civilian *gendarmerie*—many of whom were now members of *interahamwe* death squads—had left blood on French hands and prevented France from acting unilaterally. Eager for any help in halting the killing, however, the U.N. approved Opération Turquoise in mid-June.

The French entered southwestern Rwanda on June 21 and quickly moved to the eastern flank of the Nyungwe Forest. They selected the southwest because it was not yet taken by the RPF. Politically, the French also believed that the region offered the best chance for positive television footage showing Tutsi civilians saved by French soldiers. Some within the French military assumed their humanitarian orders would be revised to call for an occupation of Kigali and perhaps even a restoration of the Hutu government-in-retreat. Other French soldiers were appalled that their arrival was cheered by Hutu killers. If the French ever entertained the idea of moving on Kigali, it was ended by the RPF capture of Butare. Blocked along the only major route north, the French drew a line along the Nyungwe Forest's eastern border, behind which they declared a *zone de sécurité*.

Meanwhile, the RPF captured Kigali on July 4, Ruhengeri on July 13, and Gisenyi on July 18. Hutu political and military leaders escaped into neighboring Congo, followed by 800,000 terrified and confused supporters. Kagame declared victory and formed a new government. To their credit, the French rescued at least fifteen thousand Tutsi within their security zone. They might have saved many more if they had brought more trucks instead of armored personnel carriers. Otherwise, the French turned a blind eye to the scorched-earth policy of the Hutu military and political leaders behind their line. Every building and piece of equipment that couldn't be moved was destroyed, as were crops and permanent water sources. After Gisenyi fell and the end was clear, every vehicle, weapon, and portable machine in southwestern Rwanda was transported across the border into Congo with the final surge of another 400,000 refugees out of Rwanda. The French made no effort to arrest any of the known leaders of the genocide during this retreat, nor did they attempt to remove any of their military weapons. On August 21, they declared their mission a success and departed Rwanda.

The new government in Rwanda was carefully balanced, with a majority of Hutu officials including President Pasteur Bizimungu and Prime Minister Faustin Twagiramungu. The RPF leader Paul Kagame became minister of defense and vice president. But no one questioned that Kagame and the Tutsi were in charge. They presided over a country with a devastated infrastructure and whose foreign and domestic finances had been looted by the outgoing government. Crops lay rotting in untended fields. In a period of only a few months, the population of Rwanda had dropped from 7.6 million at the start of 1994 to no more than 4.8 million: the same number as when we first arrived sixteen years earlier. Eight hundred thousand Rwandan Tutsi were dead, from an original population of perhaps 930,000. Thousands suffered physical and psychological wounds. Two million Hutu had fled the country in fear of Tutsi retribution, regardless of their personal guilt. The remaining Hutu, one third of them refugees inside their own country, were in shock at the speed and totality of their subjugation.

The task before the new government was Herculean. But Kagame and his RPF associates were not encumbered by second thoughts about seizing power by force. They had acted decisively, overcome staggering odds, and put an end to one of the twentieth century's most horrific genocides. They could move forward in good conscience, knowing that had they not responded quickly, the genocide would still have occurred, many more would have been killed—and the *génocidaires* would still be in power. The consciences of the Western nations and institutions that failed to intervene and almost allowed this crime to succeed should have been more troubled.

Chapter Thirty-two

Aftermath

A MY RETURNED TO RWANDA on February 6, 1995. Ten months had passed since the outbreak of the genocide. The first rape babies were being born. Tutsi women were making grim choices. Along with whatever guilt they felt for surviving the genocide that killed more than 80 percent of their people, many endured the humiliation of being raped by Hutu *interahamwe* and their opportunistic followers. Now they were forced to decide whether to keep the children born of their suffering. Some killed the babies at birth. Many turned the newborns over to churches and orphanages, already overwhelmed by responsibility for more than 100,000 children who had lost their parents to death or flight in the preceding year. Many other Tutsi women kept their babies to raise as they would have raised their other children, had they not been killed in the conflagration.

Rwanda in early 1995 was no longer a boiling cauldron. Yet powerful memories and emotions still simmered.

A MY'S HELICOPTER FLIGHT over the countryside between Kigali and Cyangugu confirmed the extent of the killing and destruction, far beyond the cities and towns. Empty houses stood out across the rural landscape, their thatched roofs burned away to reveal charred interiors. Perhaps one in ten houses was burned, most of their former Tutsi occupants now dead. A greater number of houses were abandoned by Hutu who had fled with their government into nearby Congo. These houses were generally intact, but many of the surrounding fields lay fallow or untended—an exceptional sight in land-hungry Rwanda. Driving from Cyangugu up to the Nyungwe Forest, Amy could see firsthand the scorched-earth practices of the retreating Hutu forces. Vehicles were burned, windows broken, tin roofs removed and carried away to Congo. Almost every large building had been hit with grenades or rockets. Braced for the worst, she pulled in to inspect our Nyungwe project headquarters at Gisakura. But aside from a few broken windows and other incidental damage, the main office, residences, and dormitory remained remarkably intact. A U.N. officer explained that the French had found our accommodations to their liking and their occupation prevented any looting or destruction during Opération Turquoise. Following the French evacuation, U.N. peacekeeping troops from Eritrea moved in.

During the three months before the French occupation, the Nyungwe project office had been maintained by local staff. Since all were Hutu, they were left in relative peace as the killing continued around them. Their treatment would have been very different had the *interahamwe* learned that they were hiding and feeding the project's Tutsi secretary, Assumpta Mukeshimana, in a locked, shuttered building behind the office. Those who saved and sheltered her were not the only Good Samaritans to emerge during the genocide, but they belonged to an exceptionally small group. Meanwhile, the project's senior Rwandan staff joined the flood of refugees that fled into Congo when the Hutu government fell.

One Hutu official who didn't flee was Shaban Turykunkiko. Shaban had worked with Kurt Kristensen on the study of gold mining impacts and had helped with other project activities. When ORTPN sent its first delegation of guards to Nyungwe, he served as their hardworking chief warden. He was shot once while on patrol in the forest in the early 1990s, yet he continued to monitor the forest to the best of his abilities throughout *les troubles*. A few days after the French departure in August 1994, Shaban tried to resume his responsibilities. He drove to Butare with a businesswoman friend, hoping to purchase a spare part for his own disabled vehicle. On their return, the two were stopped, robbed, and executed by unknown gunmen as they approached the forest from the east. The killers might have been *interahamwe* who stayed behind to spread fear and disrupt the RPF's ability to govern. Or they might have been common thieves taking advantage of the disorder and wide availability of weapons.

In Butare, Amy stopped to see who was left at the National University. So many had been killed in the blind eradication of Hutu intellectuals that accompanied the Tutsi genocide. A few, like Jean-Baptiste Barabwiliza, were arrested by the RPF and held in prison under suspicion of complicity in the killing. Amy had first known Barabwiliza as the botanist who identified her gorilla food samples in the late 1970s. He then obtained a doctorate in Belgium and returned to rise rapidly in the university—and ruling party—hierarchy. Two other former colleagues, Samuel Kanyamibwa and Isaac Kabera, had survived the Butare bloodshed but feared for their lives if they remained in the region. Samuel fled to Bukavu, where he sold his car to buy a plane ticket to France, then went on to England. He recently moved to Kenya. Tim Moermond intervened on behalf of Isaac Kabera to secure a place for him as a graduate student at the University of Wisconsin. It was a painful choice for Isaac as he was required to temporarily leave his family behind. Amy visited his wife, delivering a small package of gifts from Isaac and asking if she wished to send any messages to him. She seemed reluctant to talk. Soon after the meeting, she was visited by RPF representatives who questioned her about the meeting. Isaac, like almost all Hutu, was a potential suspect in the genocide, even though he, too, had in fact been a target of the *génocidaires.* Amy was reminded that foreign visitors can bring unwanted attention to their Rwandan hosts in times of trouble.

OUTSIDE RWANDA, the cataclysmic events of 1994 reverberate to the present day. Of the two million Rwandan Hutu cajoled, cowed, and intimidated to follow their defeated government into exile, almost 600,000 fled east to Tanzania; another quarter million went south into Burundi. More than 1.2 million landed in eastern Congo. At least 800,000 of these refugees surged out of northwestern Rwanda over a single twenty-four-hour period: the most intensive border crossing in history. The international relief community was not prepared to address the needs of so many people in such a short time. The relief agencies also made a series of extremely poor decisions. First, they took pity on the refugees and allowed them to establish camps near the Congolese border town of Goma, rather than walk another fifteen or twenty miles inland. This violated international guidelines that call for the removal of refugees from border areas of the countries from which they have fled, for obvious security reasons. It also resulted in a series of camps being built on the lava flows that dominate the land between Lake Kivu and the Virunga volcanoes. The lava rock was too hard to cut latrines, yet too porous to hold fresh water. Combined with extreme overcrowding, the result was a sanitary disaster on an unprecedented scale. Cholera raced through the camps surrounding Goma, killing thousands within the first few weeks. More than thirty thousand refugees suffered gruesome deaths in July and August alone.

As if to vent its displeasure, the Nyiragongo volcano burst into activity, lighting the evening sky and threatening to add a natural disaster to the human calamity below. It was a field day for the international press corps. Denied access to Rwanda during the genocide, they could now fill the global airwaves with powerful images of individual suffering and mass death, set against a stunning backdrop of Lake Kivu and Nyiragongo. Americans and Europeans watching from the comfort of their homes could also feel that their governments, international agencies, and nongovernmental organizations were finally helping. Sympathy—and contributions—poured in. Although not one government or agency had been able to muster the will or means to intervene during the genocide, groups competed to help the refugees in Goma. The U.S. government had no trouble mobilizing its vast military machine to send troops and supplies to aid the refugees in Congo. The French found trucks to transport the dead that had been unavailable to save the living during Opération Turquoise. The U.N. High Commission on Refugees worked with dozens of nongovernmental organizations, such as the Red Cross, CARE, Oxfam, Médecins sans Frontières, and others. This outpouring of assistance helped save the lives of many innocent Hutu who fled their country in panic. But it also helped the *génocidaires:* the Rwandan army, *interahamwe,* and others who had organized and carried out the killing in Rwanda. It was too late to help their Tutsi victims. This critical fact was lost on most Westerners, though, who never really grasped who was killing whom in that faraway little country. CNN couldn't open its big eye on the big killing, but now it could show Rwandans suffering, and many in the viewing public just assumed that the people on TV were the survivors of the genocide.

We knew better, but we, too, had mixed feelings. Most of the refugees in the Goma camps came from northern Rwanda, including many people that we knew. All of the staff from Karisoke were in those camps. Many Parc des Volcans guides and guards were in the camps. Clementine was in the camps. We had stayed in touch with her since we left Rwanda and knew that she had married an army mechanic and had a child. When Amy returned in February of 1995, though, no one knew of her whereabouts. We were shocked and relieved to receive a letter from Clementine a few months later. She described her flight from Rwanda and gave some indication of the difficulties of life in one of the camps near Bukavu, south of Goma. We wrote back, but her next letter gave no indication that she had received our response. We wrote again and asked her to make contact through the Red Cross with another Rwandan through whom we thought we could send money. Her last letter told of a miscarriage and described a bleak existence. Her camp was shut down during a wave of violence in 1996, but she was not among those who returned to Ruhengeri.

Although we couldn't make direct contact, we were relieved that Clementine benefited from the international relief effort in Congo. We knew her well enough to be convinced that she was innocent of the killing, a wife and mother caught up in a terrible human tragedy, forced perhaps by her husband or the power of fear to flee her country. It was more difficult to be certain of the innocence of others who had fled or were missing. Big Nemeye loved to sing songs of ethnic harmony on census treks, but he was also a proud mountain Hutu. Who knows how he and other men from Karisoke or the park acted during the dark days of 1994. We hoped for the best for all of them.

An even bigger mistake than the poor location of the camps was the failure of the UNHCR to segregate and disarm the military, paramilitary, and extremist elements within the camps. Again, this is standard practice in refugee situations. Instead the innocent, the frightened, and peasant farmers swept up by events were hopelessly mixed in with the most guilty of the *génocidaires*. Once in the camps, the defeated government leaders reasserted their authority and the hundreds of thousands of refugees were quickly reorganized by their former prefecture, commune, sector, and cell. The refugees welcomed these traditional political structures that also helped the relief agencies to organize the delivery of aid. Mostly, however, they served to secure the absolute power—reinforced by propaganda and armed coercion—of the Hutu Power extremists. Food and medicine were distributed first to the soldiers and *interahamwe* militias. Next came their supporters. Any who wavered in their loyalties were denied their most basic needs. Many items were sold to local Congolese citizens in illegal markets, further enriching the Hutu leaders.

The Hutu masses served as a shield for their leaders, too. This was evident after the cholera epidemic in August 1994, when more than 100,000 refugees left the camps and walked back to Rwanda. The Hutu Power leaders quickly realized that if the exodus continued, they would be left exposed for what they were: the most notorious organizers and killers of the recent genocide. Among those who tried to return in August were several of the workers from Karisoke, including Little Nemeye, Rukera, Kana, and a recent hire named Jean-Bosco. Pascale Sicotte, acting director of Karisoke at that time, traveled at considerable personal risk to the Goma camp to arrange for the men's return. While waiting for the arranged transportation back to Rwanda, however, they were attacked by armed Hutu from the camp. Accused of not being "real Hutu," they were severely beaten. Jean-Bosco, the youngest of all, was struck on the head with a machete and had a bayonet thrust in his mouth. Only the fact that his attackers proceeded to fight over his money allowed him to escape with the others. A few days later, the Karisoke staff made it safely onto a U.N. truck for repatriation: a testament to their determination—and apparent innocence. Others were not so fortunate. The flow of returnees to Rwanda was stanched, and Hutu Power forces intensified the level of coercion to keep others in the camps. Increasingly, relief workers felt used and confused. They weren't supposed to deal with political matters, but they couldn't help noticing the number of strapping young men—often in uniform—strutting through crowds of otherwise weakened and emaciated refugees. Nor did they detect any sign of remorse for their past or present actions on the part of the leaders of the genocide. Over time, open military drills and arms practice took place within the camps. Médecins sans Frontières refused to participate in the sham and removed its personnel. Most other groups swallowed hard and carried on.

Throughout 1995 and 1996, the archipelago of refugee camps along its borders was a constant threat to Rwanda's internal security. Armed bands of young men would cross Lake Kivu or infiltrate through the Virunga and Nyungwe forests, then attack RPF government installations such as military outposts or prisons, or simply kill Tutsi civilians. Many then retreated back to the camps, while others sought refuge within the forests or among the still mostly Hutu population. Those who didn't support the renewed *interahamwe* campaigns risked their lives, regardless of their ethnicity. Jonas Gwitonda, the second guide hired by Bill to start the gorilla tourism program, was killed for failing to share his profits from trafficking cattle through the park. Two former Karisoke workers, Kanyarogano and Nshogoza, were executed for unknown reasons by *interahamwe*. Several of the men we worked with— Kana, Little Nemeye, Semitoba—saw most of their family killed. Philip Gourevitch ends his remarkable book, *We Wish to Inform You That Tomorrow We Will Be Killed with Our Families,* with a profound tale of resistance to this continued cycle of violence. In 1997, a girls' school was attacked in northwestern Rwanda and the girls taken outside. There, the *interahamwe* insisted that the Hutu girls identify themselves so the attackers could kill their Tutsi classmates. The Hutu refused, and girls of both groups were beaten and killed. Although a tragic story, it contains a powerful seed of hope for ethnic reconciliation.

Hutu death squads were not the only ones with bloody hands in Rwanda. Their attacks led to counterattacks by RPF forces in which some innocent civilians paid with their lives. Hutu accused of supporting insurgents saw their homes burned and men taken away. One of those men was the Karisoke tracker Vatiri. Another Karisoke worker, apparently jealous of Vatiri for some reason, approached the RPF and denounced him as a member of the *interahamwe*. Although this claim appears to be unfounded, the RPF arrested Vatiri and took him to a detention camp. He was never seen again.

The RPF is generally given considerable credit for its discipline, but the refugee army changed dramatically as the fighting, occupation, and conquest of Rwanda progressed. Tutsi from neighboring Burundi and Congo, as well as native Rwandan Tutsi, joined the army without any serious training. They went unpaid until early 1995. With victory, some sought spoils. Others sought revenge. At the same time, 700,000 Tutsi poured into Rwanda after a generation of exile in neighboring countries. Some found unoccupied homes and lands, others hired Tutsi soldiers to take what they wanted at gunpoint. Thousands of Hutu from northern Rwanda were killed during the first six months of RPF rule. The RPF leader, Paul Kagame, acknowledged the breakdown in discipline and took some strong countermeasures. He also implied that it would be impossible to completely restrain his troops and ignore the Tutsi desire for revenge.

KAGAME AND HIS RPF COLLEAGUES sat on a powderkeg. No matter how many top positions were occupied by Hutu, the new government was widely—and correctly—perceived to be controlled by the minority Tutsi. The former Hutu government had escaped largely intact and was now threatening a forcible return to power. Armed insurgents undermined public security from within. The prisons bulged with accused participants in the genocide. The country's infrastructure was in ruins and the national treasury was bare. Yet, incredibly, Rwandan requests for international assistance went unanswered. The French took a particularly active role in frustrating the new regime, which they regarded as a band of English-speaking refugees from Uganda. To highlight their annoyance, the French did not invite Rwanda to the biennial conference they hosted for francophone African leaders in November 1994. The French further retaliated by blocking a $200 million economic assistance package from the European Union. Then the World Bank announced that a $140 million loan to Rwanda could not be processed until a $4.5 million payment was made on a previous debt. One can only imagine what the RPF leadership thought of Western nations that withheld critical assistance needed for postwar recovery due to a $4.5 million debt *owed by the former government*—at the same time that we poured $1.5 billion into disastrously mismanaged refugee camps along their border. This after failing to act in any meaningful way to halt the brutal genocide that ravaged the Rwandan people nation, creating the very need for recovery assistance.

Meanwhile, the situation within Rwanda festered. More killers infiltrated the country and guns were more widely dispersed. Westerners became targets. Four tourists were killed in the Congolese section of the Virungas, less than a week after gorilla tourism reopened. Three Red Cross workers and a Spanish priest were murdered in Ruhengeri and two U.N. human rights workers were killed in Cyangugu. Temporarily avoiding the violence in Rwanda, Amy was working across the Congolese border in the Kahuzi-Biega National Park in June and July of 1996. She had joined a Wildlife Conservation Society team led by Jefferson Hall and Inogwabini Bila Isia, conducting the first survey since 1959 of the eastern lowland, or Grauer's gorilla. Near the end of Amy's eight weeks in the field, her mother called Bill to ask if he had heard from her lately. Bill said he didn't expect to hear anything until she left Congo but asked if there was a problem. Marion Vedder then read from a newspaper account that reported thousand-dollar rewards for white people's heads in eastern Congo. Bill confirmed the story on the Internet, where at least one article attributed the rewards to militant Hutu within the region's many refugee camps. Amy and her team were oblivious to the threat until two weeks before they were to leave, when a Congolese associate who had been in Bukavu reported the news. Jefferson joked that the five white heads in the room were worth several years' salary in the impoverished Congo. Another Congolese contact reported that unspecified "troubles" were brewing in the region and that the "foreigners should get out soon." To be safe, the work schedule was accelerated so Amy and the others could leave a week early. The group exited without incident through Rwanda. But the "heads for bounty" threat indicated that the cancer of the refugee camps was spreading to the surrounding region.

Two months after Amy left Kahuzi-Biega, Hutu extremists in the refugee camps turned their hatred and firepower on the Banyamulenge, an ethnic group related to the Tutsi but native to eastern Congo. The Banyamulenge fought back, attacking the camps with the clear support of the RPF. The international relief community, depressed and fatigued by its compromised role, stood aside and allowed the Tutsi and their allies to end the deadly charade of the camps. Nearly one million Hutu streamed back into Rwanda, creating more stress on the government and requiring further international assistance. The military wing of the RPF ignored the refugees, pursuing instead the twenty to thirty thousand hard-core *génocidaires*—and whoever else they could coerce to go with them—as they fled west into the dense Congo Basin.

As the RPF pushed farther into Congo, they encountered no resistance from local Congolese. In fact, many encouraged them to continue west—not to hunt down the fleeing Hutu, but to overthrow their own president and dictator for life, Mobutu Sese Seko. In his thirty years in power, Mobutu had been responsible for theft on such a grand scale that the term *kleptocracy* was coined to describe his distinctive form of governance. Now, his regime was collapsing, and the RPF saw an opportunity to remove an old friend of the Habyarimana government and its extremist successors. With support from Congolese allies and opportunists, the RPF won a stunning series of rapid victories. Laurent Kabila, a longtime opponent of Mobutu, assumed leadership of the expanding coalition of opposition groups. Mobutu left for medical treatment in Switzerland in the fall of 1997, and Kinshasa fell to Kabila and his allies. Shortly thereafter, Mobutu died in exile in Morocco.

Laurent Kabila proved no more willing to control the *interahamwe* in his country than had Mobutu. Rwanda responded by forming an alliance with Uganda and aggressively expanding their combined military presence in Congo. Kabila, in turn, invited Zimbabwe, Angola, and Sudan to intervene in what quickly became Africa's first regional war. At risk were sovereignty, security, and, above all, resources. Zimbabwe alone lost thousands of men in the first year of combat. Yet each of the occupying forces soon learned there was much more to be gained from carving up Congo's vast mineral wealth than in fighting each other. Rwanda retained a single-minded focus on hunting down remnant Hutu bands, but even the Rwandans increasingly turned to exploitation of diamonds, gold, and coltan. The latter mineral proved especially lucrative, as it was used to make a key component in cell phones and other high-tech devices in great demand. The West had few sources of coltan and was willing to pay top prices to suppliers, without regard for the mineral's origin or rightful ownership. Nor did Western buyers care about the devastating effect of coltan mining operations on Congo's Kahuzi-Biega National Park, where thousands of elephants and gorillas were slaughtered by miners.

As the twenty-first century dawned on central Africa, Congo lay in ruins. An estimated 1.3 million of its people had died and another two million were displaced during six years of fighting since the end of the genocide in Rwanda. Ordinary Congolese hated their foreign occupiers. The Rwandan presence, in particular, fueled strong anti-Tutsi sentiments throughout eastern Congo. The assassination of Laurent Kabila in January 2001 and rapid succession of his son, Joseph, provided an opportunity to end the Congo quagmire. Joseph Kabila reached out to the various antagonists occupying his country and reopened negotiations with the United Nations. While early signs are encouraging, the end game remains to be played. Rwanda, for one, has signaled its willingness to remove its forces from Congo—but only if its internal security can be guaranteed.

~~~~~~~~~~~~

A MY RETURNED TO RWANDA AGAIN in 1999 to find that the country had made surprising domestic progress. Its costly foreign entanglement had created breathing room along its western border and much greater internal security. In many ways it looked the same as before the terrible events of the 1990s. In the countryside, farmers produced food according to ancient cycles and rituals. Women and young girls carried water and wood on their heads. The cries of goats and babies projected far through the thin mountain air with little competition from motors and machines. Individual *rugos* still dotted the hillsides, but more and more people lived in villages. *Villagisation* was in fact a policy of the new government, intended as much to improve security and control the population as to provide better access to services. There were certainly more cows than before. The returning Tutsi had brought their cultural icons from surrounding lands and now the limited land base was forced to absorb them. It was common to see young boys herding cattle through an intersection in Kigali while cars stopped at the light. The capital was crowded with more cars than before, many with foreign license plates. The Tutsi raised in Uganda spoke English, and Rwanda was now officially trilingual, with meetings, radio broadcasts, and road signs in French, English, and Kinyarwanda.

Though greatly outnumbered by repatriated Tutsi, the native-born Rwandan Tutsi and their suffering were central to the identity of the new country. More than 80 percent of the pre-1994 Rwandan Tutsi population had been killed and many survivors were maimed or victims of rape. Everyone had lost relatives and loved ones. Memorials to the genocide were maintained at numerous sites. Ghostly white corpses filled the classrooms of Murambi. At Ntarama the bodies were left where they fell, slaughtered by the hundreds among the pews and crucifixes of the violated church. Other memorials were equally macabre, yet the Tutsi believed strongly they must be maintained to bear witness.

The new government of Rwanda also believed that justice was required before the country could move forward. And by justice they meant the death of those responsible for the genocide. The most guilty of the *génocidaires* had escaped to Europe or to other African nations. A few had been extradited to be tried by a specially created international tribunal convened in Arusha, Tanzania. The RPF was dismissive of this court, however, because it could not apply the death sentence, even for the most monstrous crimes. Several dozen senior officials had already been executed in Rwanda. But 125,000 suspects remained in prison. This was a sore point with Western donors. Many Western nations had opened their checkbooks to support democratic and judicial reforms. But European nations expressed concern that their funds would go toward executions. This perceived squeamishness over killings in retribution for the genocide was difficult for the Tutsi to comprehend, particularly on the part of nations that had failed to react to the slaughter of innocent civilians in the first place. The course of justice in Rwanda remained a serious barrier to progress with Western nations and to reconciliation among the Rwandan people.

The prison population included many former staff from Karisoke. After voluntarily leaving the refugee camps as soon as possible, they returned to their homes or a displaced persons camp near Kinigi. Karisoke was off-limits for security reasons, but all of the men continued to be employed by the Dian Fossey Gorilla Fund (formerly the Digit Fund) to help monitor the research gorillas. In April 1997, however, six staff members were arrested. They were charged with killing members of Tutsi families fleeing through the forest in February 1991—years before the organized genocide began and a few months after the initial RPF invasion of Rwanda. The details are sketchy and hotly disputed. Despite evidence that some of the men actually intervened to save lives, Little Nemeye, Kana, Rukera, and three others were all sentenced to life in prison in a series of trials that left much to be desired.

WITH FEWER SECURITY CONCERNS, Amy focused much of her time and energy on conservation issues in 1999. For some areas, it was too late. What little remained of the Gishwati Forest was converted to farmland by returnees from Congo. Gishwati's rare golden monkeys and chimpanzees were eliminated. More than seven hundred square miles, or two thirds, of the vast Akagera Park was converted to pastureland for 86,000 cattle and their Tutsi owners. Only a small fraction of the formerly abundant savanna wildlife still survived in the southern third of the park. In stark comparison, the Nyungwe and Virunga forests remained intact and in surprisingly good condition.

Near Bweyeye, in southern Nyungwe, an old woman stopped hoeing her field. She seemed relieved to rest her body, yet anxious about her discussion with the unexpected visitors. Amy and Michel Masozera, the Nyungwe project director, had stopped their Isuzu jeep along a nearby road and walked to her fields. Did she know that she was farming illegally on forest reserve land? *Yes*. Did she know she would have to leave? *Yes, but tell me where can an old widow go?* Amy was impressed with Michel's manner as he asked the questions: direct, but sympathetic. She had wondered how Michel, a repatriated Tutsi from Congo whose parents had fled Rwanda in 1959, would interact with the predominantly Hutu local population. He had no answer to the woman's question, but told her and others nearby to finish their harvest cycle that year, then not return again to farm in the forest reserve. It was the law. The woman nodded and thanked him. *Urakoze.*

Despite numerous small-scale encroachments, Nyungwe survived largely intact. Trapping had increased in some areas, but the removal of thousands of gold miners by the new government ended the most serious threat to wildlife. The cultural taboo on eating primates provided protection for the forest's monkeys and chimpanzees, and its rich bird population continued to thrive. As important as the habitat and wildlife, though, were the Rwandan people who worked for the Nyungwe project. Throughout the fighting and terrorism of the early 1990s, local staff continued to monitor the animals, trails, and various research plots of the Projet Conservation de la Forêt de Nyungwe. Immediately after the French withdrawal from Nyungwe, local personnel renewed their efforts. For nearly five months they worked without pay, until the Wildlife Conservation Society could again establish a mechanism to deliver funds to Rwanda.

Several years later, Andrew Plumptre, WCS Assistant Director, conducted a study of the effects of conflict on conservation and he asked the Nyungwe staff why they had continued to work in spite of the risks and lack of pay. They responded that the project had existed for a long time and that they trusted WCS to return and treat them in a proper manner. It was the highest compliment possible. It was also a testament to the importance of hiring truly local staff with a vested interest in the long-term success of the project and its conservation program. PCFN was also fortunate in its new leadership. Eugène Rutagarama had worked for the project in its early years, then moved to the Parc des Volcans. He was imprisoned along with thousands of other Rwandan Tutsi after the initial RPF attack in 1990, freed in the raid on the Ruhengeri prison, and then escaped to Burundi. After the RPF victory, he returned to serve as a very capable PCFN director for nearly two years. When Eugène moved into another position, the project hired Michel Masozera to replace him. Michel earned the respect of local project staff for his hard work in the forest, while also representing Nyungwe's interests within both ORTPN and MINAGRI. His exceptional skills as a biodiplomat were evident in late 1999, when he helped convince the government to upgrade Nyungwe's status to that of a national park.

Status was never an issue for the Parc des Volcans, or at least for its most famous residents. The efforts of Akeley, Schaller, Fossey, and the many people who worked on the Mountain Gorilla Project had created an unparalleled global constituency for its gorillas. The MGP added a critical Rwandan support base through its tourism and education programs. The result of this combination was seen in the exceptional declaration—by both sides—after the outbreak of hostilities in and around the park in 1990, that *everything will be done to protect the mountain gorillas.* We know of no other military conflict where the combatants proclaimed to the outside world that they would not harm wildlife. Remarkably, they kept their word. Only one gorilla is known to have been killed in Rwanda during the civil war or genocide that followed. In the predawn hours of a May morning in 1992, Rwandans living near the park border heard automatic fire from the forest. Later that morning, park personnel discovered the bullet-riddled body of Mrithi. Mrithi had led Group 13 for thirteen years, since the killing of his predecessor silverback in November of 1979. There was no obvious reason for his death. His head and other body parts were intact. No infants were discovered missing in the following days. It is likely that a night patrol by either the Rwandan army or the RPF had walked into the nest site of Group 13 by accident, then fired in fear when greeted by Mrithi's roar.

Between 1995 and 1998, another sixteen mountain gorillas are known to have been killed in three incidents in the Congolese sector of the Virungas, including ten slaughtered in a single mindless attack by Congolese soldiers. Karisoke staff entered the park two or three days per week to monitor known gorilla groups. Only two gorillas were missing, according to their counts, both of whom could have died from natural causes. At least eighty other named gorillas were accounted for.

Despite the good news about known groups, the status of the entire Virunga gorilla population was still uncertain. Amy discussed plans for a complete census—the first since 1989—but government officials decided it was too dangerous. Bands of *interahamwe* continued to penetrate the park, gorilla tourists had recently been kidnapped and murdered in Uganda's nearby Bwindi Impenetrable National Park, and land mines were believed to be present in some areas of the Parc des Volcans. But on September 4, 1999, Amy joined a small group of tourists to visit a gorilla family that stayed close to the forest edge. Their guide was Augustin, a young Rwandan whom Amy had met during an earlier gorilla census in Uganda. He spoke with remarkable passion about the gorillas he felt privileged to know. Amy also noticed a second guide smiling broadly in the background, then recognized his face. He was François Bigirimana, whom Bill had first hired as a watchman, who then worked with Mark Condiotti, and was now an official park guide.

There were other changes in the Virungas, however: the fifteen soldiers armed with AK-47s and grenades who were there to protect Amy and the other visitors from a possible terrorist attack added a disturbing, but necessary, new element.

Amy and the others slipped into the park to the south of Mt. Visoke. The immediate smell of bamboo and the sight of *Hagenia* were powerful stimulants, but nothing compared to her first sight of the gorillas. The Susa Group had grown from the four individuals first identified by Bill in 1978 to a superfamily of more than thirty. Seven or eight family members moved into a clearing, close enough for Amy to see the sunlight reflect on their faces. She looked for Poppy or Muraha, two Group 5 natives who had transferred to Susa, but they weren't in sight. The other gorillas acted just like those she remembered: calm, accepting, familiar. Still, Amy missed knowing the personal histories and relationships of the individuals she was watching. It was like walking into a roomful of people that she didn't know. Yet, knowing that they were all survivors, it was wonderful to see them alive and well.

Myths and Realities

Chapter Thirty-three

Tarnished Notions

WE BEGAN OUR WORK IN RWANDA with the belief that greater understanding could help us solve serious conservation problems. In many important ways, we were right. Yet recent events have shown that complete understanding is a myth—one of many notions tarnished, if not destroyed, by the cataclysm in Rwanda. In a single decade, Rwanda was transformed from a model of cooperative development to a case study of genocidal madness. Democracy may be a pillar of modern civilization, but it was a volatile element that contributed to the Rwandan breakdown. The same Western countries and institutions that promoted democratization failed to intervene in an unambiguous genocide. Western-funded development activities first helped to meet basic human needs, then increasingly fed the cravings of corrupt elites. Ironically, renewed economic assistance to the Tutsi-led government today is subject to more conditions than were ever applied to the discredited former regime. The perception that mountain gorilla and forest conservation are good for Rwanda and its people seems to have been tempered and hardened in the crucible of recent experience. Yet we need to understand clearly why this is so and recognize that conservation continues to occupy precarious ground in Rwanda.

In 1959, Rwanda's Hutu majority overthrew the Tutsi monarchy. Three years later, they replaced the Belgians as leaders of their own nation. Over the next thirty years, successive Hutu governments equated democracy with demography. With an 85 percent majority and one of the world's highest birth rates, it seemed an equation for perpetual rule in a single-party Hutu state. During the Cold War, Western democracies weren't particular about the governments they supported in Africa. They asked only that their leaders oppose the spread of communism. Absolute dictators like Mobutu Sese Seko in Congo met our perceived needs, as did far more moderate leaders like Juvénal Habyarimana of Rwanda. With the fall of the Berlin Wall in 1989, the West rediscovered its democratic values and pressured client states like Rwanda to adopt political reforms and embrace multiparty government. We Westerners like democracy, but we don't think very much about the conditions required to make democracy work. We take it for granted that democracy is representative, that minority and individual rights are protected, and that a free press will inform an educated public. Yet none of these conditions existed in Rwanda.

Multiparty "democracy" in Rwanda exposed fissures in the façade of Hutu unity. Southerners, resentful of the unequal benefits of development that the Habyarimana government had steered toward its northern core constituencies, created new parties to redress the inequities of the regional patronage system. If southern Hutu had little power, the minority Tutsi had none. It is widely believed that the RPF attack in late 1990 was not intended to achieve a surprise military victory, but rather to assure Tutsi participation in the new representative government. Yet the attack diverted attention from serious intra-Hutu conflicts and rekindled old ethnic tensions. Calls for Hutu unity in the face of a renewed Tutsi threat neutralized key opposition figures while raising the profile of Hutu supremacists.

Democratization in Rwanda was accompanied by an explosion of party-supported newspapers. Rather than inform the public—or even highlight party differences—the press fanned the flames of ethnic division and hatred. Rabid editorials and grotesque political cartoons fueled intense anti-Tutsi sentiment. The creation of *Radio Mille Collines* in 1993 ignited this potent mix with false reports of RPF atrocities and claims of a Tutsi campaign to restore the hated monarchy. The racist mantra of the Hutu *Ten Commandments* played daily. Radio broadcasters prepared their Hutu listeners for the genocide with a simple choice: kill or be killed. For an uneducated Hutu population, still sensitive to the abuses of the past and susceptible to new myths packaged by manipulative media, it was a credible message. When called to act on this belief, they responded with a frightening intensity that left 800,000 Tutsi dead. In the rush to kill their Tutsi neighbors, however, they left the door open to a Tutsi military victory.

Will the RPF be any more sensitive to the nuances of democratic government? At first glance, it might seem unlikely. Even with the influx of returnees from abroad, the Tutsi remain a minority of only 15 to 20 percent.

Yet Paul Kagame and many of his RPF colleagues developed a strong sense of minority rights as a result of the discrimination they experienced during their long years of exile in Uganda. The challenge for the RPF in Rwanda is to protect the rights of the *majority* Hutu while retaining sufficient Tutsi control to avoid any repetition of past abuses. The fact that Hutu outnumber Tutsi in the RPF coalition government is an auspicious start, even though Tutsi retain control of the military and the most important ministries. More promising is the gradual implementation of representative government. Rwandans have already selected their local representatives and communal *bourgmestres* in a series of open elections. Hutu have won most of these posts. National elections for parliament and president scheduled for 2003 will provide a more difficult test for the RPF. With the horror of the recent genocide too fresh to allow absolute majority rule, most observers believe that the RPF will retain some control over the selection of national leaders, both Hutu and Tutsi. It is a political formula that permits the release of certain pressures, yet guarantees underlying tensions will continue to rise. At this time, there is no evident alternative.

IN 2000, THE UNITED NATIONS published a critical assessment entitled *Rwanda: The Preventable Genocide.* This report correctly castigates the Hutu perpetrators who bear primary responsibility for the premeditated slaughter. It condemns the failure to intervene by its own U.N. Security Council, the obstruction of the United States, and the vacillation of the Europeans. The French are rebuked for first arming, then failing to restrain the Habyarimana government. The Catholic church is criticized for its failure to exert moral authority in the conflict.

It is worth asking whether Western countries would have known how to intervene in Rwanda, had they been so motivated. Would they have concentrated on urban centers? How would they have protected the hundreds of thousands of Tutsi living in dispersed rural areas? How would they have distinguished between Hutu and Tutsi? If force were required, whom would they attack? Western military technology has demonstrated its superiority in Iraq and Serbia. Yet smart bombs are useless against a small force of fifteen men with machetes in their hands and hatred in their eyes. Still, any level of intervention would have helped. Even a few thousand well-armed soldiers with an aggressive mandate could have protected the tens of thousands of Tutsi who sought refuge in churches, stadiums, and other public places in urban areas—perhaps 100,000 in all. An armed presence undoubtedly would have dampened the willingness of others to join in the killing. It might have emboldened some to speak out in opposition to the madness. We'll never know, because no one came to help.

A few years after the Rwandan genocide, the world's attention was gripped by an outbreak of ethnic killing in the Balkans. Thousands of Kosovars were killed by Serbian military forces and militias in the former nation of Yugoslavia. Led by the United States, the West committed vast military resources and billions of dollars in a timely and laudable effort to protect the Kosovars. We were left to wonder why no comparable effort was ever launched to protect the Tutsi in Rwanda.

A T THE END OF WORLD WAR II, the victorious allies tried and executed the top German leaders for their role in the Holocaust. French vigilantes took matters into their own hands, killing thousands of their countrymen suspected of collaboration with the Nazis. Many in the RPF have wondered if high-profile trials and executions—followed by a general pardon or a public "truth commission," as in post-apartheid South Africa— might not provide an essential first step toward healing and reconciliation in Rwanda. Thwarted by foreign pressure, however, they have turned to a traditional form of justice to process most of the 125,000 genocide suspects still held in Rwandan prisons. In precolonial times, the *gacaca* served as a local tribunal in which residents were empowered to pass collective judgment on their peers. Today, the government has revived the *gacaca* system to judge the vast majority of those charged with lesser genocide crimes. There is a danger that Hutu will be unwilling to convict or harshly sentence members of their own communities. There is a greater danger that the continued imprisonment without trial of more than 100,000 people will further erode the RPF's tenuous base of support. With or without trials, adult Hutu today in Rwanda bear a suspicion of collective guilt similar to that borne by postwar Germans. Did they know of the killing? Did they try to stop it? Did they kill? These are questions that wear on the soul, whether the body is free or in prison.

W ESTERN DEVELOPMENT ASSISTANCE helped Rwanda improve its basic economic and social indicators through the mid-1980s. Per capita GNP, agricultural productivity, access to clean water, infant mortality, and life expectancy all improved—until the latter was offset by AIDS-related mortality. With the collapse of coffee prices in the late 1980s, the mini-miracle ended in Rwanda. Foreign debt and inflation rose rapidly, the currency was devalued, and the Rwandan people were forced to tighten already snug belts. But not their leaders. Senior members of the Habyarimana government increasingly diverted funds to serve their personal and political needs. Most donor agencies stood by and watched as their assistance nourished the spread of corruption.

Foreign assistance is again flowing into Rwanda to meet a broad array of basic needs facing the RPF government. Between 1996 and 1998, bilateral and multilateral agencies contributed more than $1 billion to help restore the nation's infrastructure, promote economic recovery, defray its energy costs, and combat AIDS. These are all essential needs, but less than 5 percent of foreign assistance has gone into education. We would argue that education is a special case, deserving of far greater attention. Democracy requires an educated public. An educated public is far more likely to resist the kind of simplistic propaganda that triggered the Rwandan genocide. Educated farmers are more likely to adopt new technologies, or to seek employment outside of the agricultural sector. Educated women are more likely to practice some form of family planning. Support for gorilla and park conservation was significantly higher among Rwandans who attended at least three years of primary school. One can only wonder how much more progress Africa might have made if the West had invested as much in basic education as it did on military assistance and political intrigue during three decades of the Cold War. Finally, the West must act quickly and decisively on the issue of debt relief. Accumulated debt is crushing any hope for real development in dozens of the world's poorest countries. In Rwanda, this injury is aggravated by the insult of asking the current government to pay off the debts of its corrupt and murderous predecessors.

DEMOCRACY AND DEVELOPMENT are essential elements in building cohesive nations: a very difficult process of balancing multiple interests and antagonisms. After centuries of concerted effort—and countless wars, civil wars, and revolutions—the most stable of the Western democracies are still working out the details of respect and representation for minorities within national structures. Some of the younger nations of Eastern Europe continue to struggle, violently at times, to sort out who wants to live with whom and what their appropriate boundaries should be. Most Latin American nations have suffered the self-inflicted wounds of serious human rights abuses and only now are entering a fragile phase of democratic government.

We hold African nations to an unreasonable standard on this issue of nation-building. Modern African states were created by the European colonial powers with virtually no regard for traditional boundaries. The resulting bundles of unrelated and sometimes antagonistic ethnic groups were held together by force, then set free with little serious preparation as supposedly independent nations barely forty years ago. Most have spent the intervening time trying to develop a sense of national identity and linguistic unity for dozens, sometimes hundreds, of disparate ethnic constituencies. Rwanda has several advantages over most other African nations in that its current borders roughly coincide with the traditional limits of the Tutsi kingdom, and its single Bantu language is rare among its African sister states. But Rwanda has formidable problems, too. Poverty, population pressure, and AIDS remain a potent triple threat, to which must now be added civil war, genocide, recovery, and reconciliation. Europe suffered through comparable scourges of warfare, plague, urbanization, and industrialization. But we should recognize that modern European nations struggled with mixed success to surmount these and other challenges over the course of centuries, an effort which continues to this day. We can only hope that most African countries can shorten the apparently long time frame required to build coherent nations, confident in their unifying identity and tolerant of their internal diversity.

CONSERVATION IS AN ESSENTIAL part of the development process. The conservation experience in Rwanda reveals two distinct models at work in the Virunga and Nyungwe forests. Mountain gorillas have star quality. They command attention. They have a global constituency that is prepared to act when they are threatened. The Nyungwe Forest is much larger and more diverse than the Virunga ecosystem, with 270 bird species and thirteen different kinds of primates. Yet not even Nyungwe's chimpanzees can capture attention in the manner of the mountain gorillas. This is reflected in the amount that tourists will pay for visits to the two reserves: $250 for an hour-long visit to the gorillas, versus $20 for a guided primate tour of Nyungwe. With twice as many visitors each year, this means that the Parc des Volcans earns twenty-five times the direct revenue of Nyungwe. The park garners political support in line with recognition that its mountain gorillas contribute to national economic growth. Nyungwe, too, will play a growing role in tourism development, as a satellite attraction that will keep visitors spending money in Rwanda for another two or three days. But Nyungwe has survived thus far because of the project's emphasis on local hiring and involvement. Throughout the worst of the early 1990s, the local staff continued to work without pay and at considerable personal risk. Today, the same local staff plays a central role as the Nyungwe Forest project operates under proud, capable, and ethnically mixed Rwandan management.

According to many conservation theorists, the greater size and biological richness of Nyungwe should merit greater conservation attention than the Virungas. Yet the presence of mountain gorillas has attracted a half-dozen international conservation organizations to focus on the Parc des Volcans, while only one is dedicated to the fate of the Nyungwe Forest. This outside attention is central to the survival of both sites, however. When the proposal was made to clear one third of the Parc des Volcans for cattle in late 1978, we were able to help mobilize several major groups to support the MGP as an alternative to the scheme. The Gishwati Forest was three times the size of the Parc des Volcans and more biologically diverse. But no conservation group was there to defend the interests of its plants and animals when the Habyarimana government and the World Bank decided to clear the forest for cattle. The return of nearly one million Tutsi exiles to Rwanda turned covetous eyes toward the vast grasslands of the Akagera Park. With few in-country advocates, two thirds of the striking savanna complex was converted to pastureland and much of its wildlife lost. Through it all, Nyungwe and the Parc des Volcans have escaped conversion, and there is strong government support for their continued protection. The active presence, assistance, and advocacy of conservation organizations at both sites have been essential factors in their survival.

Western organizations should be contributing much more to help conservation in Rwanda. We ask the Rwandan people and government to protect mountain gorillas and tropical rain forests as part of our shared global heritage. This notion of a global heritage is a noble concept, but we in the world's most developed countries are the ones who benefit most from wildlife conservation in much poorer countries. Western audiences have an insatiable appetite for exotic wildlife in films, magazines, and books. The nature film industry alone is a $100 million annual business dominated by National Geographic, BBC, and Discovery, among others. Countries such as Rwanda that provide the raw material for these films—the animals and natural areas—earn a pittance in permit fees, usually only a few thousand dollars per film. Westerners are also primary consumers of gorillas and other tropical attractions through the burgeoning ecotourism industry. And while Rwanda can make several million dollars each year from gorilla tourism, the international airlines that transport gorilla visitors make millions more. Most hotel and rental-car chains are also foreign owned. So we should pay the lion's share of the costs of conservation because we are the primary benefici- aries.

Rwandans have other more pressing priorities. Economic development, AIDS, ethnic reconciliation, peace, and security all rank higher at the na- tional level; food, land, health, education, and safety are more important for individuals. It's not that Rwandans don't care about conservation; our work showed that Rwandans at all levels of society were intrigued by the gorillas' personalities and family life. They created art and music to celebrate the go- rillas. But Rwandans changed their attitudes to favor conservation of the Parc des Volcans only when they saw the benefits of tourism revenues and employment offsetting the perceived loss of land and wood in parks. The fact that we insisted on hiring people from the local region greatly reinforced this economic argument. National political support for mountain gorilla conservation is directly linked to gorilla economics.

We do not believe in the popular concept that wildlife must "pay its way" to survive—a dangerous trend that ultimately allows conservation values to be superseded or trumped by alternative economic uses. But money helps. And when it is so clearly in our interest to save the mountain gorillas and other forest wildlife of Rwanda, we should pay accordingly. Yet the United States contributes pathetically little foreign aid to developing countries and far less for conservation. We rank seventeenth—just behind Austria—in overall foreign assistance as a percentage of our gross national product. The U.S. should set the standard for the world's richest countries by dedicating a percentage of its GNP to international conservation and environmental mit- igation and an even higher percentage of our annual income to development assistance.

~~~~~~~~~~~~~~

MANY WOULD ARGUE the most fundamental reason to save gorillas and other wildlife is that animals have a basic right to exist. This pow- erful argument is gaining currency and support in many quarters. Some ad- vocate the extension of rights to all living beings, based on their capacity to feel pain and suffer. Others prefer to start with creatures such as dolphins, whales, and primates because of their higher intelligence. Mountain gorillas and chimpanzees are at the top of every candidate list. The world will be a much better place the day we learn to treat other species with respect and rec- ognize that they, too, have certain rights.

We have a well-developed notion of human rights that has evolved over time and is embodied in the Global Charter on Human Rights, a document signed by every member nation of the United Nations. Yet human rights are abused to varying degrees in every one of those member states. The rights of children, women, and national minorities are ignored or trampled on a regular basis. The most basic rights—to life and liberty—are violated in many countries where individuals are imprisoned or killed because of their beliefs or ethnicity. In Rwanda, the recent genocide was the ultimate violation of human rights. We have so far to go in recognizing and respecting human rights that it is difficult to be hopeful for the adoption of animal rights any time soon. Those who are passionate about the subject should act in accordance with their beliefs. With regard to mountain gorillas, however, animal rights advocates should recognize that their concerns might be seen as irrelevant—or even offensive—in the Rwandan cultural context, where the most extreme violations of human rights remain deeply etched in the national consciousness. There are more than a few Rwandans who wonder if the Western world would have intervened more quickly and forcefully if mountain gorillas, rather than Africans, were being slaughtered in 1994.

THERE IS NO QUESTION that greater knowledge helped our conservation efforts. Applied research on mountain gorilla demography and ecology provided critical information for park management and helped to fight off the extreme threat of the cattle-raising project proposed in the late 1970s. Studies of social and economic factors in gorilla conservation provided baseline information for a targeted education program and the rationale for a new kind of tourism. We then expanded this multidisciplinary approach to the entire Virunga watershed and adapted it to a very different set of conditions in the Nyungwe Forest. Yet applied research alone did not carry the day. Our knowledge of French and Swahili helped as much as our scientific knowledge in opening lines of communication with local populations, government officials, and foreign advisors. Diplomacy was needed to maneuver among competing personalities and agendas. Our commitment to apply knowledge to conservation action kept us from retreating into academic irrelevancy. And listening to local voices taught us about a world beyond the reach of our studies.

From our experiences in Rwanda over the past twenty-three years, we have gained understanding in ways we never imagined on issues we never considered. There are many lessons we wish we hadn't learned. But understanding is essential to conservation. We can't love animals or save wildlife without understanding the social, economic, and political context in which conservation occurs. We must extend the relevant landscape beyond park boundaries and the forest's edge. To achieve our goals, we must act on our understanding at local, national, and international levels. It is pure hubris to believe that we can fully appreciate what is happening in a markedly different culture, like that of Rwanda. But if we make the effort, we can open a window to better comprehension of its people, its politics, and its priorities.

Chapter Thirty-four

Pablo

THE TRAIL UP MT. VISOKE wound past familiar markers. Fallen trees, exposed rocks, and boot-sucking mud all brought back memories of our early years in the Virungas. The pungent, almost bitter odor of abundant wild celery spiked the air around Return Ravine. Yet the mountain slopes appeared more open, with less bamboo and fewer *Hagenia* trees than before. Maybe fewer signs of Cape buffalo, too. Poaching, wood-cutting, and bamboo harvesting had certainly increased during the genocide and its aftermath. Cleanup patrols were still finding fields, huts, dishes, clothing—and skeletons—within the Parc des Volcans. Halfway up the old trail to Karisoke, fresh gorilla sign led straight up the mountain. The thought of seeing the gorillas gave Amy a second wind that carried her past the treeline and into the more open subalpine zone. Still, as they neared twelve thousand feet, she was relieved when her two women companions, each half her age, suggested a rest break. Amy sat on an exposed shelf of lava rock and looked out over the moorland vegetation that surrounded her. She still marveled at the ability of the grasses and shrubs to extract life's essentials from a thin crust of rocky soil and the nearly perpetual mist in which they bathed. They were the rugged ascetics of the rich tropical rain forest community.

Amy reflected on the past few weeks leading up to this extraordinary day. She was back in Rwanda in the summer of 2000 at the request of the government to design a major new conservation initiative. The source of support was a global environmental fund managed by the United Nations Development Program. The goal was to secure financing for a package of activities that would restore and reenergize Rwanda's protected areas. She first met with officials from the key government agencies with jurisdiction over forestry, parks, and environment. She then worked closely with the major conservation groups concerned with various sites and issues—three of them with a singular focus on mountain gorillas. In the end, she negotiated among the various parties and forged a collaborative approach to the serious problems at hand. Before leaving Kigali, she submitted an $8 million proposal for a seven-year action program for the Virunga and Nyungwe forests. It was a strong proposal, but she worried it might not be approved—not because it wasn't good enough, but because decision-makers in the UNDP might not see conservation as a high priority for Rwanda at that time.

Bolstering Amy's spirits was a concurrent government announcement that legislation was being drafted to make Nyungwe a national park. This was a dream that went back to our first visit to Nyungwe with Jean Pierre Vande weghe in 1983. It had taken persistent effort by many individuals and consistent support from the Wildlife Conservation Society. But the final breakthrough was a tribute to the Rwandans who had guided the project through a grueling decade of civil war, genocide, and recovery, men such as Félix Mulindihabi, Martin Sindikubwabo, Eugène Rutagarama, and Michel Masozera. Many important decisions remained before Nyungwe could become a national park. Could local people continue to harvest honey, bamboo, or medicinal plants from a national park? If not, should a buffer zone be created for controlled subsistence use? Who would monitor such use? These were tough issues for the future. For now, Amy felt satisfied that Nyungwe might take its rightful place among the world's great protected areas.

A T VISOKE'S SUMMIT, Amy watched a bank of low clouds roll over the far rim and flow like a waterfall to the crater lake below. Moments earlier, the gorillas had passed by the spot where she was standing. Now the sounds of cracking *Senecio* announced their presence just below the outer rim. Amy worked her way through the dense tangle of a *Senecio* thicket, breathing more deeply than she wanted to admit. She emerged to peer down on a scene she had longed to see for ten years. Set against a backdrop of Mts. Mikeno and Karisimbi in the distance and the Karisoke saddle two thousand feet below, dozens of mountain gorillas were feeding peacefully on the exposed ridge. Limbs of giant *Senecio* shrubs and broken spikes of alpine *Lobelia* littered the ground after yielding their pith to the gorillas' powerful jaws. Family clusters gathered around as adult females cracked open fresh stems or dug up the delicate roots that were a gorilla favorite. Off to the side, two silverbacks staked out their solitary claims.

One of the silverbacks was Pablo, the abandoned four-year-old who had briefly turned to Amy as a surrogate mother twenty-two years earlier. Leaving him to grow up on his own and watching his adolescent antics as the clown prince of Group 5, she could not have imagined him as the patriarch of his own clan. Yet now he was the dominant silverback of the largest group of mountain gorillas ever seen: a stable family of forty-four individuals. Gorillas from Group 5, the Susa Group, and others had merged under the leadership of one individual. Now it was Pablo's Group. His alpha status had not made Pablo any more attractive than he had been as an infant. His huge body accommodated his large head a bit better, but he was still slightly cross-eyed, and his mouth hung partially open in a perpetual slack-jaw pose. Some of his offspring shared these physical traits, as well as his rambunctious personality. More troubling was the fact that many of the gorillas showed signs of syndactyly, or fused fingers and toes—another trait of Pablo's believed to be linked to inbreeding. A younger silverback was challenging his group leadership by strutting and posturing in front of Pablo. He even mated with a young female in full view of Amy, while Pablo was out of sight. This twenty-two-year-old silverback was Cantsbee, named by Dian to reflect her shock when the supposedly male Puck gave birth. Amy had last known him as a wide-eyed eighteen-month-old toddler.

Puck approached to within three feet of Amy and sat down to feed. If Puck had any recollection of the white ape who long ago had followed her and her family around, scribbling on little pieces of paper for almost two thousand hours, she gave no sign. Or at least none that Amy could discern. She found herself observing the quality of the pith that Puck was eating and fighting the urge to take notes. In fact, the two energetic women who accompanied Amy that day were building on her earlier feeding ecology study. Jess Cantlon and Chloe Wilson worked for the Dian Fossey Gorilla Fund, which continued to support monitoring of the main Karisoke research groups. Both women were as eager to learn the early history of the group's individuals and lineages from Amy as she was to learn recent news of her former family and their new relations. Most remarkable was the continued influence of Effie. The dominant matriarch of Group 5, Effie had at least seven surviving offspring before she died of old age, including Puck. Each bore the distinguishing mark of a white blaze around the ears; each had an attractive face and walked with confidence. Puck alone added another five offspring to the maternal line, the most recent of whom began to nurse as Puck continued feeding. One notable change for a study group was that the names of the younger generation were now all African. Western audiences might not find Umwana, Makuba, and Isabukuru as appealing as

Puck, Tuck, and Effie, but the future of these gorillas was now in Rwandan hands.

Pablo's Group continued feeding on alpine delicacies for the next two hours. Puck and two other females then began to move downslope, toward new feeding areas and less exposed nesting sites for the evening. Cantsbee moved off next, strutting toward the front of the group. Pablo jostled with the younger silverback for position. Cantsbee countered each move, before finally taking the lead. A brief scream erupted as the two disappeared over the ridge. Family clusters followed, infants rolling and twirling purposefully down the grassy slope, until the entire clan moved out of sight. Amy froze the moment in her mind. The forest had enveloped the gorillas in its life-giving caress. This was Eden as it should be, a place where mountain gorillas moved freely about, innocent of the fray beyond their forest borders, where a bounty of food could be had for the hiking, and every log and tree was a playground.

The forest was a refuge for Amy, too. It had always been so, but now it was also a haven from persistent concerns. From worries about grants and raising money, to fears for the safety of project personnel. From memories of the nation's horrific recent past, to deep-seated anxiety about its uncertain future. All such thoughts dissipated in the thin mountain air while Amy watched the gorillas. She felt privileged to see the continuity and change within a family she had known for more than a generation: Pablo shoulder-ing the responsibility of leadership, Puck a successful mother, and Effie liv-ing on through a line that now included two great-grandchildren. She felt a profound peace as she watched them slip away and then stared off through the afternoon haze at the massive forms of Karisimbi and Mikeno.

Within minutes of the gorillas' departure, reality intruded in the form of twenty heavily armed guards. Each carried a rifle and multiple ammo belts slung over his shoulder. One hefted a grenade launcher. Many smiled broadly in appreciation of the gorilla spectacle they had watched in hiding. But their task was deadly serious. They were present to protect the re-searchers and their visitor from any possible attack by *interahamwe* or other rogue forces. The sanctuary could be violated. This was the reason that the offer to coordinate a much needed gorilla census was again politely deferred by the government. The need was real, but so were the *interahamwe* and the possibility of land mines. The presence of foreigners would attract greater attention to the operation and require even more soldiers to guard the survey teams. Work would be especially risky in the Congolese sector of the Virun-gas under incomplete Rwandan control. We would have to wait. Meanwhile, it was reassuring to know that births apparently exceeded deaths in those groups that were regularly monitored, and that research and tourist groups, like Pablo's and Susa's, were thriving.

Hiking down the mountain, the group came to the abandoned ruins of Karisoke. The research station had been looted and destroyed seven years earlier. Amy located the site of our old cabin and that of Dian's, but most of the meadow complex was completely overgrown. Once again the giant *Hagenia* trees reigned supreme, their contorted forms towering over thousands of young *Hypericum* that now competed for space and sunlight in the former clearing. It was beautiful, perhaps the way a young Dian Fossey had first seen the site more than thirty years before. Karisoke had returned to the earth where Dian herself lay buried—though her gravestone had been removed for safekeeping. Up on the slopes, her legacy lived on in the gorillas we knew by name, by personality, and by family history.

A T THE BASE OF VISOKE, Semitoba was waiting. The metal hut where he had first worked for us as a housekeeper still stood nearby. Semitoba had heard that Amy was in the area and he came to say hello and send his greetings to Bill. He confirmed earlier news that his family had been killed by *interahamwe*. He also told Amy that Big Nemeye had returned from Congo, but then died of a disease contracted in the refugee camps. Semitoba recounted his losses and those of others in a matter-of-fact manner, as Amy listened and nodded. He then smiled broadly as they discussed past times and struggled to recall the names of others who had worked with Karisoke or the Mountain Gorilla Project. Semitoba said he would like to work for us again, but any job would help. With a parting hug, Amy left for Ruhengeri.

Leaving the parking area, Amy was again struck by the fact that all the houses near the park had been razed. Well-tended farm fields still stretched to the park boundary, but all the people had been relocated. The government claimed that this was done to protect residents from the *interahamwe,* but it also removed potential sympathizers from direct contact with *interahamwe* operating from within the park. Farther along, Amy passed miles of houses clustered on either side of the road. Brick houses with tin roofs, mud houses with thatch, and even stick frame structures covered with blue U.N. tarps were all crowded together. This was where Bidele and many others who were removed from the park boundary now lived in concentrated settlements under the government's policy of *villagisation*.

Just outside of Ruhengeri, Amy stopped at the family home of Clementine, the boys' former nanny. Rust covered the roof, and the door needed paint. No one was home in the dark house. The arrival of an *umuzungu* quickly attracted a crowd, from which a woman stepped forward in a bright *kanga* with a matching scarf. She said that Clementine's mother and brother had been killed a few years earlier, but she took Amy to where a sister lived. The sister's beautiful face lit up when Amy showed her a picture of Clementine holding Ethan as a baby. Amy quickly added that the picture was old, fearing her sister would take it as recent proof that Clementine was alive. In fact, the family had never heard from her since the last refugee camps were hurriedly shut down in 1996. She could have been killed then, she might have died moving deeper into Congo. No one knew, but the family believed that Clementine was dead. On saying this, her sister broke down in silent tears. Amy gave her the picture to keep, and the two women hugged as Clementine's mother-in-law watched in silence.

In Ruhengeri, heavy foot and bicycle traffic like Amy had never seen before almost brought her car to a halt. Freshly painted storefronts and the percussion of carpenters' hammers attested to dynamic growth. Across from our former home, in an open lot where Noah and Ethan once played, cement and wooden crosses marked the graves of genocide victims, recent wounds on the body of a recovering nation.

On the outskirts of Ruhengeri, Karisoke-in-exile operated from a rented house, surrounded by beautiful flowers and fruit trees. Karisoke's director, Liz Williamson, had invited Amy to stay at the house. Liz was a former director of our Nyungwe project and a good friend. At times, her lanky frame and long, dark hair were eerily reminiscent of Dian, with whom she shared a headstrong determination and singular dedication to the gorillas. Yet Liz had a very different personality from Dian's and a balanced understanding of the complexities of modern conservation. She maintained regular contact with key government officials and worked cooperatively with other conservation groups. She believed in the importance of sharing information and worked hard to maintain open lines of communication with multiple groups. But Liz was tightly wound and the demands of her work were excessive.

On her return from seeing Pablo's Group, Amy found an empty house and a note marked "urgent," asking her to join Liz at a local restaurant. Liz had just received confirmation of a road to be built through the heart of the Parc des Volcans, continuing through the forest to the Congolese side of the range. The road undoubtedly made sense to Rwandan military leaders, who wished to increase their mobility and strengthen their control over occupied areas of Congo. For the gorillas and the park, it was a pending disaster. Tony Mudakikwa from the Mountain Gorilla Veterinary Project and Anecto Kayitare from the International Gorilla Conservation Project joined Amy and Liz to discuss the situation with the park warden, François Bizimungu. All committed to write letters of concern, but it was agreed that a foreign-based group would have the greatest freedom to speak its mind. So Amy drafted a separate letter, complementary to the others, addressed to Paul Kagame, president of Rwanda and minister of defense. The letters were delivered in Kigali, where Amy and Liz also discussed the matter with the U.S. ambassador, who said he would raise the matter with Kagame at their next meeting.

Back in the U.S., Amy unpacked her bags. Her day with Pablo came to life in the sweet smell of everlasting flowers and *Lobelia* on her field shirt. A few days later, we watched slides of her trip, stirring in deep-seated memories with old pictures of Pablo, Puck, and Cantsbee as youngsters. Noah was home from college so we even dug out the fourteen-year-old video of his visit to Group 5.

A few weeks after Amy's return, we received the excellent news that President Kagame had canceled plans to build the road through the Virunga forest. Then came word that the protected areas proposal had passed its first-stage review and that a more detailed design phase would be funded through the U.N. We both agreed to help with this multidisciplinary effort, setting the stage to work together in Rwanda for the first time in a dozen years.

In January of 2001, we received even better news from Liz. With support and training from the International Gorilla Conservation Project—the direct descendant of the Mountain Gorilla Project—a combination of park guards and military personnel had been counting as many of the Virunga gorilla groups as possible. The results of this "ranger monitoring" program indicated a preliminary count of 355 mountain gorillas—a 10 percent increase since the last census in 1989. Liz noted that this was a conservative count since some areas around Mt. Mikeno were still too dangerous to survey. Amy confirmed the number with IGCP and everyone agreed that a full census and population analysis were still needed. But for now, it was a bracing dose of good news to start the new century.

When we first came to Rwanda, gorillas were dying and the Parc des Volcans was a battleground. Outside the park, Rwanda was at peace with itself and its neighbors. Today, more than a million Rwandans are dead from genocide and warfare; several hundred thousand are in exile in neighboring countries. AIDS continues to take a deadly toll. Still, there are several million more people living in this small country than when we first arrived in 1978. Terrorism threatens the nation's internal security, while its armed forces are embroiled in a high-stakes regional conflict with no end in sight. Ironically, it is the mountain gorillas who now enjoy a fragile peace. If there is symbolic value in this fact, it may be that twenty-five years ago few believed that Rwanda's gorillas could survive the deadly forces arrayed against them. The fact that so many continue to survive—and even thrive—within the confined space of the Virunga forest might offer hope. Hope not only for the mountain gorillas, but also for the millions of Rwandans who must now find a way to reconcile a brutal past and bring lasting peace to the beautiful land they call home.

Index